F.G.R. FOWKES (Ed.)

Epidemiology of Peripheral Vascular Disease

With 51 Figures

Springer-Verlag
London Berlin Heidelberg New York
Paris Tokyo Hong Kong
Barcelona Budapest

F.G.R. Fowkes, PhD, FRCPEd, FFPHM
Director, Wolfson Unit for Prevention of Peripheral Vascular
Diseases, and Reader in Epidemiology, Department of Community
Medicine, University of Edinburgh Medical School, Teviot Place,
Edinburgh EH8 9AG, UK

ISBN-13: 978-1-4471-1891-6   e-ISBN-13: 978-1-4471-1889-3
DOI: 10.1007/978-1-4471-1889-3

British Library Cataloguing-in-Publication Data
Fowkes, F.G.R.
  Epidemiology of peripheral vascular disease.
  I. Title
  616.131
  ISBN-13: 978-1-4471-1891-6

Library of Congress Cataloguing-in-Publication Data
Epidemiology of peripheral vascular disease/F.G.R. Fowkes (editor).
  p.   cm.
  Includes index.
  ISBN-13: 978-1-4471-1891-6
  1. Peripheral vascular diseases—Epidemiology. I. Fowkes, F.G.R., 1946–
  [DNLM: 1. Vascular Diseases—epidemiology. 2. Vascular Diseases—prevention
  and control. WG 500–E64]
RA645.P48E65   1991
614.5'9131—dc20
DNLM/DLC
for Library of Congress   91–4899
                                                                        CIP

Typeset by Wilmaset, Birkenhead, Wirral
Printed by The Alden Press, Osney Mead, Oxford. Bound at the Bath Press, Bath
28/3830–543210 Printed on acid-free paper

Preface

Atherosclerotic disease of arteries in the lower limbs is a major cause of morbidity in developed countries. In many Western populations approximately 2% of adults in late middle age have intermittent claudication. Among some patients the disease is severe enough to warrant hospital admission. Each year, for example, around 50 000 patients in England and Wales are admitted with a diagnosis of peripheral vascular disease and of these, over 15 000 have major surgery, including amputations.

Since the mid-1950s, epidemiological research into circulatory diseases has quite rightly concentrated on ischaemic heart disease and stroke because of their high mortality and morbidity, but consequently peripheral vascular disease in the lower limbs has been overshadowed and somewhat ignored for epidemiological study. Most population-based research into peripheral vascular disease has been an adjunct to major studies of ischaemic heart disease in which questions on intermittent claudication have been included in cardiovascular questionnaires. Few studies have attempted to employ other measurement techniques used commonly in clinical practice to diagnose peripheral vascular disease.

It is probably true that the cause and pathogenesis of atheroma may be investigated equally well in the legs as in the coronary and cerebral arteries. Narrowing of the iliac or femoral vessels is almost invariably due to atherosclerosis so that imaging or non-invasive physiological measurements may be good proxy measures of disease. The arteries in the legs are accessible to simple non-invasive assessment by measuring the ankle-brachial systolic pressure index or other haemodynamic features. However, an important difference between research into atherosclerosis in the peripheral vessels compared with elsewhere is that the epidemiologist is usually concerned with the chronic effects of disease in the legs because few patients have acute complications. This contrasts with coronary and cerebrovascular disease in which much of the research has focused on the acute events of myocardial infarction, stroke and cardiovascular death. The claudicant is in fact at great risk, not from an acute embolism or even amputation, but from a heart attack. Thus, despite studying a chronic non-fatal disease in the legs, epidemiologists may still be frustrated by a high level of wastage in their study population!

This book is an attempt to redress the balance between peripheral and coronary artery disease. It is the first to provide a review of the epidemiology of atherosclerotic disease affecting the legs. Professor Widmer in Basle provided the groundwork in an important book describing his study in pharmaceutical workers. In this present book, the purpose is to provide both review and research material concerned with epidemiological measurement, descriptive epidemiology, risk factors and prevention. Conventional cardiovascular risk factors and more recent aspects of epidemiology and pathogenesis are included. The risk factors are considered separately but because they are usually interrelated in real life, there is by necessity some overlap between chapters. Also, some factors such as diet, personality, and social deprivation have been so little studied in peripheral vascular disease, that we have to use predominantly our own work in the Edinburgh Artery Study. In some instances, the lack of material on peripheral vascular disease has meant that authors have had to draw parallels from research on heart disease and stroke.

Nevertheless, this book does provide a reasonably comprehensive overview of the subject at this time.

Each year in Edinburgh, the Pfizer Foundation generously supports a research symposium held under the auspices of the Edinburgh Post-Graduate Board for Medicine (under the aegis of The University of Edinburgh, The Royal College of Physicians of Edinburgh, and The Royal College of Surgeons of Edinburgh). The authors of this book will each give a presentation at the 1991 symposium on the epidemiology of peripheral vascular disease. Contrary to normal practice, each contributor produced their chapter six months ahead of time so that the book could be available at the symposium. Thus, there was a tight schedule resulting in a very up-to-date publication. However, readers should be aware that this book is not a record of the proceedings of the symposium because it contains review material in addition to presented research.

This book should appeal to a wide range of experts in the field of vascular disease – cardiovascular epidemiologists, angiologists and vascular surgeons. Also, family doctors who are becoming increasingly concerned about prevention may like to find out more about the cause of disease. On the other hand, cardiovascular researchers may appreciate the review material and be made aware of opportunities for research in this field. For many clinicians, however, there is simply the opportunity to "complete the clinical picture" of disease beyond the narrow spectrum observed in the medical setting.

April 1991   Gerry Fowkes

Acknowledgements

The production of this book has come about because of the efforts of those who contributed to the Pfizer Foundation symposium on the epidemiology of peripheral vascular disease held in Edinburgh in 1991. I am grateful to Dr Alexander Muir, Dean of Post-Graduate Medicine, University of Edinburgh, for taking a major role in organising the symposium, ably assisted by Sheila Edwards, Deputy Director of the Pfizer Institute, and Clare Perry. Also, valuable advice was given by Dr E.A. Stevens of the Pfizer Foundation. I am also grateful to Professor Michael Oliver for setting the scene for the symposium and to the chairmen: Professors Roger Greenhalgh, Gerry Shaper, Michael Garraway, Michael Marmot, Charles Forbes, and Hugh Tunstall-Pedoe. Also I should like to give special thanks to the individual authors who were required to produce their chapters in a relatively short period of time.

I am grateful to publishers and authors who have agreed to the reproduction of various tables, figures and illustrations in the text. In particular, the editor of the *International Journal of Epidemiology* gave permission to reproduce much of the material in Chapters 1 and 8.

Contents

Contributors

P.L.P. Allan, BSc, DMRD, FRCPE, FRCR
Senior Lecturer in Medical Radiology, University of Edinburgh,
X-ray Department, Royal Infirmary, Edinburgh, EH3 9YW, UK

D.J.P. Barker, MD, PhD, FRCP
Professor and Director, MRC Environmental Epidemiology Unit,
University of Southampton, Southampton General Hospital, South-
ampton SO9 4XY, UK

C.J. Bulpitt, MSc, FRCP
Professor of Geriatric Medicine, Royal Postgraduate Medical School,
Hammersmith Hospital, Du Cane Road, London, W12 0NN, UK

V.D.L. Carstairs, BSc, Hon MFPHM
National Co-ordinator of Health Services Research Networks,
University of Edinburgh, Buccleuch Place, Edinburgh EH8 9JT, UK

J.M. Connor, MD, BSc, FRCP
Professor of Medical Genetics, Duncan Guthrie Institute of Medical
Genetics, Yorkhill, Glasgow G3 8SJ, UK

M.H. Criqui MD, MPH
Professor, Departments of Community & Family Medicine and
Medicine, Director, Preventive Cardiology Academic Award Pro-
gram, University of California at San Diego, School of Medicine, 9500
Gilman Drive, La Jolla, California 92093–0607, USA

R. Cuming, BSc
Research Technician, Department of Surgery, Charing Cross Hospital, Fulham Palace Road, London, W6 8RF, UK

A. da Silva, MD
Assistant of the Medical School of Lisbon, Servico de Cirurgia
Vascular piso 4, Hospital de Santa Maria, Lisbon, Portugal

G. Davey Smith, MD, MSc, MA
Lecturer in Epidemiology, Department of Epidemiology and Popu-
lation Sciences, London School of Hygiene and Tropical Medicine,
Keppel Street, London WC1E 7HT, UK

I.J. Deary, BSc, MBChB, MRCPsych
Senior Lecturer in Psychology, Department of Psychology, University
of Edinburgh, 7 George Square, Edinburgh, EH8 9JZ, UK

P.T. Donnan, BA, BSc, MSc
Research Statistician, Wolfson Unit for Prevention of Peripheral
Vascular Diseases, Medical School, Edinburgh University, Teviot
Place, Edinburgh, EH8 9AG, UK

J. A. Dormandy, DSc, FRCS
Consultant Vascular Surgeon, Department of Surgery, St George's
Hospital, Blackshaw Road, London SW17 0QT, UK

M.R. Ellis, BSc
Chief Technician, Professorial Department of Surgery, Charing Cross
and Westminster Medical School, St. Dunstan's Road, London, W6
8RP, UK

P.C. Elwood, MD, FRCP, FFCM
Director, MRC Epidemiology Unit (South Wales), Llandough Hospital, Penarth, South Glamorgan CF6 1XX, UK

H.S. Fiegelson
Department of Community and Family Medicine, University of
California at San Diego, School of Medicine, La Jolla, California
92093–0607, USA

J.F. Forbes, BA, MSc, PhD
Senior Lecturer in Health Economics, Department of Public Health
Sciences, Usher Institute, Medical School, Teviot Place, Edinburgh
EH8 9AG, UK

F.G.R. Fowkes, PhD, FRCPEd, FFPHM
Director/Reader in Epidemiology, Wolfson Unit for Prevention of
Peripheral Vascular Diseases, University of Edinburgh, Teviot Place,
Edinburgh, EH8 9AG, UK

P.J. Franks, PhD
Lecturer, Department of Surgery, Charing Cross and Westminster
Medical School, St. Dunstan's Road, London, W6 8RF, UK

A. Fronek, PhD
Department of Surgery and Department of Bioengineering, Univer-
sity of California at San Diego, School of Medicine, La Jolla,
California 92093–0607, USA

R.M. Greenhalgh, MA, MD, MChir, FRCS
Professor of Surgery, Department of Surgery, Charing Cross and
Westminster Medical School, St. Dunstan's Road, London, W6 8RF,
UK

M. Horrocks, MSc, FRCS
Consultant Surgeon, Department of Surgery, Bristol Royal Infirmary, Bristol BS2 8HW, UK

E. Housley, FRCPE, FRCP
Consultant Physician, Honorary Senior Lecturer in Medicine, Peripheral Vascular Clinic, Royal Infirmary, Edinburgh, EH3 9YW, UK

L.U.CH. Janzon, MD
Professor in Epidemiology, Department of Community Health Sciences, Malmo General Hospital, S–Z1401, Malmo, Sweden

R.J. Jarrett, MD, FFCM
Professor of Clinical Epidemiology, Department of Public Health Medicine, U.M.D.S. (Guy's Campus), London, SE1 9RT, UK

V. Kaiser, MD
General Practitioner/Investigator, Department of General Practice, University of Limburg, P.O. Box 616, NL–6200 MD, Maastricht, Netherlands

M.R. Klauber
Department of Community and Family Medicine, University of California at San Diego, School of Medicine, La Jolla, California 92093–0607, USA

J.A. Knottnerus, MD, PhD
Professor of General Practice Research, Department of General Practice, University of Limburg, P.O. Box 616, NL-6200 MD Maastricht, Netherlands

R.D. Langer
Department of Community and Family Medicine, University of California at San Diego, School of Medicine, La Jolla, California 92093–0607, USA

G.C. Leng, BSc, MBChB
Clinical Research Fellow, Wolfson Unit for Prevention of Peripheral Vascular Diseases, University of Edinburgh, Teviot Place, Edinburgh, EH8 9AG, UK

G.D.O. Lowe, MD, FRCP (Edin, Glasg, Lond)
Senior Lecturer and Consultant Physician, University Department of Medicine, Royal Infirmary, 10 Alexandra Parade, Glasgow G31 2ER, UK

C.C.A. Macintyre, MSc
Lecturer in Medical Statistics, Medical Statistics Unit, Department of Public Health Sciences, Edinburgh University, Teviot Place, Edinburgh, EH8 9AG, UK

M.G. Marmot, MB, MPH, PhD, FFPHM
Professor of Epidemiology and Public Health, Department of Epi-
demiology and Population Sciences, London School of Hygiene and
Tropical Medicine, Keppel Street, London WC1E 7HT, UK

T.W. Meade, DM, FRCP
Director, MRC Epidemiology and Medical Care Unit, Northwick
Park Hospital, Watford Road, Harrow, Middlesex HA1 3UJ, UK

R.H. Mohiaddin, MD, MSc
Senior Research Fellow, Magnetic Resonance Unit, Royal Brompton
National Heart and Lung Hospital, Sydney Street, London SW3 6NP,
UK

M.F. Oliver, MD, FRCP, FRSE
Director, Wynn Institute for Metabolic Research, 21 Wellington
Road, St. John's Wood, London NW8 9SQ, UK

J.T. Powell, MD, PhD
Senior Lecturer, Department of Surgery, Charing Cross and West-
minster Medical School, St. Dunstan's Road, London, W6 8RF,
UK

C.V. Ruckley, ChM, FRCSEd
Reader in Surgery, University of Edinburgh Department of Surgery,
The Royal Infirmary, Edinburgh EH3 9YW, UK

D.J.A. Scott MD, FRCS
Senior Registrar in Surgery, Department of Surgery, Bristol Royal
Infirmary, Bristol BS2 8HW, UK

A.G. Shaper, FRCP, FRCPath, FFPHM
Professor of Clinical Epidemiology, Department of Public Health and
Primary Care, Royal Free Hospital School of Medicine, Rowland Hill
Street, London, NW3 2PF, UK

M.J. Shipley, BA, MSc
Lecturer in Medical Statistics, Department of Epidemiology and
Population Sciences, London School of Hygiene and Tropical Medi-
cine, Keppel Street, London WC1E 7HT, UK

F.B. Smith, HND
Research Associate, Wolfson Unit for Prevention of Peripheral
Vascular Diseases, University of Edinburgh, Teviot Place, Edinburgh
EH8 9AG, UK

W.C.S. Smith, MD, MPH, FFPHM
Leprosy Mission, Katong, PO Box 149, Singapore 9143 (Formerly
Deputy Director, Cardiovascular Epidemiology Unit, Dundee,
UK)

H.E.J.H. Stoffers, MD
General Practitioner/Investigator, Department of General Practice,
University of Limburg, P.O. Box 616, NL–6200 MD, Maastricht,
Netherlands

M. Thomson, MPhil, SRD
Chief Dietitian, Department of Nutrition and Dietetics, Eastern
General Hospital, Seafield Street, Edinburgh, EH8 9AG, UK

H. Tunstall-Pedoe, MA, MD, FRCP, FFPHM
Professor and Director, Cardiovascular Epidemiology Unit, Nine-
wells Hospital and Medical School, Dundee DD1 9SY, UK

S.R. Underwood, MA, MRCP
Senior Lecturer in Cardiac Imaging, Royal Brompton National Heart
and Lung Hospital, Sydney Street, London SW3 6NP, UK

M.K. Walker, MA, RGN, RM
Research Administrator, Department of Public Health and Primary
Care, Royal Free Hospital School of Medicine, Rowland Hill Street,
London, NW3 2PF, UK

S.G. Wannamethee, MSc
Research Statistician, Department of Public Heath and Primary Care,
Royal Free Hospital School of Medicine, Rowland Hill Street,
London NW3 2PF, UK

M.R. Whyman, MBBS, FRCS
Research Fellow, Wolfson Unit for Prevention of Peripheral Vascular
Diseases, University of Edinburgh, Teviot Place, Edinburgh, EH8
9AG, UK

L.K. Widmer, MD, PhD
Professor, Division of Angiology, Department of Medicine, University, CH 4031, Basle, Switzerland

M. Woodward, PhD
Department of Applied Statistics, The University, PO Box 217,
Whiteknights, Reading RG6 2AN, UK
(Formerly Senior Statistician, Cardiovascular Epidemiology Unit,
Dundee, UK)

SECTION I

Measurement in Communities

1 Review of Simple Measuring Techniques

F.G.R. FOWKES

In measuring disease in epidemiological surveys, the techniques usually need to be quick, easy, safe and inexpensive and the equipment must often be mobile. Thus diagnostic techniques used in clinical practice are not always appropriate for use in the field. Often in epidemiological surveys, it is only possible to use one or two tests to measure disease, and so we must be sure that these tests are reasonably valid in correctly classifying individuals as diseased or healthy. Also, the test may be applied only once to each individual and so we must be sure that the results are reasonably reliable and would not differ greatly if taken by another observer or on another day. The purpose of this chapter is to provide an overview of those techniques which might be used in epidemiological surveys to measure peripheral vascular disease and, in particular, to assess their validity and reliability.

Definition of Atherosclerotic Peripheral Vascular Disease

In determining the validity of a diagnostic test, the disease which the test is attempting to measure requires definition. Although the term "peripheral vascular disease" is sometimes used to refer to disease affecting arteries other than the coronary arteries, in clinical practice the term is often restricted, as in this review, to disease affecting arteries in the upper limb and in the lower limb distal to the aortic bifurcation. But since atherosclerosis rarely affects arteries in the upper limb, or the smaller arteries in the lower limb, the diagnosis of atherosclerotic peripheral vascular disease in practice is limited to measurement of disease in the major arteries of the lower limb.

The term "atherosclerotic peripheral vascular disease" implies that the definition of disease be based on pathological findings. But in Western countries almost every adult has pathological evidence of atherosclerosis affecting their peripheral arteries. A post-mortem study in Oxford, for example, found that in 15% of men and 5% of women more than half the diameter of the lumen of a major peripheral artery was occluded (Mitchell and Schwartz 1965). Since almost every adult has atherosclerosis to varying degrees, it is difficult to classify individuals meaningfully as simply diseased or non-diseased on the basis of pathological findings.

A definition of disease based on the degree of atherosclerosis is more appropriate. Angiography has traditionally been used as the best method of assessing this in living subjects (Thiele and Strandness 1983), but the results of one study examining inter-observer variability indicates that angiography is far from being a perfect arbiter of disease (Bruins Slot et al. 1981). Twenty-one translumbar aortograms of patients with intermittent claudication were presented to 11 observers who were asked to categorise each of 27 segments of the arterial tree as: "normal"; "plaques without narrowing"; "stenosis less then 50%"; "stenosis greater than 50%"; "total occlusion"; or "no judgement possible". Overall the inter-observer agreement on the degree of atherosclerosis was very poor, (except for "total occlusion") and for some arterial segments was little better than would occur by chance.

Thus, there are considerable difficulties in defining precisely a universal gold standard of disease against which a non-invasive measurement technique may be assessed for epidemiological purposes. Judgements of validity have to be based on comparisons with angiography and other non-invasive measurement techniques. Unfortunately, such studies have been conducted mostly on selected groups of patients who were not representative of the population as a whole.

Questionnaire for Intermittent Claudication

The questionnaire used commonly for the assessment of symptoms in cardiovascular surveys was developed at the London School of Hygiene and Tropical Medicine (Rose 1962). In comparison with doctors' assessment of symptoms the questionnaire was shown in one survey to be highly specific (99.8%) in excluding healthy individuals but only moderately sensitive (67.5%) in detecting those with claudication (Richard et al. 1972). Repeatability is good (Reunanen et al. 1982), and the questionnaire has been adapted for self-administration (Van Ganse et al. 1972; Rose et al. 1977).

The severity of intermittent claudication can be graded using the questionnaire: Grade 1, symptoms occur while walking "at an ordinary pace on the level"; and Grade 2, symptoms occur while walking "uphill" or in a "hurry". Grades of severity could also be assessed by asking about the walking distance before onset of symptoms but the variability between such repeat assessments is likely to be considerable. The London School of Hygiene questionnaire can also be used to identify "possible" intermittent claudication by including subjects who do not have all the necessary criteria but who have calf pain on exercise and not at rest (Criqui et al. 1985a). However, it is not certain whether these individuals have less severe disease than those with the full criteria for intermittent claudication.

The degree of claudication can also be assessed by means of a standard exercise test in which patients walk on a treadmill at a fixed speed and continue until the onset of calf pain. But, if such a test is repeated after 1 week, patients will be able to walk for an average of more than 20 seconds longer than on the first occasion

(Clyne et al. 1979). When the test is repeated on several occasions the reproducibility is extremely poor (Ouriel et al. 1982). Variability in the time until the patient has to stop walking is even greater than the time until onset of claudication (Hillestadt 1963).

Although the intermittent claudication questionnaire has been used widely in population surveys, the validity of intermittent claudication as a measure of underlying atherosclerotic peripheral vascular disease is doubtful. The sensitivity in detecting subjects with a disruption to arterial blood flow assessed by non-invasive techniques is less than 20% (Schroll and Munck 1981; Criqui et al. 1985a), but is more than 50% in detecting severe arterial occlusion demonstrated by angiography (Widmer et al. 1964). On the other hand, the questionnaire is highly specific – over 98% of healthy individuals are correctly identified (Widmer et al. 1964; Isacsson 1972; Schroll and Munck 1981; Criqui et al. 1985a). False positives may be due to musculoskeletal disease and venous disorders. False negatives may be due to asymptomatic disease, a sedentary existence, or other physical conditions limiting exercise and preventing the onset of symptoms. Errors may also occur due to difficulties in understanding the questionnaire. In Edinburgh, we have attempted to overcome these problems by developing a new questionnaire, the "Edinburgh Claudication Questionnaire", described in Chapter 3.

Palpation of Peripheral Pulses

In around 20% of the adult population, at least one pulse is not detected when the femoral, posterior tibial and dorsalis pedis arteries are palpated (Schroll and Munck 1981; Criqui et al. 1985a). This is due not only to arterial disease but to obesity and other factors making detection difficult. Also, the dorsalis pedis artery is congenitally absent in between 4%–12% of the population; the posterior tibial artery on the other hand is invariably present (Barnhorst and Barner 1968; Ison 1968).

When evaluated against other measures of peripheral vascular disease, such as ankle-brachial systolic ratios, palpation of the peripheral pulses would appear to be more sensitive but less specific than the intermittent claudication question-naire in detecting disease. In one population-based study, an absent pulse in one or more arteries was found to be 48% sensitive and 96% specific (Schroll and Munck 1981) and, in another study, diminished or absent pulses were 77% sensitive and 86% specific (Criqui et al. 1985a).

Variability in the detection of peripheral pulses was found to be similar in two studies conducted on hospital patients (Ludbrook et al. 1962; Meade et al. 1968). Overall agreement among three observers on the presence or absence of 384 doralis pedis and posterior tibial pulses was 74% (Meade et al. 1968) and the probability that if one observer could not detect a pulse, the other two would agree was about 0.5 which was higher than would be expected by chance (Ludbrook et al. 1962; Meade et al. 1968). During repeat examinations, around 80% of observations were in agreement with those made on the first examinations (Meade et al. 1968). However, the authors of these two studies reached different conclusions. Meade et al. (1968) found that the agreement between observers improved during the study and reached "satisfactory levels", while Ludbrook et al. (1962) thought that there was a "very great observer error" which did not diminish during the study.

For the purposes of most epidemiological research, palpation of the peripheral pulses is probably too insensitive a measure of peripheral vascular disease. Its usefulness is also limited because measurement of a continuous variable with graded reponses is often preferred to a discrete variable which is simply present or absent.

Ankle-Brachial Systolic Pressure Ratio

The ankle systolic pressure can be measured by a cuff inflated proximally to the ankle and the return of blood flow detected by means of a Doppler probe placed over the posterior tibial or dorsalis pedis artery. This technique for measuring systolic pressure has been shown to correlate closely with direct intra-arterial recordings (Stegall et al. 1968; Kazamias et al. 1971). A Doppler probe is simpler to use than other techniques for detecting blood flow such as plethysmography using a mercury strain gauge (Strandness and Bell 1965; Neilsen et al. 1973).

The results of studies based on small numbers of subjects have suggested that an ankle-brachial systolic pressure ratio of 0.9 is up to 95% sensitive in detecting angiogram positive disease and a ratio of 0.9 or higher is almost 100% specific in identifying supposedly healthy subjects (Winsor 1950; Carter 1968; Yao 1973; Hummel et al. 1978; Bernstein and Fronek 1982; Laing and Greenhalgh 1983). Also, the magnitude of this ratio (the ankle-brachial pressure index) would appear to correlate with the severity of disease (Chamberlain et al. 1975; Bernstein and Fronek 1982). However, these studies have mostly used selected hospital patients who were symptomatic, and controls who were judged to be healthy because they were young, or had no symptoms nor signs. Angiogram negative limbs of subjects with vascular disease have also been used as controls. It is not known how ratios in the general population would correlate with the extent of peripheral arterial disease measured by angiography, nor with angiogram negative disease. However, it is likely that the results among those with severe disease would be similar to those found in hospital patients. In the Edinburgh Artery Study, we have found an association in the general population between the ankle-brachial pressure index and the severity of disease detected on duplex scanning (Table 1.1).

The variability in measuring ankle pressure has been examined in a few studies (Mahler et al. 1976; Schroll and Munck 1981; Baker and Dix 1981; Ouriel et al. 1982; Fowkes et al. 1988a). In one epidemiological study (Schroll and Munck 1981), 10 measurements were repeated on 10 subjects and found to vary in a similar way to measurements of arm systolic pressure (although the results were not reported in detail). In another study "good reproducibility" was obtained when two successive measures of ankle systolic pressure were made on 11 subjects (Mahler et al. 1976). Baker and Dix (1981) found that repeated measurements on 25 subjects produced an average range of 0.18 in ankle-brachial

Table 1.1. Duplex scanning results according to ankle brachial pressure index (ABPI) in sub-samples of subjects in Edinburgh Artery Study

ABPI	n	Percentage of subjects ($n=164$)			
		(a) Image D or E	(b) Waveform 4 or 5	(c) Peak systolic velocity <30% or >250%	Any of (a) (b) or (c)
≥1.1	52	6	28	8	33
0.9–1.09	49	0	22	27	39
0.7–0.89	26	54	77	58	85
<0.7	37	62	78	62	84

N.B. Duplex scanning measurements are based on the worst observed in any arterial segment between the inguinal ligament and the knee in the right or left leg and are indicative of approximately 50% stenosis or greater. ABPI is based on the lowest measurement of the left or right ankle. Source: Fowkes et al. (unpublished observations).

pressure indices (ABPI). Carter (1968) found that 95% of repeat ankle ABPIs were within 9% of the average. This variability in both ankle systolic pressures and ABPI is similar to that obtained when measuring arm systolic pressure.

The variability due to taking measurements on different occasions by different observers has only a marginal effect on between-subject variability suggesting that ankle systolic pressures may be a suitable measurement for use in some epidemiological studies (Fowkes et al. 1988a). Table 1.2, for example, shows results taken from our study in Edinburgh in which repeated ankle systolic measurements were taken on 24 patients with peripheral vascular disease by four different observers on two different days (Fowkes et al. 1988a). As expected, the greatest variability was between subjects and between a subject's left and right legs because disease does not usually affect each leg equally. Observer bias was minimal, and this may have been due partly to the use of a random zero sphygmomanometer Fig 1.1. Despite persistent controversy about the accuracy

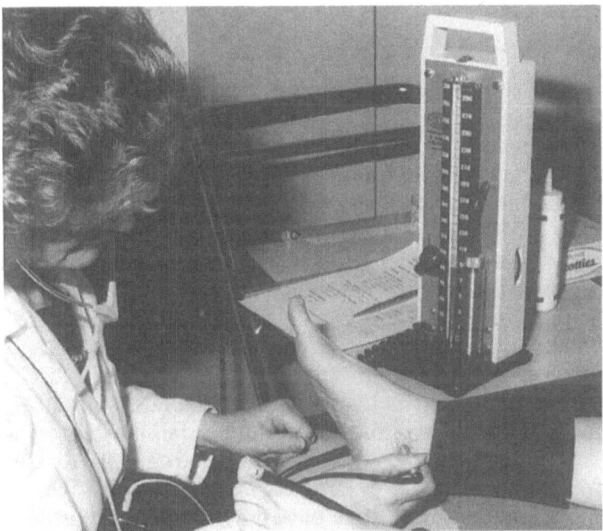

Fig. 1.1. Hawksley random zero sphygmomanometer used with a doppler probe to detect ankle systolic pressure.

Table 1.2. Components of variance (standard errors) of measurement of ankle systolic pressure and ankle brachial ratio in subjects with peripheral vascular disease

Source of variation	Ankle systolic pressure (n=24)	Ankle-brachial ratio (n=24)
Between subjects	608 (286)	259 (127)
Systematic between left and right legs	64 (125)	39 (70)
Between a subject's left and right legs	525 (169)	252 (80)
Subjects on different days	16 (26)	<10
Day-to-day differences between left and right legs	62 (27)	22 (9)
Observer bias	<10	<10
Interaction between subjects and observers	21 (12)	19 (7)
Differences in way observers measured each leg	15 (7)	16 (5)
Repeat readings	23 (2)	17 (2)

Source: Adapted from Fowkes et al. (1988a) with permission.

of this instrument, automated blood pressure measuring devices also have their problems, and the random zero sphygmomanometer is probably still the method of choice for epidemiological studies.

Stress Tests

In some patients known to have significant peripheral vascular disease, blood flows and pressures may be normal at rest but may become abnormal after exercise or an equivalent stress such as that caused by obstructing blood flow (reactive hyperaemia test).

Exercise

When normal subjects undergo moderate exercise the ankle systolic pressure will remain the same or be slightly elevated (Strandness and Bell 1964; Yao 1970; Laing and Greenhalgh 1983), but may fall after severe exercise (Stahler and Strandness 1967). In patients with peripheral vascular disease demonstrated by angiography, a standard amount of exercise will produce a significantly greater fall in ankle pressure and longer recovery time than in normal subjects (Carter 1972; Thulesius 1978; Osmundson et al. 1981; Laing and Greenhalgh 1983). Furthermore the reduction in systolic ankle pressure is related to the severity of disease determined by angiography (Strandness and Sumner 1969). The exercise is usually performed on a treadmill and it has been shown that 1 minute of exercise at 4 km/hour on a 1 in 10 slope will produce a similar fall in pressure to higher work loads (Laing and Greenhalgh 1980). A standardised exercise test has been shown in one study on a small number of subjects to be 97% sensitive and 96% specific in differentiating angiogram positive patients from healthy subjects (Ouriel et al. 1982). In the same study, repeated measurements on 10 subjects on five occasions by one observer indicated that 95% of observations of the post-exercise ankle-brachial pressure index would occur within 15.6% of the mean value.

It is not known the extent to which an exercise test when used in an epidemiological study in the general population will improve the detection of peripheral vascular disease. Nearly all the research has been conducted on hospital patients with severe disease. However, Carter (1972) conducted an exercise test on 18 patients with "questionable" peripheral vascular disease and found that 11 patients had a fall in pressure equivalent to patients with known mild peripheral vascular disease suggesting that the exercise test was detecting disease in individuals who on other medical grounds might be classified as normal.

Reactive Hyperaemia

In the reactive hyperaemia test, blood flow is occluded at the upper thigh or above the knee, and the ankle systolic pressure measured on releasing the obstruction. The systolic pressure in patients with peripheral vascular disease is reduced two or three times further and takes longer to return to pre-occlusion levels than in healthy subjects (Johnson 1975; Baker 1978; Hummel et al. 1978; Verhagen et al. 1983). An occlusion time of 3 minutes has been shown to produce similar effects to that occuring after longer occlusion times (Gunderson 1973; Dedichen and Myre 1975). Measuring the blood pressure at between 15 and 30 seconds after the cuff has been released would appear to be the optimum time to distinguish between subjects with angiogram positive disease and controls (Johnson 1975; Hummel et al. 1978).

Occlusion of blood flow in the limbs for a short duration of time is not known to be hazardous although may be undesirable in patients with severe peripheral vascular disease where compression could conceivably cause damage to the arterial wall. During some studies in which blood flow was occluded for between 5 and 7 minutes in the thigh, some patients experienced discomfort underneath the cuff but this was not sufficient to discontinue the compression (Johnson 1975; Baker 1978). If occlusion takes place above the knee as opposed to the thigh, and the occlusion is maintained for only 3 to 4 minutes, the discomfort is minimal (Hummel et al. 1978).

There have been few reports in the literature of the variability of the reactive hyperaemia test. In one study, the reproducibility of the ankle-brachial pressure index after occlusion when measured in 10 patients by one observer was similar to that after exercise and marginally greater than the variation at rest (Ouriel et al. 1982). When 36 subjects had repeated measurements taken by four observers on two visits in a study we conducted in Edinburgh, the variability was slightly higher than for resting ankle-brachial pressure indices taken prior to occlusion (Fowkes et al. 1988b). Table 1.3 shows how the 95% confidence intervals are reduced when a mean of several readings is taken under different circumstances. Also, the confidence interval of the difference between two readings indicates the variability that can occur without a change in the subject's clinical condition; or conversely, a difference between two clinical readings greater than that stated in the table indicates that the subject's clinical condition is likely to have changed.

Clinicians have traditionally used exercise as a means of stressing the peripheral arterial system. But during recent years, the reactive hyperaemia test

Table 1.3. Ninety-five per cent confidence intervals of reduction in ankle systolic pressures and post-occlusion ankle-brachial ratio

Mean of	Pre-minus post-occlusion ankle systolic pressure (+ mmHg)		Post-occlusion ankle-brachial ratio (+ %)	
	Normal	Arterial disease	Normal	Arterial disease
One observer, one measurement	22.9	28.6	20.2	19.2
One observer, two repeat measurements	20.2	26.4	17.7	18.1
Two observers, one measurement each (same day)	20.0	21.6	16.4	15.0
Two observers, two measurements each (all on same day)	18.5	20.2	14.9	14.3
Two observers, two measurements each (one on one day and another on a different day)	15.3	18.9	12.8	13.1
Differences between two occasions:				
Same observer	31.9	26.9	25.2	18.1
Different observers	32.4	40.4	28.6	27.1

Source: Fowkes et al. (1988b) reprinted with permission.

has been advocated because it is simpler, cheaper and quicker to perform (Johnson 1975; Baker 1978; Hummel et al. 1978; Bernstein and Fronek 1982) and the results obtained show good correlation with exercise tests (Mahler et al. 1976; Fox et al. 1977; Baker 1978; Hummel et al. 1978; Keagy et al. 1981). Furthermore, the test does not put cardiac patients at risk of developing ischaemia or arrhythmias and is less prone to subjects affecting the degree of stress applied, particularly if they are old and have orthopaedic or neurological abnormalities. Another advantage of the reactive hyperaemia test is that the stress is applied when the patient is lying down and does not require any movement of the patient after the cessation of the stress with the possibility that the ankle blood pressure will be measured too late to detect the optimum difference from the resting state (Baker 1978). For these reasons, the reactive hyperaemia test might be the most suitable stress test to apply in large scale epidemiological investigations.

However, as with the exercise test, the extent to which the reactive hyperaemia test improves diagnostic sensitivity in the general population is not clearly known. In a study on hospital patients comparing exercise and reactive hyperaemia tests, the exercise test had marginally better sensitivity, but the patients in each group were not comparable (Ouriel et al. 1982). When a reactive hyperaemia test was used in combination with a test of pulse reappearance time (see below) on a non-hospital population of over 600 subjects, an additional 16% of subjects (who had normal ankle pressures at rest) were thought to have disease (Criqui et al. 1985). The hyperaemic response, however, was measured by means of the Doppler flow velocity in the femoral artery and not pressure measurements at the ankle.

Toe Pulse Reappearance Time

A test which has not been used widely in clinical practice, but has the potential for use in epidemiological studies, is the toe pulse reappearance time. In a similar way to the reactive hyperaemia test, blood flow is occluded in the leg and, on releasing the occlusion, the time taken for pulsation to return in the toe is measured using a strain gauge or photo-plethysmograph (Guttierrez et al. 1981). In normal individuals, the pulse will reappear in less than 1 second, but in patients with known peripheral arterial disease the return of pulsation will take longer – up to 2 minutes in one study (Fronek et al. 1977). The time taken for the pulse volume to return to 50% of the pre-occlusion amplitude (PRT/2) may be prone to less measurement error and may be better at identifying diseased subjects than the pulse reapperance time (Fronek et al. 1977). The PRT/2 has been shown to be of value in predicting the outcome of aortofemoral bypass surgery (Bernstein et al. 1981). Since the toe pulse reappearance time can be measured easily using fairly inexpensive equipment, it may be suitable for use in epidemiological surveys and can be conducted simultaneously with a reactive hyperaemia test (Criqui et al. 1985b). But as with the exercise and hyperaemia tests, it is not known what the distribution of results would be in individuals with mild to moderate disease.

Conclusion: Measurement in Epidemiological Research

During recent years, many sophisticated non-invasive tests have been developed for the assessment of peripheral vascular disease in clinical practice (Atkinson and Woodcock 1982; Bernstein and Fronek 1982; Budinger 1983). These include the use of Doppler wave form analysis (Chapter 2), Duplex ultrasound scanning (Chapter 4) and nuclear magnetic resonance (Chapter 5). Duplex scanning offers considerable potential because of the ability to visualise peripheral vessels by means of pulse echo imaging and to simultaneously measure blood flow by means of Doppler waveform analysis (Atkinson and Woodcock 1982). This has been used with some success in the assessment of carotid artery stenosis (Wetzner et al. 1984), but has proved difficult in assessing disease of the lower limbs because of the more diffuse nature of the atherosclerotic lesions. The epidemiologist should be aware, however, that such techniques currently require cumbersome and expensive equipment, are difficult and time consuming to use, are of unknown reliability and do not offer substantial advantages over the simpler non-invasive tests described above. But, with further innovation and evaluation, these more sophisticated tests may well prove to be useful in the not too distant future in large epidemiological studies. Duplex scanning of the popliteal artery, for example, is being used in the Atherosclerosis Risks in Communities Study in the United States (ARIC Investigators 1989) and was used in a subsample of subjects participating in the Edinburgh Artery Study (Fowkes et al. 1991).

Important characteristics of the simpler measurement tools described in this chapter are summarised in Table 1.4. The WHO intermittent claudication

Table 1.4. Non-invasive instruments for measuring peripheral vascular disease in epidemiological surveys

	Sensitivity[a]	Variability	Advantages	Disadvantages
WHO intermittent claudication questionnaire	50%	80%–90% repeatable	Self-administration feasible	Only symptomatic disease detected
Palpation of peripheral pulses	Not reported	70%–80% repeatable	Simple to perform	Discrete variable – could not relate to grade of diseases
Ankle-brachial pressure ratio	90+%	95% confidence interval of one reading = ±16%	Detect asymptomatic disease	Doppler probe required
Treadmill exercise test	95+%	95% confidence interval of one reading = ±22%	Detect asymptomatic disease with normal resting pressures disease	Time consuming. Expensive heavy equipment. Hazardous if co-existing coronary
Reactive hyperaemia test	95+%	95% confidence interval of one reading = ±20%	Detect asymptomatic disease with normal resting pressures. Simple routine equipment	Time consuming. Some discomfort
Toe pulse reappearance time	95+%	Not reported	Detect wide range of asymptomatic disease	Time consuming. Non-atherosclerotic disease of distal arteries gives abnormal results. Strain gauge or photo-plethysmograph required

[a] Sensitivity in detecting severe grades of disease (angiogram positive). Source: Fowkes (1988) reprinted with permission.

questionnaire lacks sensitivity, but is highly specific; palpation of pulses is more sensitive, but less specific. Care must be taken in the use and interpretation of the ankle-branchial pressure ratio in epidemiological studies because of doubts about validity in the general population. In population-based case–control studies, some masking of the importance of risk factors may occur due to overlap between test results and the true extent of disease. On the other hand, in longitudinal studies examining the value of the test in identifying groups at risk of future cardiovascular events, doubts about validity are less important as it is the value of the test and not peripheral vascular disease that is being studied.

The advantage of a stress test, such as the exercise or reactive hyperaemia tests, is that the sensitivities associated with an ankle-brachial pressure ratio at rest will be increased further. This may be important in epidemiological field surveys in which a considerable proportion of the population may have mild to moderate asymptomatic disease which would not be detected by other means. The variabilities of both these tests are comparable to that for ankle-brachial pressure ratios at rest, but the reactive hyperaemia test may offer some advantage over the treadmill exercise test in being easier and cheaper to perform. The variability of ankle pressure measurements at rest or during stress tests is adequate for most research purposes.

This review of measurement techniques would indicate that, in addition to questionnaires on intermittent claudication, non-invasive tests should be used in epidemiological surveys. The development of a standard investigative protocol of questionnaire, measurement of the ankle-brachial ratio, and a stress test, would allow better comparisons to be made between populations (Prineas et al. 1982).

References

ARIC Investigators (1989) The Atherosclerosis Risk in Communities (ARIC) Study: design and objectives. Am J Epidemiol 129:687–702

Atkinson P, Woodcock J (1982) Doppler ultrasound and its use in clinical measurement. Academic Press, London

Baker JD (1978) Post stress Doppler ankle pressures. Arch Surg 113:1171–3

Baker JD, Dix D (1981) Variability of Doppler ankle pressures with arterial occlusive disease: an evaluation of ankle index and brachial-ankle pressure gradient. Surgery 89:134–7

Barnhorst DA, Barner HB (1968) Prevalence of congenitally absent pedal pulses. N Engl J Med 278:264–5

Bernstein EF, Fronek A (1982) Current status of non-invasive tests in the diagnosis of peripheral arterial disease. Surg Clin North Am 62:475–87

Bernstein EF, Rhodes GA, Stuart SH, Coel MN, Fronek A (1981) Toe pulse re-appearance time in prediction of aorto-femoral bypass success. Ann Surg 20:1–54

Bruins Slot H, Strijbosch L, Greep JM (1981) Inter-observer variability in single plane aortography. Surgery 90:497–503

Budinger TF (1983) Overview of imaging technologies: present and future trends. Atherosclerosis Rev 10:7–12

Carter SA (1968) Indirect systolic pressures and pulse waves in arterial occlusive disease of the lower extremities. Circulation 37:624–38

Carter SA (1972) Response of ankle systolic pressure to leg exercise in mild or questionable arterial disease. N Engl J Med 287:578–82

Chamberlain J, Housley E, Macpherson AIS (1975) The relationship between ultrasound assessment and angiography in occlusive arterial disease of the lower limb. Br J Surg 62:64–7

Clyne CAC, Tripolitis A, Jamieson CW et al. (1979) The reproducibility of the treadmill walking test for claudication. Surg Gynecol Obstet 149:727–8

Criqui MH, Fronek A, Klauber MR et al. (1985a) The sensitivity, specificity and predictive value of traditional clinical evaluation of peripheral arterial disease: results from non-invasive testing in a defined population. Circulation 71:516–22

Criqui MH, Fronek A, Barrett-Connor E et al. (1985b) The prevalence of peripheral arterial disease in a defined population. Circulation 71:510–15

Dedichen H, Myre HO (1975) Reactive hyperaemia of the human lower limb. Acta Chir Scand 141:517–21

Fowkes FGR (1988) The measurement of atherosclerotic peripheral arterial disease in epidemiological surveys. Int J Epidemiol 17:248–54

Fowkes FGR, Housley E, Macintyre CCA, Prescott RJ, Ruckley CV (1988a) Variability of ankle and brachial systolic pressures in the measurement of atherosclerotic peripheral arterial disease. J Epidemiol Comm Health 42:128–33

Fowkes FGR, Housley E, Macintyre CCA, Prescott RJ, Ruckley CV (1988b) Reproducibility of reactive hyperaemia test in the measurement of peripheral arterial disease. Br J Surg 75:743–6

Fowkes FGR, Housley E, Cawood EHH et al. (1991) Edinburgh Artery Study: prevalence of asymptomatic and symptomatic peripheral arterial disease in the general population. Int J Epidemiol 20:384–92

Fox MJ, Tripolitis A, Kirby S, Jamieson CW (1977) A comparison of reactive hyperaemia test with a standard exercise test in the evaluation of peripheral arterial disease. Br J Surg 64:290 (abstract)

Fronek A, Coel M, Bernstein EF (1977) The pulse reappearance time – an index of overall blood flow impairment in the ischaemic extremity. Surgery 81:376–81

Gunderson J (1973) Standardised multisegmental measurements of blood pressure for quantitative evaluation of the circulation in the limbs. Scand Clin Lab Invest 31 (Suppl 128):111–15

Guttierrez JZ, Gage AA, Makula PA (1981) Toe pulse reappearance study in ischaemic arterial disease of the legs. Surg Gynecol Obstet 153:889

Hillestadt LK (1963) The peripheral blood flow in intermittent claudication. The significance of the claudication distance. Acta Med Scand 173:467

Hummel BW, Hummel BA, Mowbry A et al. (1978) Reactive hyperemia vs. treadmill exercise testing in arterial disease. Arch Surg 113:95–8

Isacsson S (1972) Venous occlusion plethysmography in 55-year-old men: a population study in Malmo, Sweden. Acta Med Scand 537 (Suppl):1–62

Ison JW (1968) Palpation of dorsalis pedis pulse. JAMA 206:2745

Johnson WC (1975) Doppler ankle pressure and reactive hyperaemia in the diagnosis of arterial insufficiency. J Surg Res 18:177–80

Kazamias TM, Gander MP, Franklin DL et al. (1971) Blood pressure measurement with Doppler ultrasonic flowmeter. J Appl Physiol 30:585–8

Keagy BA, Pharr WF, Thomas D et al. (1981) Comparison of reactive hyperaemia and treadmill tests in the evaluation of peripheral vascular disease. Am J Surg 142:158–61

Laing SP, Greenhalgh RM (1980) Standard exercise test to assess peripheral arterial disease. Br Med J 280:13–16

Laing S, Greenhalgh RM (1983) The detection and progression of asymptomatic peripheral arterial disease. Br J Surg 70:628–30

Ludbrook J, Clarke AM, McKenzie JK (1962) Significance of absent ankle pulse. Br Med J i:1724–6

Mahler F, Koen L, Jahnasen KH et al. (1976) Post-occlusion and post-exercise flow velocity and ankle pressures in normal and marathon runners. Angiology 72:721–9

Meade TW, Gardner MJ, Cannon P et al. (1968) Observer variability in reading the peripheral pulses. Br Heart J 30:661–5

Mitchell JRA, Schwartz CJ (1965) Arterial disease. Blackwell, Oxford

Neilsen PE, Bell G, Lassen NA (1973) Strain gauge studies of distal blood pressure in normal subjects and in patients with peripheral arterial disease in analysis of normal variation and reproducibility and comparison to intra-arterial measurements. Scand J Clin Lab Invest 31 (Suppl 128):103–9

Osmundson PJ, Chesebro JH, O'Fallon WM et al. (1981) A prospective study of peripheral occlusive arterial disease in diabetes. II. Vascular laboratory assessment. Mayo Clin Proc 56:223–32

Ouriel K, McDonnell AE, Metz CE et al. (1982) A critical evaluation of stress testing in the diagnosis of peripheral vascular disease. Surgery 91:686–93

Prineas RJ, Harland WR, Janzon L et al. (1982) American Heart Association Council on Epidemiology. Recommendations for use of non-invasive methods to detect atherosclerotic peripheral arterial disease in population studies. Circulation 65:1561A–6A

Reunanen A, Takkunen H, Aromaa A (1982) Prevalence of intermittent claudication and its effect on mortality. Acta Med Scand 211:249–56

Richard JL, Ducimétière P, Elgrishi I et al. (1972) Dépistage par questionnaire de l'insuffisance coronarienne et de la claudication intermittente. Rev Epidemiol Med Soc Sante Publ 20:735–55

Rose GA (1962) The diagnosis of ischaemic heart pain and intermittent claudication in field surveys. Bull WHO 27:645–58

Rose G, McCartney P, Reid DD et al. (1977) Self administration of a questionnaire on chest pain and intermittent claudication. Br J Prev Soc Med 31:42–8

Schroll M, Munck O (1981) Estimation of peripheral arteriosclerotic disease by ankle blood pressure measurements in a population study of 60-year-old men and women. J Chron Dis 34:261–9

Stahler C, Strandness DE (1967) Ankle blood pressure response to graded treadmill exercise. Angiology 18:237–41

Stegall HF, Kardon MB, Kemmerer WT (1968) Indirect measurement of arterial blood pressure by Doppler ultrasound sphygmomanometry. J Appl Physiol 25:793–8

Strandness DE, Bell JW (1964) An evaluation of the haemodynamic response of the claudicating extremity to exercise. Surg Gynecol Obstet 119:1237–42

Strandness DE, Bell JW (1965) Peripheral vascular disease: diagnosis and objective evaluation using a mercury strain gauge. Am Surg 161 (Suppl):1–35

Strandness DE, Sumner DS (1969) The relationship between calf blood flow and ankle blood pressure in patients with intermittent claudication. Surgery 65:763–7

Thiele BL, Strandness DE (1983) Accuracy of angiographic quantification of peripheral atherosclerosis. Prog Cardiovasc Dis 26:223–35

Thulesius O (1978) Systemic and ankle blood pressure before and after exercise in patients with arterial insufficiency. Angiology 29:274–8

Van Ganse W, Van Hoorne N, De Backer G et al. (1972) L'interview et le questionnaire auto-administré de Rose dans une étude pilote d'atherosclerosie. Rev Epidemiol Med Soc Sante Publ 20:7

Verhagen PF, de Jong TJ, Van Vroonhoven Th J (1983) Ankle pressure changes during reactive hyperaemia in peripheral arterial disease. VASA 12:29–34

Wetzner SM, Kiser JC, Bezreh JS (1984) Doppler ultrasound imaging: vascular applications. Radiology 150:507–14

Widmer LK, Greensher A, Kannel WB (1964) Occlusion of peripheral arteries: a study of 6400 working subjects. Circulation 30:836–42

Winsor T (1950) Influence of arterial disease on the systolic blood pressure gradients of the extremity. Am J Med Sci 220:117–26

Yao ST (1970) Haemodynamic studies in peripheral arterial disease. Br J Surg 57:761–6

Yao ST (1973) New techniques of objective arterial evaluation. Arch Surg 106:600–4

2 Non-invasive Tests

M. HORROCKS and D.J.A. SCOTT

Clinical examination is reported to underestimate the prevalence and progression of arterial disease. Criqui et al. (1985) reported that 1.9% of patients between 38 and 82 years complained of claudication. By contrast non-invasive testing in the same population reported an 11.7% prevalence of large vessel arterial disease. Strandness and Stahler (1966) reported that over a 3-year period 52% of patients had documented disease progression on non-invasive testing. Of these only 40% reported a change in symptoms. In addition non-invasive tests may identify those patients likely to develop loss of limb and life. Sumner (1989) has reported that an ankle index of <0.3 is associated with an amputation rate of 32% and a mortality of 60% over a 6-year period. This contrasts sharply with a 13% amputation rate and 33% mortality rate in patients with ankle indices between 0.3 and 0.5.

The assessment of peripheral vascular disease in large scale community programmes can only be investigated by non-invasive, cheap and reproducible techniques. The aim of testing should be to identify the incidence, prevalence and high risk groups in the community and also to locate and quantify the severity of arterial disease. In those patients with multi-level disease, non-invasive testing may only identify the most haemodynamically significant lesion.

The purpose of adding a stress testing is to help identify significant arterial occlusive lesions which may have been missed under resting circumstances. It may also provide some degree of qualitative assessment of the lesion and allow the clinician the opportunity to observe the relevance of these lesions in relation to other pathological processes including orthopaedic, neurological and cardio-vascular disease.

The simple non-invasive tests which are used commonly in clinical practice to assess peripheral vascular disease are the resting ankle systolic pressure and resting ankle-brachial pressure ratio or index (ABPI). Ankle pressure measurement was first described by Yao et al. (1969) and is available in most vascular laboratories. The test has a low coefficient of variation (less than 10%) and changes of >0.15 are indicative of progression of atherosclerosis.

The ratio between the ankle and branchial systolic blood pressure gives an objective measure of the severity of peripheral vascular disease. The ankle systolic pressure may be affected by the central aortic pressure and can therefore be normalised by comparison with the brachial pressure thus producing a ratio of 1.0 at rest in normals (Yao et al. 1969). In general, patients with claudication have an ABPI of <0.95 and those with critical ischaemia of <0.4.

The ABPI is however subject to variation including change in sympathetic activity, length of rest before the study and the effect of smoking. In general a serial fall in the ABPI of >0.15 indicates progression of arterial disease (Mozersky et al. 1972a,b) and a rise in the development of a collateral circulation (Skinner and Strandness 1967a,b).

Exercise Testing

In normal individuals the post-exercise ankle-brachial ratio will rise with exercise and fall in patients with peripheral vascular disease. Simple treadmill tests have been used to grade the severity of disease (Fig. 2.1). Laing and Greenhalgh (1983) reported that of 100 limbs assessed, 78% had a normal resting pressure. After a 1-minute exercise test at 4 km/hr at 10° incline, 41% of the normal limbs had an abnormal pressure. This test is unfortunately not applicable to the majority of patients with claudication who are unable to complete the test (Wyatt et al. 1990). In addition it only provides information about the worst leg and may be influenced by the presence of orthopaedic and neurological diseases. Furthermore, patients with peripheral vascular disease undergoing treadmill tests are at risk of cardiac arrhythmias and even cardiac arrest. Alternative methods have been evaluated including occlusive calf hyperaemia (Baker 1978) and ankle flexion (Wyatt et al. 1990). These tests allow each leg to be assessed in turn and do not require cardiac monitoring.

Segmental Pressures

It is possible to assess the severity of arterial disease by indirect measurement of arterial pressures down the leg. Four pneumatic cuffs are applied to the leg; high

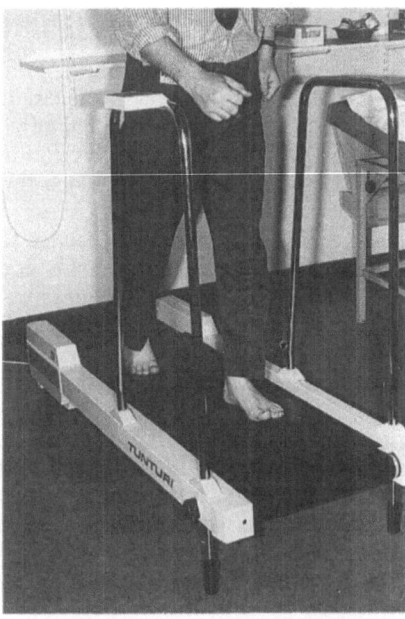

Fig. 2.1. Standard exercise test.

thigh, above knee, below knee and ankle. The ankle systolic pressure is recorded at different levels of inflation for each cuff and expressed as a ratio to the brachial systolic pressure. In general a thigh-brachial index of >1.2 excludes significant aorto-iliac disease, between 0.8 and 1.2 indicates the presence of stenotic disease and less than 0.8 total occlusion (Skinner and Strandness 1967b). False positives may occur when the superficial femoral artery is occluded distal to an iliac stenosis. Fronek et al. (1973) have reported that a vertical gradient of >40 mmHg or a horizontal difference of >20 mmHg can be used as an indicator of significant disease. By contrast Rutherford and colleagues (1979) have shown that segmental limb pressures are of limited value in the assessment of combined aorto-iliac and femoro-popliteal disease. Segmental pressures may be affected by many other factors including poor cardiac output, hypertension and incompressible blood vessels. A normal result provides functional rather than anatomical information about the leg. This method has not found favour as a screening technique because it is time consuming and has poor accuracy and reproducibility.

In assessing toe pressures, a small pneumatic cuff is placed around the proximal phalanx and a flow sensor, e.g. photoplethysmograph applied distally. In normal elderly patients the toe pressure is 9.8 ± 10.7 mmHg below that of the brachial pressure (Nielsen et al. 1972). In claudication the majority (89%) of patients will have a toe pressure of >30 mmHg. By contrast a toe pressure of less than 30 mmHg is a reliable index of critical ischaemia (Ramsey et al. 1983). This is a simple and reproducible technique which is suitable for widespread screening provided the equipment is available.

Doppler Ultrasonography

Satomura (1959) first described the use of a Doppler ultrasound probe to detect blood flow in arteries and suggested that it may be used in the detection of atherosclerosis. Over the last 30 years Doppler ultrasound has become the mainstay of non-invasive investigation of lower limb arterial disease. Ultrasound velocimeters are readily available and can cost as little as £300 (US $500) (Fig. 2.2). Superficial vessels are best insonated using a 10 MHz pencil probe and deeper vessels a 5 or 3 MHz probe. The output from a Doppler velocimeter can be analysed simply by listening to the signal or by a variety of complex mathematical techniques, e.g. Laplace transform damping. Simple audible interpretation of the Doppler signal is easy to perform but is operator dependent. Normal Doppler signals are biphasic or triphasic (Fig. 2.3). In diseased vessels the Doppler signal will vary according to its sample site. For example, at a stenosis there is a high-pitched monophasic signal suggestive of a jet; distal to the stenosis the signal is low-pitched and monophasic. By contrast Doppler signals obtained proximal to an occluded vessel have a "thumping" quality.

Qualitative Waveform Analysis

Audible interpretation can be limited and for this reason the waveform can be further analysed with respect to its contour and frequency spectrum. Although the latter has been shown to accurately predict the presence of haemodynamically

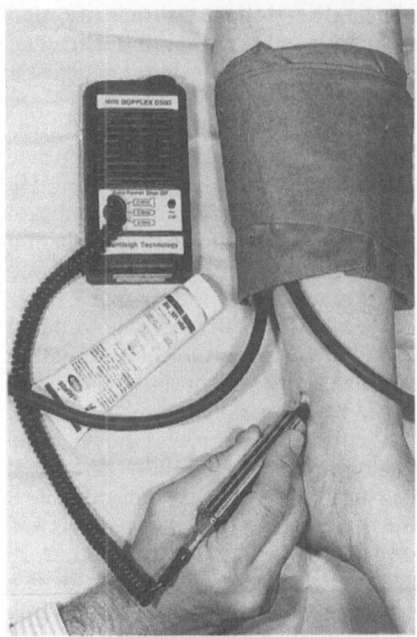

Fig. 2.2. Doppler velocimeter.

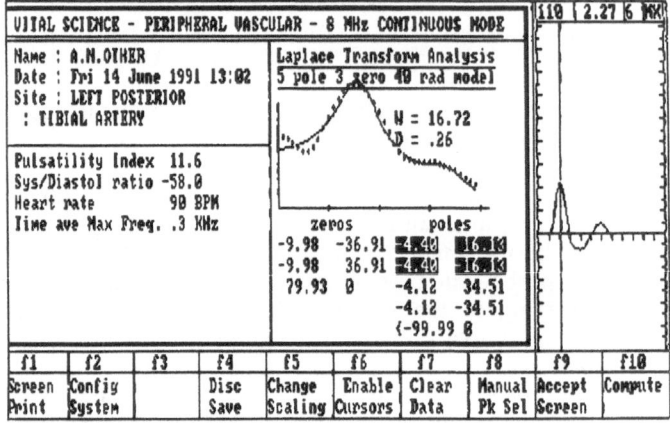

Fig. 2.3. Triphasic Doppler signal.

significant lesions, it does require a skilled technician and takes at least 1 hour to assess each leg.

Quantitative Waveform Analysis

Simple Doppler waveforms can be acquired and analysed using a number of mathematical techniques. These include pulsatility index (Gosling et al. 1971),

Laplace transform damping (Skidmore et al. 1980) and principal component analysis (Evans et al. 1981; Macpherson et al. 1984). Their role lies in the detection of significant aorto-iliac inflow disease. By contrast, there is little value in the assessment of severe distal disease as the waveforms are often severely damped and not suitable for analysis.

In theory, pulsatility indices should be relatively consistent as they are independent of heart rate and the angle of insonation of the Doppler probe. However, there are numerous reports highlighting the variability of the technique particularly in the presence of multilevel disease. In general, a femoral pulsatility index of >4.0 excludes significant iliac disease, but values less than 4.0 should be interpreted with caution (Thiele et al. 1983). The calculation of inverse damping factors (DF^{-1}) has been reported to improve the discrimination between iliac and superficial femoral artery disease (Johnston et al. 1978). A damping factor greater than 1 identifies most disease in the superficial femoral artery (Campbell et al. 1985).

Laplace Transform Damping

There are several studies to support the use of Laplace transform damping in the assessment of presymptomatic atherosclerosis. Campbell et al. (1985) reported that high values of Laplace δ and ω_0 were associated with a significantly higher risk of arterial disease. Theoretically Laplace δ should identify proximal luminal stenoses and ω_0 arterial wall elasticity but because of the inter- and intra-observer variability the technique has not gained favour and remains a research tool.

Principal Component Analysis

Principal component analysis is a mathematical technique where the waveforms are described in terms of derived coefficients. Thirty-two components have been described, but the majority of the information is held in the first two components. Macpherson et al. (1984) reported that principal component analysis was a more sensitive method of identifying proximal arterial disease than Laplace transform damping and pulsatility index. The disadvantage of this technique is that it does require a microcomputer to provide on-line analysis and has therefore not gained favour.

Velocity Measurements

These have been assessed in several clinical situations by Fronek et al. (1976). A number of measurements have been made including peak forward velocity, deceleration and peak velocity/mean velocity. Although there was a difference between normals and patients with arterial disease, a significant overlap occurred between the various clinical conditions which made the test unreliable and of little practical help. Similar findings were reported when velocity measurements were combined with a reactive hyperaemia test (Hirai and Schoop 1984).

Several other measurements have been utilised including pulse transit time and rise time (Humphries et al. 1980; Aukland and Hurlow 1982). Both are reported to identify patients with occlusions or severe stenoses but this information is often readily available from clinical examination.

Plethysmography

Pulse Volume Recorder (PVR)

This is a semi-quantitative segmental air plethysmograph which gives a reproducible non-invasive pressure wave trace (Kaufman et al. 1989) (Fig. 2.4). Air-filled cuffs can be applied to the thigh, calf, ankle and toe. The normal PVR trace is characterised by a rapid upstroke, a sharp systolic peak, a downslope and a dicrotic notch. In proximal disease there is a decreased upstroke time and flattening of the downslope with loss of the dicrotic notch. Amplitude and upstroke times can be calculated from a series of five waveforms. The pulse amplitude may vary and is dependent upon cardiac function, vasomotor tone and the size and position of the limb. Rutherford et al. (1979) reported that PVR correctly identified normal limbs in 97% of cases. In patients with isolated or combined disease affecting the aorto-iliac and femoro-popliteal segment PVR had a 70% accuracy. These figures can be improved if PVR is combined with segmental pressure measurements. Finally, Kaufman et al. (1989) reported a 35% mortality rate for patients with a flat PVR trace. In those patients with no symptoms of ischaemia and a flat PVR, 72.4% eventually required some form of reconstruction.

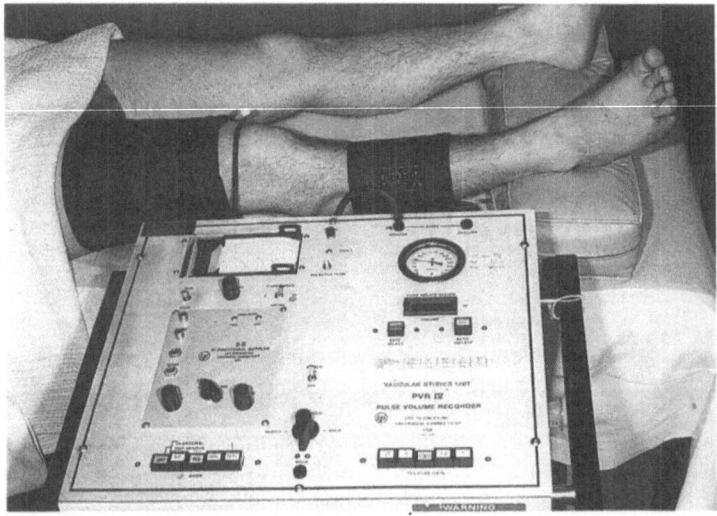

Fig. 2.4. Pulse volume recorder (PVR)

Digital Plethysmography

This form of plethysmography represents the state of the arterial tree between the heart and the arterioles. Three instruments have been devised to measure digital pulse volume: the air plethysmograph, mercury strain gauges and photoplethysmographs. Patients are placed in a warm environment with a humidity of 40%, which avoids vasospasm. The role of digital plethysmography lies in the evaluation of patients with digital ischaemia and normal ankle systolic pressures (Sumner 1981). Used alone it is not a good indicator of arterial disease and is affected by sympathetic tone. The combination of digital pulse volume and reactive hyperaemia is a good indicator of arterial disease (Strandness and Bell 1965).

Transcutaneous Oxygen Tension

Transcutaneous oxygen tension ($TCPO_2$) is a function of cutaneous blood flow, metabolic activity, oxyhaemoglobin dissociation and oxygen diffusion through tissues. Clearly systemic and local factors will affect the validity of $TCPO_2$ measurements, e.g. systemic factors will be affected by oxygen content and blood flow, and local factors by increased skin thickness. At rest a value of 55 mmHg is considered to be normal, but there is considerable overlap between normals and patients with peripheral vascular disease (Cina et al. 1984). In an attempt to improve the reliability of $TCPO_2$ it can be expressed as a regional perfusion index or foot to chest $TCPO_2$. This is ratio of limb $TCPO_2$ to subclavicular $TCPO_2$ and is usually in the range of 90% (Hauser and Shoemaker 1983). Further attempts to improve the accuracy $TCPO_2$ have included measurements during exercise (Matsen et al. 1984), reperfusion (Kram et al. 1984), after oxygen breathing (Feenstra et al. 1988) and in various leg positions (Franzeck et al. 1982). The test involves expensive equipment and is time consuming. It may well have a role in diabetics who have calcified blood vessels but is not ideal as a screening test.

Laser Doppler

Laser Doppler velocimetry measures blood flow velocity in superficial arterioles and capillaries. Monochromatic light generated from a standard helium-neon laser Doppler is conducted to the skin by a series of fibreoptic cables. The light penetrates the skin to a depth of 1.5 mm and is reflected back to a second photodiode by contact with flowing red blood cells. The Doppler shift is processed and expressed as millivolts. Three measurements are obtained: (i) pulse wave amplitude which reflects velocity changes in arterioles; (ii) vasomotor waves which reflect variations in sympathetic innervation; and (iii) skin blood flow velocity. This technique has been used to measure skin blood flow in patients with peripheral vascular disease (Karanfilian et al. 1984, 1986). However, in its present form it is an expensive piece of equipment which remains in the domain of research institutions offering no advantage over $TCPO_2$ measurements.

Magnetic Resonance Spectroscopy

Phosphorus magnetic resonance spectroscopy is a non-invasive technique of measuring muscle metabolism, but poor signal quality appears to be the important factor limiting its current application in clinical evaluation. The role of other nuclei, e.g. sodium and potassium, still have to be developed. The introduction of smaller cheaper magnets may make this technology more widely available for patients with peripheral vascular disease (Hands 1988), but currently the technique is too expensive and time consuming for screening.

Biochemical Parameters

Microalbuminuria

There is increasing evidence to suggest that during periods of relative ischaemia, (i.e. claudication) there is release of various inflammatory mediators and lipid peroxides (Di Perri et al. 1988, Shearman et al. 1988). These cause endothelial cell damage and increased vascular permeability to macromolecules (e.g. albumin). Hickey et al. (1990) have reported an increased albumin/creatinine ratio in patients with claudication. In addition, Stringer and Kakkar (1990) have reported that lipid peroxide/total lipid ratio was the best index of atherosclerotic disease severity, but this has not been applied as a screening test.

White Cell Count

Dormandy and Murray (1991) have recently reported that a raised white cell count is a significant predictor of myocardial infarction, stroke and vascular mortality.

In conclusion, the identification of presymptomatic arterial disease remains an important challenge. Early recognition of arterial disease may allow the clinician and patient to modify the known risk factors of arterial disease including smoking, hyperlipidaemia, hypertension and diabetes.

The measurement of Doppler ankle pressures at rest and after exercise is a simple, cheap and reproducible test of arterial insufficiency and of long-term mortality. Where there is doubt about the ABPI result, toe pressure indices provide an excellent measure of the severity of occlusive arterial disease. Theoretically Doppler waveform analysis should hold the key to the early detection of arterial disease, but in practice the technique is operator dependent, time consuming, expensive and inaccurate.

An alternative approach would be to assess the impedance of the limb non-invasively (Fig. 2.5). This technique has been successively applied in the detection of post-operative graft stenoses with a high degree of accuracy and could easily be applied for screening (Wyatt et al. 1991).

There is increasing evidence that biochemical markers, e.g. raised white cell count, microalbuminuria and lipid peroxide/total lipid content ratio may identify

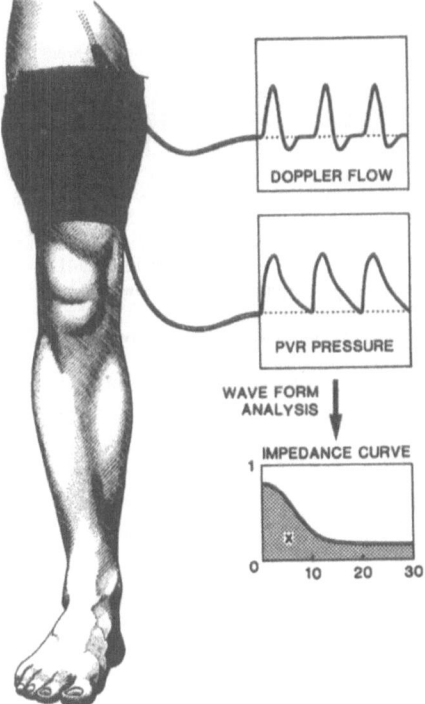

Fig. 2.5. Measurement of thigh impedance. A Doppler velocimeter measures pulsatile flow from the common femoral artery. A pulse volume recorder measures pulsatile pressure from a cuff around the thigh. An impedance curve is produced by performing waveform analysis on the paired flow/pressure signals. The area (x) under the curve represents the mean thigh impedance and is expressed as a fraction of the total area (x/30).

individuals at high risk of developing peripheral vascular disease but further research is required.

References

Aukland A, Hurlow RA (1982) Spectral analysis of Doppler ultrasound: Its clinical applications in lower limb ischaemia. Br J Surg 69:539–42

Baker AR, Evans DH, Prytherch DR, Bell PRF (1986) Haemodynamic assessment of the femoropopliteal segment: Comparison of pressure and Doppler methods using ROC curve analysis. Br J Surg 73:559–62

Baker JD (1978) Post-stress Doppler ankle pressures. A comparison of treadmill exercise with two other methods of induced hyperemia. Arch Surg 113:1171–3

Campbell WB, Skidmore R, Woodcock JP, Baird RN (1985) Detection of early arterial disease: a study using Doppler waveform analysis. Cardiovascular Res 19:206–11.

Cina C, Katsamouris A, Megerman J et al. (1984) Utility of transcutaneous oxygen tension measurements in peripheral arterial occlusive disease. J Vasc Surg 1:362–71

Criqui MH, Fronek A, Barrett-Connor E et al. (1985) The prevalence of peripheral arterial disease in a defined population. Circulation 71:510–15

Darling RC, Raines JK, Brener BJ, Austen WG (1972) Quantitative segmental pulse volume recorder: A clinical tool. Surgery 72:873–87

Di Perri T, Laghi Pasini F, Ralli L et al. (1988) Effect of controlled physical activity on hemorheological and metabolic changes in POAD patients. Clin Hemorheol 8:737–49

Dormandy JA, Murray GD (1991) The fate of the claudicant – A prospective study of 1969 claudicants. Eur J Vasc Surg 5:131–3

Evans DH, Macpherson DS Bentley S, Asher MJ, Bell PRF (1981) The effect of proximal stenosis on Doppler waveforms: a comparison of three methods of waveform analysis in an animal model. Clin Phys Physiol Meas 2:17–25

Feenstra BWA, Meiss L, Montauban van Swinjndregt AD, Stigter H, van Urk H (1988) Assessment of peripheral vascular obliterative disease by transcutaneous oxygen tension tests. Eur J Vasc Surg 2:19–26

Franzeck UK, Talke P, Bernstein EF, Goldbrandson FL, Fronek A (1982) Transcutaneous PO_2 measurements in health and peripheral arterial occlusive disease. Surgery 91:156–63

Fronek A, Johansen KH, Dilley RB, Bernstein EF (1973) Non-invasive physiologic tests in the diagnosis and characterization of peripheral arterial occlusive disease. Am J Surg 126:205–14.

Fronek A, Coel M, Bernstein EF (1976) Quantitative ultrasonographic studies of lower extremity flow velocities in health and disease. Circulation 53:957–60

Gosling RG, Dunbar G, King DH et al. (1971) The quantitative analysis of occlusive peripheral arterial disease by a non-invasive ultrasonic technique. Angiology 22:52–5

Hands L (1988) Is magnetic resonance spectroscopy helpful in recognizing irreversible limb ischaemia? In: Greenhalgh RM, Jamieson CW, Nicolaides AN (eds) Limb salvage and amputation for vascular disease. Saunders, London, pp 63–73

Hauser CJ, Shoemaker WC (1983) Use of transcutaneous PO_2 regional perfusion index to quantify tissue perfusion in peripheral vascular disease. Ann Surg 197:337–43

Hickey NC, Shearman CP; Gosling P, Simms MH (1990) Assessment of intermittent claudication by quantitation of exercise induced microalbuminuria. Eur J Vasc Surg 4:603–6

Hirai M, Schoop W (1984) Hemodynamic assessment of the iliac disease by proximal thigh pressure and Doppler femoral flow velocity. J Cardiovasc Surg 25:365–9

Humphries KN, Hames TK, Smith SWJ, Cannon VA, Chant ADB (1980) Quantitative assessment to the common femoral to popliteal arterial segment using continuous wave Doppler ultrasound. Ultrasound Med Biol 6:99–105

Johnston KW, Maruzzo BC, Cobbald RSC (1978) Doppler methods for quantitative measurement and localization of peripheral arterial occlusive disease by analysis of the blood velocity waveform. Ultrasound Med Biol 4:209–23

Karanfilian RG, Lynch TG, Lee BC, Long JB, Hobson RW (1984) The assessment of skin blood flow in peripheral vascular disease by Laser Doppler velocimetry. Am Surg 50:641–4

Karanfilian RG, Lynch TG, Zirl VT et al. (1986) The value of laser Doppler velocimetry and transcutaneous oxygen tension determination in predicting healing of ischaemic forefoot ulcerations and amputations in diabetic and nondiabetic patients. J Vasc Surg 4:511–16

Kaufman JL, Fitzgerald KM, Shah DM, Corson JD, Leather RP (1989) The fate of extremities with flat lower calf pulse volume recordings. J Cardiovasc Surg 30:216–19

Kram HB, Appel PL, White RA, Shoemaker WC (1984) Assessment of peripheral vascular disease by post-occlusive transcutaneous oxygen recovery time. J Vasc Surg 1:628–34

Laing S, Greenhalgh RM (1983) The detection and progression of asymptomatic peripheral arterial disease. Br J Surg 70:628–30

Macpherson DS, Evans DH, Bell PRF (1984) Common femoral artery Doppler waveforms: a comparison of three methods of objective analysis in patients with vascular disease. Br J Surg 71:46–9

Matsen FA III, Wyss CR, Simmons CW, Robertson CL, Burgess EM (1984) The effect of exercise upon cutaneous oxygen delivery in the extremities of patients with claudication and in a human model of claudication. Surg Gynecol Obstet 158:522–8

Mozersky DJ, Sumner DS, Strandness DE Jr (1972a) Long-term results of reconstructive aorto-iliac surgery. Am J Surg 123:503–9

Mozersky DJ, Sumner DS, Standness DE Jr (1972b) Disease progression after femoro-popliteal surgical procedures. Surg Gynecol Obstet 135:700–4

Nielsen PE, Bell G, Lassen NA (1972) The measurement of digital systolic blood pressure by strain gauge technique. Scand J Clin Lab Invest 29:371–9

Ramsey DE, Manke DA, Sumner DS (1983) Toe blood pressure – a valuable adjunct to ankle pressure measurement for assessing peripheral arterial disease. J Cardiovasc Surg 24:43–8

Rutherford RB, Lowenstein DH, Klein MF (1979) Combining segmental systolic pressures and plethysmography to diagnose arterial occlusive disease of the legs. Am J Surg 138:211–18

Satomura S (1959) Study of the flow patterns in peripheral arteries by ultrasonics. J Acoust Soc Japan 15:151–8

Shearman CP, Gosling P, Gwynn BR, Simms MH (1988) Systemic effects associated with intermittent claudication: a model to study biochemical aspects of vascular disease. Eur J Vasc Surg 2:401–4

Skidmore R, Woodcock JP, Wells PMT et al. (1980) Physiological interpretation of Doppler-shift waveforms. III. Clinical results. Ultrasound Med Biol 6:227–31

Skinner JS, Strandness DE Jr (1967a) Exercise and intermittent claudication. I. Effect of repetition and intensity of exercise. Circulation 36:15–22

Skinner JS, Strandness DE Jr (1967b) Exercise and intermittent claudication. II. Effect of physical training. Circulation 36:23–9

Strandness DE Jr, Bell JW (1965) Peripheral vascular disease, diagnosis and objective evaluation using a mercury strain gauge. Ann Surg (Suppl 4) 161:1–35

Strandness DE Jr, Stahler C (1966) Arteriosclerosis obliterans. Manner and rate of progression. JAMA 196:121–4

Stringer MD, Kakkar VV (1990) Markers of disease severity in peripheral atherosclerosis. Eur J Vasc Surg 4:513–18

Sumner DS (1981) Rational use of non-invasive tests in designing a therapeutic approach to severe arterial disease of the legs. In: Puel P, Boccaslon H, Enjalbert A (eds) Hemodynamics of the limbs, Vol 2. GEPESC, Toulouse, pp 369–76

Sumner DS (1989) Non-invasive assessment of peripheral arterial occlusive disease. In: Rutherford RB (ed) Vascular surgery 3rd edn. Saunders, Philadelphia, pp 61–111

Thiele BL, Bandyk DF, Zierler RE, Strandness DE Jr (1983) A systematic approach to the assessment of aorto-iliac disease. Arch Surg 118:477–81

Wyatt MG, Muir RN, Tennant WG, Scott DJA, Horrocks M (1990) An objective comparison of four stress tests in the assessment of at risk femorodistal grafts. J Cardiovasc Surg 31:340–3

Wyatt MG, Muir RN, Tennant WG, Scott DJA, Baird RN, Horrocks M (1991) Impedance analysis to identify the at risk femoro-distal graft. J Vasc Surg 13:284–93

Yao JST, Hobbs JT, Irvine WT (1969) Ankle systolic pressure measurements in arterial diseases affecting the lower extremities. Br J Surg 56:676–9

3 Questionnaires

G.C. LENG

A questionnaire can be defined as "a prepared set of written questions for purposes of compilation or comparison". The use of questionnaires in medicine has increased steadily over the past 25 years, both in epidemiological surveys and in clinical work. Their use was encouraged by the emergence of computers to code and analyse the results, making the questionnaire a fast and efficient way of collecting and collating large amounts of data.

The impetus for the development of standardised epidemiological questionnaires came with the knowledge that medical history taking is extremely variable. In the clinical setting, however, observations are interpreted in the light of other findings, and standardisation is therefore less important. In contrast, epidemiological observations are usually analysed in isolation, and thus require careful development and pre-testing to ensure both valid and repeatable results. There are now several widely-used epidemiological questionnaires, for example the MRC Respiratory Questionnaire and the WHO/Rose Cardiovascular Questionnaire (Rose et al. 1982). This chapter describes the general principles behind the design and validation of a questionnaire, methods of application and common problems associated with their use in epidemiological surveys. The WHO/Rose questionnaire on intermittent claudication will be discussed in detail, and a new alternative proposed.

Development of a Questionnaire

The development of a questionnaire is a lengthy process. It begins with defining the domain, formatting questions, designing the layout, testing for validity and reliability, and finally piloting the questionnaire. Unfortunately many research articles do not include details of questionnaire development, or even reproduce the questionnaire; consequently there can be no peer review, and the instrument cannot be used validly in different populations.

The initial step in the formulation of a questionnaire is to define the content. Obviously if more topics are included, then less information will be omitted, but a long questionnaire is difficult to process and may produce inaccurate answers due to repondent fatigue. In general, the minimum amount of information necessary to provide sufficient data about the problem under consideration is all that is required (Del Greco and Walop 1987).

Types of Question

Open questions allow the respondent to answer freely in his own words, e.g. "How do you describe the pain you feel?" This method can, therefore, be a very rich source of data, but as the responses are recorded verbatim, the results are very difficult to summarise for statistical analysis. Open questions are probably best used in preliminary research to allow closed questions to be designed.

Closed questions restrict the respondent to preselected answers: for example, "Do you have any pain or discomfort in your chest? Yes/No". This method has the advantage of acting as a memory prompt, but does mean either that the choices must be all-inclusive, or that an additional space for comments is provided. In general these questions are easy to administer and to summarise, and are therefore widely used in questionnaires.

A special form of closed question is the rating scale, where the response is one of grading. For example, in reply to the question "How would you describe the severity of your pain?" the patient assigns a grade from 1 (mild) to 5 (severe). The main problem with this technique is that subjects tend to avoid the extremes of the scale, but allowances can be made for this during analysis.

Semi-closed questions offer the subject a limited number of choices to aid recall, plus the freedom to include additional information. For example, "When you experience chest pain, which of the following words best describes it? Tight, burning, dull, sharp, other (please specify)".

Table 3.1. Comparison of closed and open questions

Closed questions	Open questions
Require respondent to recognise something No recall required, therefore more information obtained	Require respondent to recall something, therefore less information obtained

Pre-determined answers, therefore:
  Whole range of possible responses may not be covered
  Forced choice may be resented
  Long lists of choices may be tiring, and subject may not complete the questionnaire
	Subject responds in own words, therefore:
	  A rich source of information

Easy to code and analyse	Difficult to code and analyse Inferential errors may be introduced during coding
Easy to administer	Difficult to administer

Wording, Sequence Layout

A questionnaire should contain simple straightforward questions. This does not necessarily mean short questions, as certain ideas may be better expressed using more words.

The characteristics of a well-worded questionnaire are as follows:

Grammatically simple

No vague or technical terms

Leading and loaded questions avoided where possible

One idea only in each item

Where additional information is essential, it is placed in parentheses to avoid confusion

In determining the sequence of the questionnaire, general questions are usually placed first followed by more detailed and specific questions, thus reflecting the pattern of a typical clinical interview. Questions on the same topic should be grouped together to facilitate memory. Questions of particular importance to the researchers are better placed at the beginning, before the interest of the respondent begins to wane, and embarrassing or sensitive topics should be placed at the end of the questionnaire.

The layout of the form is particularly important in the mailed questionnaire, where it helps to stimulate the interest of the subject. In versions administered by an interviewer the layout should prevent confusion and misreading of instructions. Appropriately numbered boxes alongside each item of data are useful if the responses are to be coded for statistical analysis.

Evaluation of a Questionnaire

A newly-designed questionnaire must be evaluated in terms of both validity and reliability to determine how useful it will be as an instrument of measurement.

Validity

Validity, or accuracy, indicates the extent to which a method provides a true assessment of that which it purports to measure. In relation to medical questionnaires, there are several types of validity (Del Greco et al. 1987).

Content validity is the extent to which a questionnaire covers the areas it was designed to cover. This is usually tested by reference to clinical practice.

Face validity is how well a questionnaire looks as though it measures what it is supposed to measure. This is not a very good indicator of usefulness, but a more professional-looking questionnaire is likely to elicit a better response.

Construct validity is the extent to which a new questionnaire agrees with current theories concerning the relevant areas of research. Construct validity can be tested by comparing the questionnaire with current literature, or by comparison with other questions also designed to measure the same thing.

Criterion (concurrent) validity is the extent to which a questionnaire measures what it purports to measure. This can be tested by checking the questionnaire response against a "gold standard", for example, the validity of an angina questionnaire could be compared with a physician's diagnosis or electrocardiographic findings. The criterion validity of a questionnaire may be expressed either as a correlation coefficient or, more commonly, in terms of sensitivity and specificity:

$$\text{Sensitivity} = \text{proportion of true cases correctly identified}$$
$$= \frac{\text{Number of persons with positive result on questionnaire}}{\text{Number of positive cases detected by "gold standard"}}$$

$$\text{Specificity} = \text{proportion of true "normals" correctly identified}$$
$$= \frac{\text{Number of persons with negative result on questionnaire}}{\text{Number of negative cases detected by "gold standard"}}$$

The balance between sensitivity and specificity depends upon the needs of a particular study and the expected prevalence of the disease under investigation. For example, if a questionnaire is to be used as a screening tool to select possible cases for a more detailed investigation, then it might be considered most appropriate to have a high sensitivity and a relatively low specificity. This ensures that few true cases are lost, whilst the additional false positives can be identified with subsequent investigations. In contrast, if a single test, or questionnaire, is being used to produce an accurate estimation of the true prevalence, then maintaining a high specificity is essential, especially if the disease is relatively uncommon.

Reliability

Reliability, or repeatability, refers to the extent of agreement between repeated measurements. Knowledge of the repeatability of a question is essential to predict how much of an apparent change in the measured variable is true, and how much is simply due to an error of measurement.

Test–retest reliability is the ability of a questionnaire to produce similar results when administered to the same person on different occasions. The interval between applying the questionnaire is important, as it should neither be too short so that previous responses are remembered, nor too long so that the true situation may have changed.

Inter-observer reliability is the ability of a questionnaire to produce the same result when administered by different interviewers. This can be assessed by evaluating the same subject with different observers, but again the time interval is important.

Internal consistency is the extent to which the subject answers similar questions in a similar manner. This can be tested by correlating the responses to one half of the questions with the responses to the other half, but in practice it is very difficult to assess in most health or symptom questionnaires.

Pre-test

Pre-testing, or piloting, is an essential final step in the development and evaluation of a questionnaire. The subjects used for pre-testing should be similar to those who will be used in any future study (usually between 10 and 50 subjects is an adequate number (Sudman and Bradbur 1983)). All comments and problems must be critically reviewed after each testing, and the process repeated until the research is satisfied. Typical problems encountered at this stage include:

Misinterpretation of questions

Question sequences that are difficult to follow

A long completion time producing subject fatigue

The omission of certain answers, e.g. because of lack of space or because the topic is a sensitive one

Administered or Self-Administered Questionnaires

An administered questionnaire is one that is read out by an interviewer who also records the responses. There are two basic approaches:

1. Standardised using predetermined questions which are always asked in the same order and with the same wording. The interviewer is not allowed to reword the questions in any way.

2. Unstandardised using a questionnaire really as a guide to the areas that must be covered, as the interviewer is allowed to introduce probes and reword sections to ensure that the subject has understood.

A self-administered questionnaire is completed by the subject without the help of an interviewer. It may consist of the traditional printed form or be administered by computer.

Table 3.2. Comparison of Administered and Self-Administered Questionnaires

Self-administered questionnaires	Administered questionnaires
No interviewer training or supervision	Requires interviewer training and supervision
No interviewer bias	Interviewer bias
Possible problems with completion e.g. low intelligence, arthritis	Few problems with completion
More difficult to design	Easy to design
Easy to use – postal surveys possible	More difficult to use
Cheap	Expensive
Generally more standardised – higher repeatability	Generally less standardised – lower repeatability

Self-administered questionnaires have been widely used in medicine, primarily because they can be posted. This allows housebound subjects to be contacted, and follow-up information to be obtained relatively easily. The main disadvantage of postal questionnaires is the poor response rate, often only 40%–60% (Bennett and Ritchie 1975), thus if the respondents are not typical of the whole population, non-respondent bias will be introduced. The response rate may be increased to 80%–90% by sending follow-up questionnaires, but the use of various incentives such as a free dinner (Woodward et al. 1985), a lottery ticket (Rissel 1989), and money (Mortagy et al. 1985) have produced inconsistent results, suggesting that they have only a small effect.

WHO/Rose Questionnaire for Intermittent Claudication

The most widely used cardiovascular questionnaire was designed by Professor G. Rose in 1962 at the London School of Hygiene, to identify subjects with angina, myocardial infarction and intermittent claudication. It was subsequently adopted by the World Health Organisation, and has since been applied in over 20 countries to estimate the prevalence of intermittent claudication. Despite this, there has been only one independent assessment of the questionnaire's ability to confirm a physician's diagnosis of claudication (Richard et al. 1972).

Large international differences in the prevalence of intermittent claudication have been reported using the WHO/Rose questionnaire. Figures in men over 50 years of age range from 0.4% in Odense (Rose et al. 1968) to 10% in East Finland (Heliovaara et al. 1978). Some of this variation will reflect true international differences but as the prevalence is undoubtedly low, slight changes in specificity (such as might result from errors in translation or administration) will greatly affect the number of false positives, and hence the measured prevalence.

Most epidemiological surveys also include simple non-invasive tests, for example the ankle-brachial systolic pressure ratio, toe pulse reappearance time and stress tests (the treadmill exercise test and the reactive hyperaemia test). However, all of these tests have their limitations (Fowkes 1988), and a subcommittee of the American Heart Association Council on Epidemiology has stated that "no single test will describe the whole range of atherosclerotic peripheral arterial disease, and each should be employed to characterize the whole spectrum of disease" (Prineas et al. 1982). Work is therefore needed to evaluate these tests with the view to developing a standard investigative protocol. This protocol should be as simple as possible, making it easy to apply in different populations to enable valid international comparisons to be made. It would ideally contain a questionnaire to diagnose intermittent claudication, plus one or two non-invasive tests.

Development of WHO/Rose Questionnaire

A group of patients with intermittent claudication and other causes of leg pain were subjected to detailed interview, and the responses analysed to determine the precise characteristics and distinguishing features of claudication pain. Descriptions of the quality of the pain were too variable to be useful, but the pain of intermittent claudication could be fully defined by six characteristics; these features were then used to form the basis of the questionnaire:

1. Pain situated in one or both calves

2. Provoked by either hurrying or walking uphill (or by walking on the flat, for those who never attempt more)

3. Must never start at rest

4. Makes the subject either stop or slacken pace

5. Disappears on a majority of occasions in 10 minutes or less from the time when the subject stands still

6. Never disappears while walking continues

The original WHO/Rose cardiovascular questionnaire was designed for administration by an interviewer, but in view of the benefit of postal surveys, it was adapted for self-administration in 1977 (Appendix 3.1). "Good agreement" was found between the administered and self-administered versions in a group of over 1800 male civil servants in Whitehall, but unfortunately no actual data were presented for the questions on intermittent claudication (Rose et al. 1977).

In 1972, Van Ganse et al. carried out an independent comparison of the two versions, and found only 80% concordance. However, there was no significant difference between the prevalence rates, possibly due to a loss of specificity in the self-administered questionnaire.

Validation

Rose tested the interview-administered questionnaire on a group of 37 patients with claudication and 18 control patients with leg pain due to other causes. Thirty-four of the claudicants met the questionnaire criteria, but none of the others did (92% sensitivity, 100% specificity). Unfortunately, these subjects were not a population sample, and therefore the results may not accurately reflect the expected performance of the questionnaire in the field.

In 1972, Richard et al. applied the questionnaire to a large population sample in France, and compared the responses with a physician's diagnosis of intermittent claudication. The results confirmed the high specificity demonstrated by Rose, but revealed a much lower sensitivity (Table 3.3); this difference may in part be caused by errors of translation.

The Rose questionnaire has also been tested against certain non-invasive tests of peripheral vascular disease, which also show the sensitivity to be lower than that identified by Rose (Table 3.3). However, good agreement between a questionnaire designed to diagnose a symptom complex and a test of peripheral atherosclerosis could not be expected, although a knowledge of the predictive value would be useful.

There are no published reports of the accuracy of the self-administered WHO/ Rose questionnaire on intermittent claudication, but recent results from two studies in Edinburgh suggest that it has a high specificity and moderate sensitivity at predicting the diagnosis made by a physician (F.G.R. Fowkes and G.C. Leng,

Table 3.3. Accuracy of the WHO/Rose questionnaire on intermittent claudication

Author	Subjects (n)	Interviewer/ self-administered	"Gold" standard	Accuracy	
				Sensitivity (%)	Specificity (%)
Rose 1962	Patients with leg pain (55)	Interviewer	Physician's diagnosis	92	100
Richard et al. 1972	Population sample (7996)	Interviewer	Physician's diagnosis	68	100
Isacsson 1972	Population sample (651)	Interviewer	Venous occlusion plethysmography	30	98
Criqui et al. 1985	Population sample and hyperlipidemics	Interviewer	Segmental blood pressure recording	9	99
Fowkes et al. 1991[a]	Patients with leg pain (647)	Self-administered	Physician's diagnosis	60	90
Fowkes et al. 1991[a]	Population sample (1592)	Self-administered	ABPI ≤0.9	12	96

[a] Unpublished data.

unpublished work). However, as expected, it shows very poor agreement with a physiological test of peripheral blood flow (Table 3.3).

Reliability

The repeatability of the questionnaire was tested after an interval of 6 months in a large population survey in Finland (Reunanen et al. 1982). The reliability coefficient κ was calculated for both men (0.45) and women (0.38), showing reasonable, but not good, repeatability. The prevalence of symptoms was the same at both interviews, although several people with mild or atypical symptoms changed from symptom positive to symptom negative, and vice versa.

Modified Classification

In 1985, Criqui et al. extended the classification of the WHO/Rose questionnaire to include "possible claudicants", defined as those with "exercise leg pain not present at rest, but not otherwise fully concordant with the Rose criteria". This expansion of the original definition was designed to improve sensitivity and, indeed, in detecting subjects with large vessel peripheral arterial disease, the sensitivity of the questionnaire increased from 9.2% to 20.0%. Unfortunately, specificity was reduced from 99.0% to 95.9%, and as a result the proportion of "cases" which were false positives increased from 47% to 62%, although in absolute terms the identification of genuine cases more than doubled.

A similar effect on sensitivity and specificity was also found in a recent large population survey in Edinburgh (Fowkes et al. 1991). Using the presence of a low ankle-brachial systolic pressure ratio as the definition of peripheral arterial disease, the addition of possible claudicants to Rose claudicants again increased sensitivity (12% to 15%) and decreased specificity slightly (96% to 95%). These differences are less pronounced than those found by Criqui et al. (1985), possibly because the "gold standard" is not the same.

An alternative classification of a "possible claudicant" has recently been proposed by Davey Smith et al. (1990), which includes those whose pain disappears whilst walking (possibly because the subject has slowed down), but who otherwise fulfil the Rose criteria. Adding this group to the Rose positives identified in the Whitehall study in 1977 increased the prevalence of intermittent claudication from 0.8% to 1% (Davey Smith et al. 1990). After 17 years of follow-up both subjects with Rose and possible claudication showed an increase in mortality rates from all causes and cardiovascular causes, although mortality from non-cardiovascular causes was elevated in the possible group only. It can therefore be concluded that the possible claudication group contains "many genuine cases", and that because intermittent claudication appears to be independently related to increased mortality rates, a simple claudication questionnaire should be included in cardiovascular risk assessment and screening programmes.

Thus the inclusion of a possible claudication category is important to increase case detection, but the need for this refinement emphasises that there are problems associated with the original WHO/Rose questionnaire. Ideally, a claudication questionnaire should have a high sensitivity without any substantial loss of specificity in order to prevent the dilution of true cases by false positives, an inevitable consequence of using a non-specific test to identify a disease with a low prevalence.

Edinburgh Claudication Questionnaire

Following the recent application of the WHO/Rose questionnaire in two large studies in Edinburgh, the responses to each question were analysed and particular problems identified. Several new questions were then formulated and tested, and finally a new questionnaire was developed (Appendix 3.2). This new questionnaire is designed to be self-administered, and it has now been tested on 150 patients aged over 55 years, either attending their general practitioner or the Peripheral Vascular Clinic at the Royal Infirmary of Edinburgh. Preliminary results indicate that this new questionnaire shows extremely good agreement with the diagnosis of intermittent claudication as made by an experienced vascular consultant (specificity 99%, sensitivity 85%) (G.C. Leng, unpublished work). This possible improvement in accuracy suggests that the Edinburgh Claudication Questionnaire, which has been developed from use of the WHO/Rose question-naire, might be adopted for use in future epidemiological surveys of intermittent claudication.

Appendix 3.1. The WHO/Rose Questionnaire on Intermittent Claudication

(a) Do you get a pain in either leg on walking?
       1. ☐ Yes   2. ☐ No
(b) Does this pain ever begin when you are standing still or sitting?
       1. ☐ Yes   2. ☐ No
(c) Do you get this pain in your calf (or calves)?
       1. ☐ Yes   2. ☐ No
(d) Do you get it when you walk uphill or hurry?
       1. ☐ Yes   2. ☐ No
(e) Do you get it when you walk at an ordinary pace on the level?
       1. ☐ Yes   2. ☐ No
(f) Does the pain ever disappear while you are still walking?
       1. ☐ Yes   2. ☐ No
(g) What do you do if you get it when you are walking?
       1. ☐ Stop
       2. ☐ Slow down
       3. ☐ Continue at same pace

(h) What happens to it if you stand still?
       1. ☐ Usually continues more than 10 minutes
       2. ☐ Usually disappears in 10 minutes or less

Definition of positive classification: "Yes" to (a), (c) and (d); "No" to (b) and (f); "stop" or "slow down" to (g); and "usually disappears in 10 minutes or less" to (h). Grade 1 = "No" to (e); Grade 2 = "Yes" to (e). Criqui et al. (1985) defined as "possible" claudicant those with exercise calf pain not present at rest, but otherwise not fully concordant with Rose criteria.

Appendix 3.2. The Edinburgh Claudication Questionnaire

1. Do you get a pain or discomfort in   Yes ☐
   your leg(s) when you walk?   No ☐
                                       I am unable to walk ☐

If you answered YES to question 1 – please answer the following questions. Otherwise you need not continue.

2. Does this pain ever begin when you are   Yes ☐   No ☐
   standing still or sitting?

3. Do you get it if you walk uphill or hurry?   Yes ☐   No ☐

4. Do you get it when you walk at an ordinary   Yes ☐   No ☐
   pace on the level?

5. What happens to it if you stand still?

            Usually continues more than 10 minutes   ☐
            Usually disappears in 10 minutes or less   ☐

6. Where do you get this pain or discomfort?
   Mark the place(s) with X on the diagram below.

Front   Back

Definition of positive classification: "Yes" to 1, "No" to 2, "Yes" to 3, and "usually disappears in 10 minutes or less" to 5, Grade 1 = "No" to 4 and Grade 2 = "Yes" to 4. If these criteria are fulfilled a definite claudicant is one who indicates pain in the calf, and a "possible" claudicant one where pain occurs in the thigh or buttock. Subjects should be considered *not* to have claudication if crosses are placed over the joints, or arrows are used to indicate a pain which radiates.

References

Bennett AE, Ritchie K (1975) Questionnaires in medicine. A guide to their design and use. Oxford University Press, Oxford

Best WR (1962) The potential role of computers in medical practice. JAMA 182:994

Criqui MH, Fronek A, Klauber MR et al. (1985) The sensitivity, specificity and predictive value of traditional clinical evaluation of peripheral arterial disease: results from non-invasive testing in a defined population. Circulation 71:516–22

Davey Smith G, Shipley MJ, Rose G (1990) Intermittent claudication, heart disease risk factors and mortality. The Whitehall Study. Circulation 82:1925–31.

Del Greco L, Walop W (1987) Questionnaire development: formulation. Can Med Assoc J 136:583–5

Del Greco L, Walop W, McCarthy RH (1987) Questionnaire development: validity and reliability. Can Med Assoc J 136:699–700

Fowkes FGR (1988) The measurement of atherosclerotic peripheral arterial disease in epidemiological surveys. Int J Epidemiol 17(2):248–54

Fowkes FGR, Housley E, Cawood EHH et al. (1991) Edinburgh Artery Study: Prevalence of asymptomatic and symptomatic peripheral arterial disease in the general population. Int J Epidemiol 20:384–92

Heliovaara M, Karvonen MJ, Vilhunden R et al. (1978) Smoking, carbon monoxide and atherosclerotic diseases. Br Med J i:268–70

Isacsson S (1972) Venous occlusion plethysmography in 55-year-old men: a population study in Malmo, Sweden. Acta Med Scand 537 (Suppl):1–62

Mortagy AK, Howell JRL, Waters WE (1985) A useless raffle. J Epidemiol Comm Health 39:183–4

Prineas RJ, Harland WR, Janzon L et al. (1982) Recommendations for use of non-invasive methods to detect atherosclerotic peripheral arterial disease in population studies. (American Heart Association Council on Epidemiology). Circulation 65:1561A–6A

Reunanen A, Takkunen H, Aromaa A (1982) Prevalence of intermittent claudication and its effect on mortality. Acta Med Scand 211:249–56

Richard JL, Ducimetiere P, Elgrishi I et al. (1972) Dépistage par questionnaire de l'insuffisance coronarienne et de la claudication intermittente. Rev Epidemiol Med Soc Sante Publ 20:735–55

Rissel C (1989) Improving response rates with incentives (letter). Med J Aust 150:411

Rose GA (1962) The diagnosis of ischaemic heart pain and intermittent claudication in field surveys. Bull WHO 27:645–58

Rose GA, Ahmeteli M, Checgacci L et al. (1968) Ischaemic heart disease in middle-aged men. Prevalence comparisons in Europe. Bull WHO 38:885–95

Rose G, McCartney P, Reid DD (1977) Self-administration of a questionnaire on chest pain and intermittent claudication. Br J Prev Soc Med 31:42–8

Rose GA, Blackburn H, Gillum RF et al. (1982) Cardiovascular survey methods. WHO, Geneva

Sudman S, Bradbur NM (1983) Asking questions. Jossey-Bass, San Francisco, p 282

Van Ganse W, Van Hoorne N, De Backer G et al. (1972) L'interview et le questionnaire auto-administré de Rosa dans une étude pilot d'atherosclerosis. Rev Epidemiol Med Soc Sante Publ 20:7

Woodward A, Douglas B, Miles H (1985) Chance of free dinner increases response to mail questionnaire (letter). Int J Epidemiol 14:641–2.

4 Duplex Ultrasound

P.L.P. ALLAN

History

The Doppler effect was described by Christian Andreas Doppler in 1842 (Doppler 1843). It describes the effect on a waveform when there is relative movement between the transmitter and the receiver: if they are moving apart the received frequency will be lower than the transmitted frequency; conversely, if they are moving towards each other the received frequency will be higher than the transmitted frequency.

During the 1940s and 1950s several groups of workers in Europe and the United States developed techniques based on the use of ultrasound to obtain information from within the body. Initially there were relatively crude machines and the information from them was difficult to interpret, but in the 1960s static image scanners were developed and in the late 1970s real-time scanners became available which provided instantaneous moving images of structures within the body. Since then the development of more sophisticated computers, electronics and materials has resulted in increasingly powerful machines with greater data processing capability and high quality image resolution.

Ultrasound is a waveform and will therefore show the Doppler effect if it is reflected from a moving target. The difference in frequency between the transmitted and received frequencies can be analysed by the machine and information on the presence or absence of movement, its direction and its velocity can be obtained. Satomura (1957) reported the use of Doppler ultrasound in the assessment of moving cardiac structures in 1957 and 2 years later he described the use of the technique to obtain information about flow patterns in peripheral arteries (Satomura 1959). These first machines produced a continuous beam of ultrasound and continuously analysed any reflected waveforms from the blood flowing in the vessels; their descendants are still very much in use today, particularly in vascular units to detect patent vessels and facilitate limb blood pressure measurements. The disadvantage of these machines is that any vessel in the line of the ultrasound beam will produce a signal and they have no means of discriminating between different sources. The development of pulsed Doppler equipment (Baker 1970) allowed data collection to be confined to a known, controllable depth within the patient so that some of this confusion was resolved, but it was still not possible to identify individual vessels or specific regions of interest within vessels.

In order to identify particular vessels and to obtain Doppler signals from specific sites it was necessary to combine ultrasound imaging with ultrasound Doppler. The first of these "duplex" machines was developed in the mid 1970s (Barber et al. 1974). The initial machines combined Doppler with static image ultrasound machines but as real-time scanners were developed and the amount of computing power available increased, the "duplex" scanner has developed into the powerful machine which is obtainable today. In addition to allowing precise localisation of the source of the Doppler signal, the development of this technique included new ways of manipulating, displaying and recording the Doppler information with built-in spectrum analysers providing a display of the range of velocities present in the sample volume and their variation with time over the course of the cardiac cycle. More recently colour display techniques have enabled some of the Doppler data to be displayed in real-time superimposed on the imaging information.

Basic Principles

Duplex ultrasound of a vessel is a combination of high resolution imaging and information on the flow of blood in the vessel. In practice these two aspects tend to be assessed simultaneously as an examination is performed but for the sake of discussion it is easier to consider them separately.

High Resolution Imaging

Modern equipment allows high resolution imaging of superficial vessels such as the carotids, peripheral arteries and veins. Many of the deeper intra-abdominal vessels can also be visualised satisfactorily, although overlying bowel gas can be a problem in the abdomen.

The real-time image enables the operator to identify a specific vessel and to follow it along its course. It also allows the sample volume for the Doppler examination to be located precisely in particular segments of a vessel, or even at a particular point in the lumen of the vessel.

The image allows direct visualisation of the vessel walls and lumen; areas of disease can be examined and the severity of changes assessed. Early atheroma shows as thickening and irregularity of the internal aspect of the vessel wall. In more severe cases areas of plaque can be seen and direct measurement of the reduction in calibre of the vessel can be obtained if the margins of the plaque are seen clearly. In areas of very severe disease the vessel lumen can be difficult to identify and it is therefore necessary to rely on changes in the blood flow to assess the severity of a stenosis. In addition to identifying the presence of a plaque it is also possible, in some cases, to see changes compatible with plaque ulceration or thrombus on the surface of the plaque or intraplaque haemorrhage. Although these changes are not seen in every plaque, if they are identified they should be assessed carefully as they may indicate a source of emboli which are producing problems further downstream (Carroll 1991). Calcified plaque is clearly demonstrated, but this can be a problem as the calcification and its associated acoustic shadow mean that details of the vessel behind the plaque cannot be defined and no Doppler information can be obtained

The abdominal aorta can be visualised throughout most of its length in many patients, but in obese patients, or patients with large amounts of gas in the bowel, parts of the aorta may not be seen despite using several lines of approach. The iliac arteries can also be obscured through some or all of their course by gas in the bowel. However the vast majority of abdominal aortic aneurysms can be identified and measured on ultrasound, although in some cases the upper and lower ends cannot be clearly assessed and the relationship to the renal arteries may be difficult to define.

Other parts of the vascular system where ultrasound may have problems in visualising the vessels, or obtaining a Doppler signal, include high bifurcations of the carotids, the origins of the great vessels and arch of the aorta, the iliac vessels, the adductor canal in the thigh, and the calf vessels. Obesity can also be a problem but with experience, care, perseverance and suitable equipment it is often possible to obtain some information from these areas even if they are not well seen; however, it is sometimes necessary to recognise that an adequate ultrasound examination has not been performed rather then accept inadequate information.

Doppler Information

Despite the vast number of publications following Doppler's initial report, the basic information from Doppler examinations is fairly limited. The Doppler examination will detect the presence or absence of flow, the direction of flow, the velocity of flow, and there may be some characteristics of the flow, such as turbulence or increased diastolic flow, which give some clue to the nature of flow in the vessel concerned. Although this information appears rather limited, various methods of analysis and manipulation of the data allow much to be deduced about conditions in the vessel at the point of assessment, and also at other, more remote, locations.

The major arteries have characteristic, individual signatures on Doppler which depend on the conditions of blood inflow, the vessel itself and outflow from the vessel into its distal branches or vascular beds. These characteristic waveforms are altered by disease but it is important to remember that these changes are the result not only of local factors, such as atheroma at the site of assessment, but can also be due to disease elsewhere, such as stenoses or occlusions upstream or downstream, aortic valve disease, or poor cardiac output. The changes in the waveform that can be caused by disease include increased peak velocities or Doppler shifts, increased or decreased diastolic flow, spectral broadening (which is a sign of turbulence), widening of the systolic peak, etc. From these changes various criteria have been developed which attempt to correlate alterations in the waveform with the presence and severity of disease in the carotid vessels and, more recently, in the peripheral arteries. The most frequently used criteria in the carotids include peak systolic velocity, peak diastolic velocity, the ratio of peak systolic velocities in the internal and common carotid arteries and the presence or absence of spectral broadening (Bluth et al. 1988; Robinson et al 1988). In the lower limbs similar criteria have been developed based primarily on peak systolic velocity changes at and above areas of stenosis; spectral broadening is also a useful sign in some circumstances (Kohler et al. 1987; Cossman et al.1989).

Colour Doppler assesses the mean Doppler shifts at multiple points in the sample area and assigns a shade of colour to each point. This colour "map" is then projected over the image to provide a visual representation of blood flow data on the image. The great advantage of colour Doppler is that it decreases the examination time as abnormal areas are more quickly identified. Some abnormalities can be assessed from the colour changes alone but it is often necessary to use "black and white", spectral Doppler in order to quantify these flow abnormalities more precisely to assess the severity of a stenosis.

Initially Doppler information was given in terms of the size of the frequency shift in kilohertz (kHz). This was because the initial Doppler machines did not allow visualisation of the vessel so that the angle between the direction of flow and the Doppler beam could not be assessed. This is necessary if the kilohertz shift is to be converted into a value for velocity in metres per second. The advent of duplex machines allowed this angle to be assessed from the image and a correction factor to be applied to the Doppler shift which produces a velocity measurement. Velocity measurements are generally better than kilohertz shifts as the latter require knowledge of transducer frequencies, angle of insonation, etc. for comparisons to be made between different examinations. Velocity measurements, although not perfect, allow easier comparisons between examinations done on different machines.

Ultrasound as a Screening Technique

There are many techniques of screening for atherosclerosis, including questionnaires, history and examination, blood tests, ECG, ankle-brachial pressure indices (ABPI), stress tests, reactive hyperaemia tests, review of medical notes and discharge summaries, etc. Over the years many of these have produced satisfactory results in terms of identifying people with evidence of disease (Fowkes 1988). Doppler ultrasound initially provided a method for detecting arterial pulses with greater accuracy for peripheral limb blood pressure measurements and it is still used to great effect for this today. In skilled hands it also gave information on some aspects of blood flow in the region of concern but angiography was the only method of directly visualising vessels, and due to its invasive nature this could not be applied widely for screening purposes. The development of modern ultrasound machines has meant that information which could be obtained by angiography in the past, can now be obtained non-invasively by ultrasound.

Advantages of Ultrasound

When used as a screening technique ultrasound is safe, non-invasive, and does not require the use of ionising radiation or contrast agents. It has been shown to have a satisfactory sensitivity of 87% for the detection of significant stenoses in symptomatic patients when compared with arteriography (Ricotta et al. 1987; Cossman et al. 1989). In addition, with modern equipment, early changes in the arterial wall thickness and elasticity can be assessed and followed over a period of time even before significant plaque has formed (Salonen and Salonen 1990; Riley et al. 1986).

Ultrasound equipment is relatively mobile and some machines are easily portable. For screening of aortic aneurysms trained nursing staff have obtained reproducible, accurate measurements using small, relatively cheap, portable machines in the community (Grimshaw et al.1990). Larger machines with duplex or colour Doppler facilities are not yet portable but are sufficiently mobile to be moved around several screening centres over a period of time, or to be installed in a van which can be driven around as part of a screening programme.

Disadvantages of Ultrasound

The experience, expertise and enthusiasm of the sonographer is a crucial factor in obtaining reliable results and, depending on the complexity of the assessment, there is a variable learning curve. At the bottom end of the curve it is relatively easy to train a research nurse or technician in a day or two to obtain reasonable results for specific measurements of abdominal aortas. At the other end complex assessments involving several vessels, using more complicated imaging and Doppler information may take several months of training and experience before adequate reliability is achieved.

Problems with visualisation and acquisition of Doppler information from some important vascular segments have already been mentioned: high carotid bifurcations, iliac vessels, adductor canal segments and calf vessels can be problem areas in some patients. A further difficulty with duplex is that, due to basic physical principles, a good angle for imaging (90° to the line of the vessel) is a bad angle for Doppler, which needs a smaller angle, ideally less than 60°. In some cases the operator must therefore reach a compromise between image quality and Doppler information.

Applications of Duplex as a Screening Technique

Carotids

The carotid arteries are ideally situated for duplex examinations as they are superficial, and high frequency transducers with good resolution can be used. Initial studies concentrated on assessing the prevalence of atheroma in the carotids in subjects who may have been asymptomatic, have had asymptomatic bruits, symptoms of transient ischaemic attack (TIA) or reversible ischaemic neurological deficits (RINDS). Unfortunately many areas relating to the diagnosis and management of carotid and cerebrovascular disease remain unclear

(Scheinberg 1988). Whilst ultrasound has been shown to be sensitive and accurate in detecting and grading carotid stenoses (Cardullo et al. 1986; Ricotta et al. 1987) there has been much discussion on the relevance of these findings to the management and prognosis of the individual (Brown et al. 1989). Subjects with demonstrable disease have a higher incidence of cerebrovascular and often cardiovascular events (Busuttil, 1981). In addition, the likelihood of these events occuring is related to the severity of the changes demonstrated, particularly if these are producing more than 75%–80% diameter reduction (Caplin, 1986). The problem is that the location of the cerebrovascular incident is not closely related to the side of the lesion with some 40%–50% of strokes occurring in the contralateral hemisphere (Wolf et al. 1981; Brown et al. 1989). Much of the stimulus for developing non-invasive techniques for carotid assessment came from the requirement to screen patients who may have benefitted from carotid endarterectomy. This procedure was relatively common especially in the United States where 107 000 were performed in 1985, but the role of this operation in the management of patients with carotid disease is also unclear (Winslow et al. 1988). Most people agree that endarterectomy for severe stenoses is of value but the picture is much less clear for less severe stenoses, particularly in asymptomatic patients. Several large trials are currently in progress in Europe and the United States and the results of these should clarify matters, particularly in relation to the criteria which should be used for recommending surgery as opposed to conservative, medical management (O'Leary and Polak 1989).

More recently attention has been focused on duplex examinations of the carotids as a method of assessing the cardiovascular system as a whole. The appearance of the carotids is taken to reflect changes elsewhere in the arterial system so that if a subject has severe carotid disease he or she is also likely to have severe changes in cerebral, coronary and peripheral arteries (Scheinberg 1988); whereas if the changes in the carotids are minimal it is assumed that other vessels are also minimally involved. This would seem to be a reasonable proposition as there is an association between the severtity of carotid changes and the likelihood of a cardiovascular event. The value of this approach is that if early changes, such as intimal thickening (Fig. 4.1), can be detected reliably then the progression of disease with time can be followed and correlated with other risk factors; early studies suggest that this is a valid approach (Poli et al. 1988; Salonen and Salonen, 1990). In addition, it might be possible to assess the effect of drugs and other therapeutic regimes, such as alteration of diet or giving up smoking, over a period of time (O'Leary and Polak, 1989; Bond et al. 1989). If the early promise of these techniques is fulfilled then ultrasound will provide a simple, repeatable, non-invasive means of assessment of the cardiovascular system which could be applied relatively easily to population samples. This approach is being assessed in at least one large screening exercise where ultrasound of the carotids and a popliteal artery are part of the assessment in the Atherosclerosis Risk in Communities Study (ARIC 1989).

Abdominal Aortic Aneurysms

Aneurysms of the abdominal aorta are another aspect of cardiovascular disease (Fig. 4.2). They are often silent until they present with acute symptoms of leaking

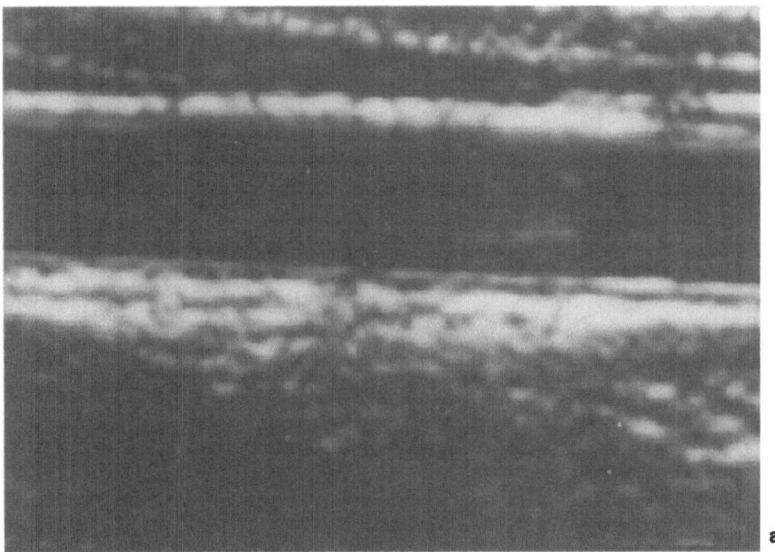

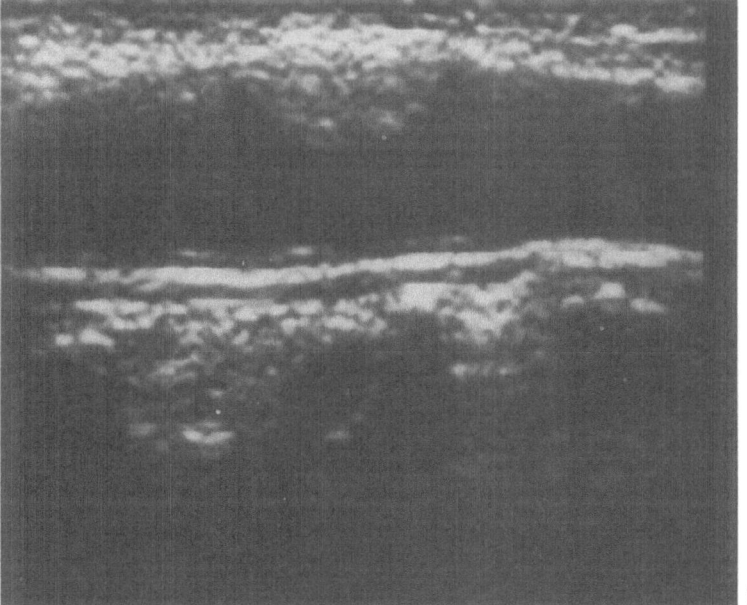

Fig. 4.1. a A normal intimal line; **b** and **c** (see over) show increasing thickness and irregularity.

or rupture. Collin et al. (1988) used ultrasound to screen a population of men aged 65–74 years and 5.5% of their sample had an abdominal aortic aneurysm. They estimated that, if this prevalence was extended to all of England and Wales they could expect to identify some 52 000 men with aortic aneurysms. If elective surgery was accepted by 60% of men with aneurysms over 4 cm diameter they

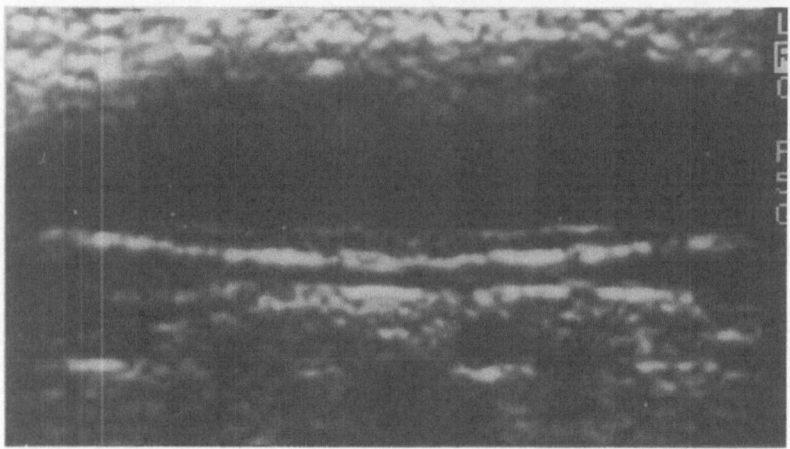

Fig. 4.1. (continued)

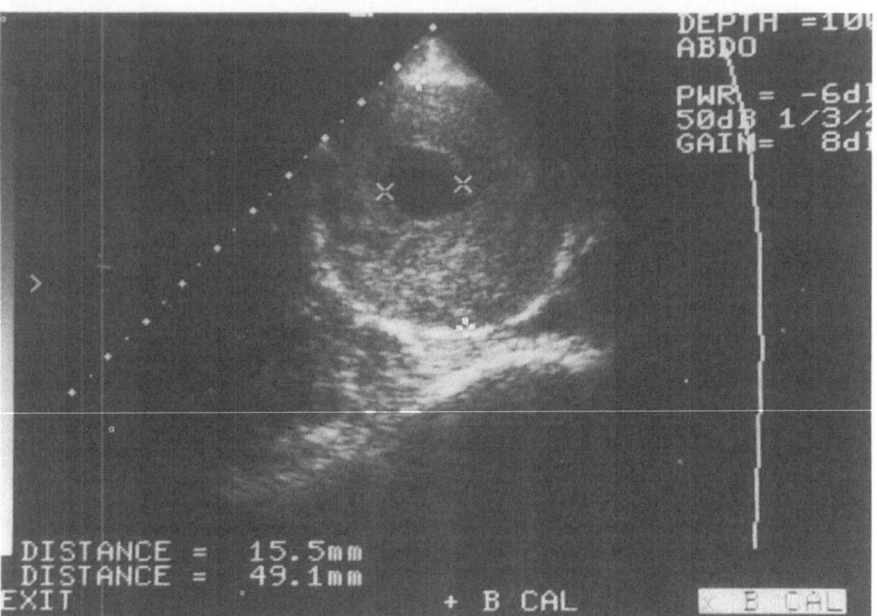

Fig. 4.2. A transverse view of a large abdominal aortic aneurysm showing the residual lumen (×–×) of 15 mm and the outer diameter (+–+) of 49 mm.

calculated that some 6000 deaths per year from ruptured aneurysms could be prevented.

In a subsequent paper these authors (Collin and Walter 1989) showed that in a small group of siblings of patients with abdominal aortic aneurysms 29% also had aneurysms, suggesting a familial link; interestingly these were all brothers; none of the small number of sisters who were examined was found to have an aneurysm although overall numbers were small and their significance uncertain.

Duplex ultrasound with Doppler is not required to diagnose abdominal aortic aneurysms so smaller, cheaper, portable real-time machines can be used. These are small enough to be carried easily in a nurse's or technician's car around screening locations such as surgeries and health centres. The performance of these machines and their operators has been shown to be satisfactory (Grimshaw et al. 1990) and larger scale screening studies to identify patients with aneurysms, particularly larger aneurysms, are being developed.

Peripheral Vascular Disease

The arteries of the limbs, particularly the legs, are also sites of atheroma (Fig. 4.3). These atheromatous changes can be localised or more generalised and vary from minor intimal irregularity to complete occlusions of varying lengths. They can occur anywhere in the vessel but there is a predilection for regions around bifurcations or in the adductor canal of the thigh. Clinically they can be asymptomatic, or symptomatic with symptoms varying from mild claudication to severe ischaemic tissue damage. Arteriography has been the main method of disease assessment in patients with significant symptoms referred to hospital, but because of the invasive nature of arteriography and the small associated morbidity it is not a practical screening tool, nor can its use be justified easily in asymptomatic patients. In addition arteriography has problems with inter- and intra-observer variability, especially if only single plane projections are available (Slot et al. 1981). Furthermore it gives anatomic rather than functional information, unless pressure gradients across stenoses are measured. This means that if there is a series of lesions along a vessel it may be difficult to identify the most significant stenosis and in these circumstances a relatively mild stenosis may have greater haemodynamic significance if there is a further stenosis upstream.

As described earlier many indirect tests of cardiovascular status have been developed over the years to try and assess the degree of involvement with atheroma (Fowkes 1988). These, together with the assessment of known risk factors, have formed the basis of many population surveys (Fowkes 1990). In comparison with arteriography these tests appear to be adequately sensitive in detecting angiographically proven disease – 95% sensitivity for ABPI <0.9 (Laing and Greenhalgh 1983) – but they provide no information on the extent or location of diseased segments.

The arteries of the lower limb from the common femoral artery at the groin to the lower popliteal artery are accessible to duplex ultrasound through most of their length. Problems may occur in the iliac and adductor canal segments and disease in these areas must often be inferred from changes in the waveform, but the precise level and severity of stenosis in these two areas may be difficult to assess. However, satisfactory sensitivity (82%–87%) and specifity (92%–96%) have been reported for ultrasound when compared with arteriography for the diagnosis of stenoses of more than 50% diameter reduction in the aorta, iliac and femoro-popliteal segments (Kohler et al. 1987; Cossman et al. 1989).

Whilst duplex performs well in comparison with angiography in symptomatic

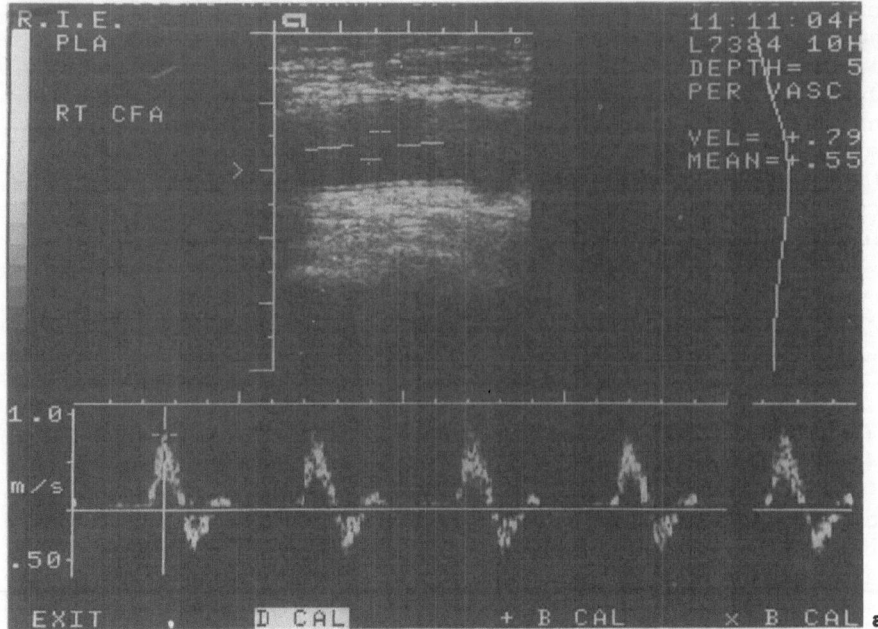

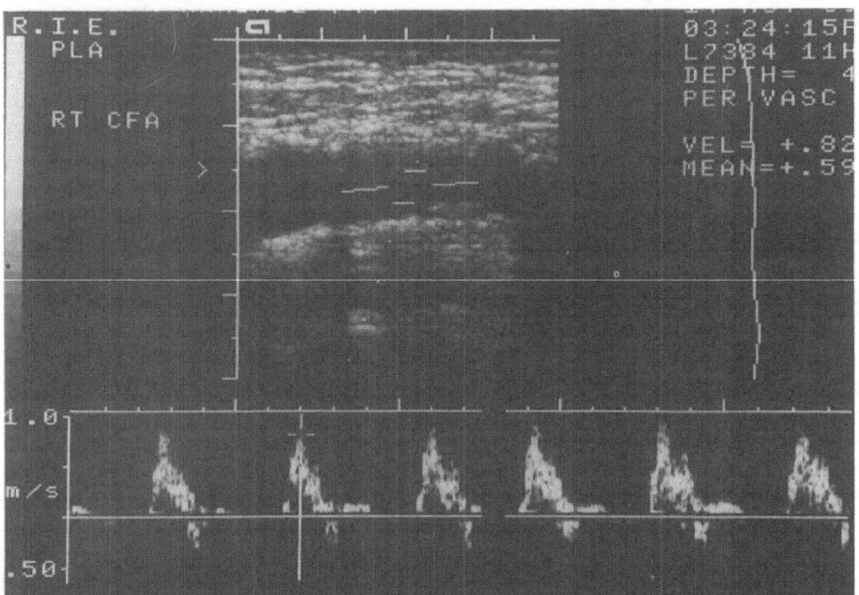

Fig. 4.3. a A normal segment of the common femoral artery showing a smooth intima. The trace below shows a normal tri-phasic waveform. **b** Moderate disease in the common femoral artery with irregular intima and a plaque. The waveform shows some spectral broadening and reduction of second and third components.

patients, there are still some problems with using it as a screening tool in the general population for the assessment of lower limb arterial disease (Fowkes 1990). In addition to the problems of equipment availability and operator expertise, which were discussed earlier, the overall validity and reliability of the technique in epidemiological studies have yet to be established. One study (Fowkes et al. unpublished observations) suggests that duplex ultrasound performs less well when trying to assess disease in the general population in comparison with ABPI and reactive hyperaemia tests. Duplex had a sensitivity of 78%, a specificity of 65% and a positive predictive value of only 19%. The best sensitivity (78%) was achieved by assessing the image, waveform and peak systolic velocity together; the best specificity (97%) and positive predictive value (67%) were obtained by assessment of the image alone. The reasons for this rather poor performance in the general population have not yet been identified. Some of the patients in whom ultrasound detected mild or moderate disease may not have had sufficient disease to show up on the ABPI and reactive hyperaemia testing. Conversely some individuals in the positive ABPI and reactive hyperaemia group may not have disease distributed in such a fashion that the scanning protocol used was able to detect it. Further experience and refinement of the technique is obviously required before the role of ultrasound as a screening tool can be defined more accurately in relation to peripheral vascular disease.

The ARIC study (ARIC 1989) is using ultrasound of the carotids and a popliteal artery as part of its assessment programme. Their methods are based on the measurements described by Pignoli et al. (1986) for assessing arterial wall thickness by measuring the intima/media thickness using computer-enhanced images. An alternative approach is suggested by Beach et al. (1989) who propose measuring the combined thickness of the superficial femoral artery and vein walls where they are in close apposition. This is also a computer-enhanced method but they feel that it overcomes some of the disadvantages of the intima/media thickness assessment.

Venous Disease

Chronic venous insufficiency is a major problem which can have significant morbidity such as varicose ulcers, swollen, painful legs, etc. Previously venography was the only way available to assess these individuals but the development of duplex ultrasound, and particularly colour techniques, allows non-invasive assessment of flow in the deep and superficial venous systems of the lower limb (Foley et al. 1989). The question of the prevalence of chronic venous insufficiency in the population is difficult to assess but it is to be hoped that ultrasound will provide an accurate method of tackling the problem. Although much work has been done on ultrasound in the diagnosis of deep vein thrombosis (Baxter et al. 1990), its role in the diagnosis and assessment of other venous abnormalities is less clear. The main factors which are responsible for chronic venous insufficiency are valvular reflux and chronic residual obstruction (Strandness 1990). The former can be well demonstrated by duplex ultrasound (Fig. 4.4) and the affected segments documented (van Bemmelen et al. 1989). Duplex ultrasound can also be used to identify signs of chronic occlusion but the effect of these changes on

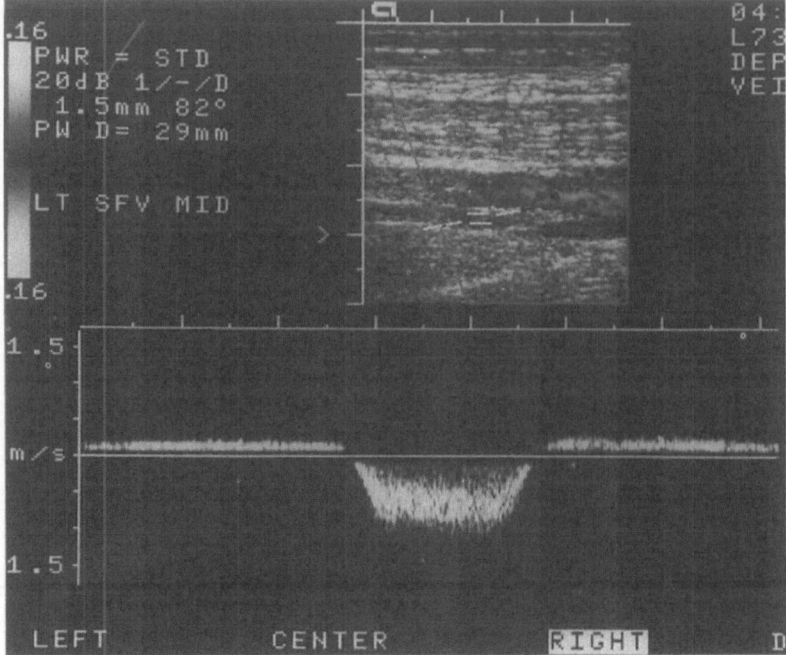

Fig. 4.4. Trace from superficial femoral vein in a patient with significant reflux (arrows). (Black and white image of colour Doppler examination.)

venous function is more difficult to assess and requires other tests such as venous outflow plethysmography for full evaluation (Strandness 1990).

The Future

Duplex ultrasound provides an excellent method of assessment of the peripheral arteries and veins. Studies comparing it with angiography in symptomatic patients have shown good sensitivity and specificity in these patients. The validity and reproducibility of results in a screening context have not yet been worked out satisfactorily but it shows great promise and it will undoubtedly have a role in the future.

In addition, improvements in equipment such as the development of smaller, lighter machines with powerful features such as colour Doppler will further enhance its value as a non-invasive vascular and blood flow imaging tool. Intravascular contrast agents are being developed and will have a major impact, initially in research and subsequently in clinical diagnosis, although it is unlikely that they will have significant applications in community screening studies in the immediate future.

References

ARIC (1989) The Atherosclerosis Risk in Communities (ARIC) study: design and objectives. Am J Epidemiol 129:687–702

Baker DW (1970) Pulsed ultrasonic blood flow sensing. IEEE Trans Biomed Eng 17:170–85

Barber FE, Baker DW, Nation AWC et al. (1974) Ultrasonic duplex echo Doppler scanner. IEEE Trans Biomed Eng 21:109–13

Baxter GM, McKechnie S, Duffy P (1990) Colour Doppler ultrasound in deep venous thrombosis: a comparison with venography. Clin Radiol 42:32–6

Beach KW, Issac CA, Phillips DJ, Strandness DE (1989) An ultrasonic measurement of superficial femoral artery wall thickness. Ultrasound Med Biol 15:723–28

Bluth E, Wetzners M, Stavros AT et al. (1988) Carotid duplex sonography: a multicenter recommendation for standardized imaging and Doppler criteria. Radiographics 8:487–506

Bond MG, Wilmoth SK, Enevold GL, Strickland HL (1989) Detectional monitoring of asymptomatic atherosclerosis in clinical trials. Am J Med 86 (Suppl 4A):33–6

Brown PB, Zwiebel WJ, Call GK (1989) Degree of cervical carotid artery stenosis and hemispheric stroke: duplex US findings. Radiology 170:541–3

Busuttil RW, Baker JD, Davidson RK et al. (1981) Carotid artery stenosis: Haemodynamic significance and clinical course. JAMA 245:1438–41

Caplan LR (1986) Carotid artery disease. N Engl J Med 315:886–8

Cardullo PA, Cutler BS, Bromwell Wheeler H (1986) Detection of carotid artery disease by Duplex ultrasound. J Diagn Med Sonogr 2:63–73

Carroll BA (1991) Carotid sonography. Radiology 178:303–13

Collin J, Walter J (1989) Is abdominal aortic aneurysm familial? Br Med J 229:493

Collin J, Walter J, Araccijo L, Lindsell D (1988) Oxford screening programme for abdominal aortic aneurysm in men aged 65–74 years. Lancet ii:613–15

Cossman DV, Ellison JE, Wagner WH et al. (1989) Comparison of contrast arteriography to arterial mapping with color-flow duplex imaging in the lower extremities. J Vasc Surg 10:522–8

Doppler CA (1843) Ueber das farbige licht der Doppelsterne und einiger anderer Gestirne des Himmels. Abhandlungen Koniglichbohmische Gesellschaft, Ser 2, 465–85

Foley WD, Middleton WD, Lawson TL, Erickson S, Quiroz FA, Macrander S (1989) Color Doppler ultrasound imaging of lower extremity venous disease. AJR 152: 371–6

Fowkes FGR (1988) The measurement of atherosclerotic peripheral arterial disease in epidemiological surveys. Int J Clin Epidemiol 17:248–54

Fowkes FGR (1990) Peripheral vascular disease: a public health perspective. J Public Health Med 12:152–9

Grimshaw GM, Hammer JD, Gannon MX (1990) Accuracy and reproducibility of screening in the community for aortic aneurysm in the elderly male using a portable ultrasound scanner. Presented at 22nd Annual Meeting of the British Medical Ultrasound Society, Harrogate, December 1990

Kohler TR, Nance DR, Cramer NM, Vandenburghe N, Strandness DE (1987) Duplex scanning for aorto-iliac and femoro-popliteal disease: a prospective study. Circulation 76:1074–80

Laing S, Greenhalgh RM (1983) The detection and progression of asymptomatic peripheral arterial disease. Br J Surg 70:628–30

O'Leary DH, Polak JF (1989) High resolution carotid sonography: past, present and future. AJR 153:699–704

Pignoli P, Tremoli E, Poli A et al. (1986) Intimal plus medial thickness of the arterial wall: a direct measurement with ultrasound imaging. Circulation 74:1399–1406

Poli A, Tremole E, Colombo A, Sirtori M et al. (1988) Ultrasonographic measurement of the common carotid arterial wall thickness in hypercholesterolaemic patients. Atherosclerosis 70:253–61

Ricotta JJ, Bryan FA, Bond MG et al (1987) Multicentre validation study of real-time (B-mode) ultrasound, arteriography and pathologic examination. J Vasc Surg 6:512–20

Riley WA, Freedman DS, Higgs NA et al. (1986) Decreased arterial elasticity associated with cardiovascular risk factors in the young: the Bogalusa Heart Study. Arteriosclerosis 6:378–86

Robinson ML, Sucles D, Perlmutter GS, Marinelli DL (1988) Diagnostic criteria for carotid duplex sonography. AJR 151:1045–9

Salonen R, Salonen JT (1990) Progression of carotid atherosclerosis and its determinants: a population-based ultrasonography study. Atherosclerosis 81:33–40

Satomura S (1957) Ultrasonic Doppler method for the inspection of cardiac functions. J Acoust Soc Am 29:1181–5

Satomura S (1959) Study of flow patterns in peripheral arteries by ultrasonics. J Acoust Soc Jpn 15:151–8

Scheinberg P (1988) Controversies in the management of cerebrovascular disease. Neurology 38:1609–16

Slot HB, Strijbosch L, Greep JM (1981) Intra-observer variability in single-plane aortography. Surgery 90:447–503

Strandness DE (1990) Duplex scanning in vascular disorders. Raven Press, New York, pp 175–83

van Bemmelen PA, Bedford G, Strandness DE (1989) Quantitative segmental evaluation of venous valvular reflux with ultrasonic duplex scanning. J Vasc Surg 10:425–31

Winslow CM, Solomon DH, Chassin MR et al. (1988) The appropriateness of carotid endarterectomy. N Engl J Med 318:721–7

Wolf PA, Kannel WB, Sorlie P, McNamara P (1981) Asymptomatic carotid bruits and risk of stroke. (The Framingham Study). JAMA 245:1442–5

Zwiebel WJ (1987) A primer of cerebrovascular ultrasound. Semin US, CT, MR 8:2–57

5 Magnetic Resonance Imaging

S.R. UNDERWOOD and R.H. MOHIADDIN

Atherosclerotic vascular disese is the commonest cause of death and disability in the Western world and so the potential contributions of magnetic resonance to its detection and assessment are important. The pathogenesis of the disease is controversial but the hypothesis that it is a response to injury, dating from the pioneering work of Virchow (1856) and recently reviewed by Ross (1986), has gained widespread acceptance. A variety of lesions affects the arterial wall, from non-protruding fatty streaks to more complex lesions consisting of lipid, smooth muscle, fibroblasts, and calcification. In addition to discrete atherotic lesions a localised and generalised sclerosis occurs. Sclerosis, or stiffness, can be demonstrated both in experimental disease in animals (Band et al. 1973) and in man (Banga and Balo 1961), and regression leads to reduced stiffness (Farrar et al. 1980).

Magnetic resonance provides techniques which allow all the aspects of atherosclerosis to be studied. Atheroma can be imaged directly, its size can be measured, its shape can be described, its lipid content can be assessed and its effects upon vascular haemodynamics can be studied (Mohiaddin et al. 1989a,b). In addition, arterial compliance, pulse-wave velocity, and the pattern of flow within the aorta can be studied (Bogren et al. 1989a; Mohiaddin et al. 1989c). It is thus a potential tool not only for the detection of disease but also for studying its natural history, risk factors and the effects of pharmacological or surgical interventions. In this chapter we review these techniques which, although at an early stage of development, promise an important role for magnetic resonance in the assessment of arterial disease.

The Assessment of Atherosis

Chemical Shift Imaging

Because signal is absent from flowing blood, high natural contrast exists between the arterial lumen and the wall, providing ideal circumstances for the demonstration of vascular anatomy and of atherosclerotic plaque. Magnetic resonance angiography and conventional imaging have been used to study atheromatous lesions in post-mortem human arteries, in experimental animal models, and in patients with atherosclerosis (Herfkens et al. 1983; Wesby et al. 1985). Most of these studies, however, have relied upon distortion of the arterial lumen and have not involved direct visualisation of the atheroma. Those that have, have used conventional techniques which image hydrogen irrespective of chemical environment, and have not exploited the chemical shift of resonant frequency between the hydrogen in water and fat (3.3 parts per million). Imaging nuclei with only a particular resonant frequency rather than the entire spectrum of resonant frequencies is called chemical shift imaging (Pykett and Rosen 1983). Several techniques of proton chemical shift imaging have been described (Brateman 1986). The original method of Dixon (1984) results in an image where pure water and pure fat produce no signal and tissues with a mixture of water and fat have higher signal. The method of Hinks and Quencer (1988) is a true water or fat imaging technique where only signal from water or from fat contributes to the image.

Fig. 5.1 shows the Dixon technique applied to imaging a lipid rich atheromatous plaque in a post-mortem human aorta. The in-phase image shows both the plaque and an area of extravascular fat without information on lipid concentration, but the subtraction image shows that the plaque has a high lipid content. In contrast, Fig. 5.2 shows a lipid deficient plaque with very low signal in the subtraction image. Patients with documented peripheral vascular disease have also been studied. Fig. 5.3 shows an image of the descending abdominal aorta in which the circular lumen is distorted by thickening of the wall with an atheromatous plaque. The Dixon subtraction image shows that this plaque has low lipid content. Fig. 5.4 shows another patient with a high lipid plaque.

The morphology and composition of arterial segments containing atheroma is of considerable importance. Plaques of different morphology (concentric or eccentric for example) have different effects on the arterial wall, such as the potential for thrombosis and the effect of arterial spasm (Wissler et al. 1985). The lipid content may also affect the propensity for fissuring, ulceration and thrombosis (Smith 1986). Other properties of atheroma that may be affected by the lipid content are the short- and long-term outcome of angioplasty, and the potential for regression (Potkin and Roberts 1988). Although little is known of these areas it is possible that there is a link between the composition of atheroma and its susceptibility to chemical or mechanical intervention. There is certainly evidence that regression can occur in experimental animals and that it may be possible to alter the rate of progression in man (Gotto 1983).

Velocity Mapping

Magnetic resonance velocity mapping is a technique in which the phase of the magnetic resonance signal is encoded with velocity in a chosen direction with respect to the imaging plane (Nayler et al. 1986; Underwood et al. 1987). Velocity profiles along an atheromatous vessel are particularly useful to localise disease and to measure the degree of stenosis. Blood flowing into a stenosed region must accelerate and the mean velocity increase is proportional to the reduction of cross sectional area (Mohiaddin et al. 1989d). This principle is illustrated in Fig. 5.5

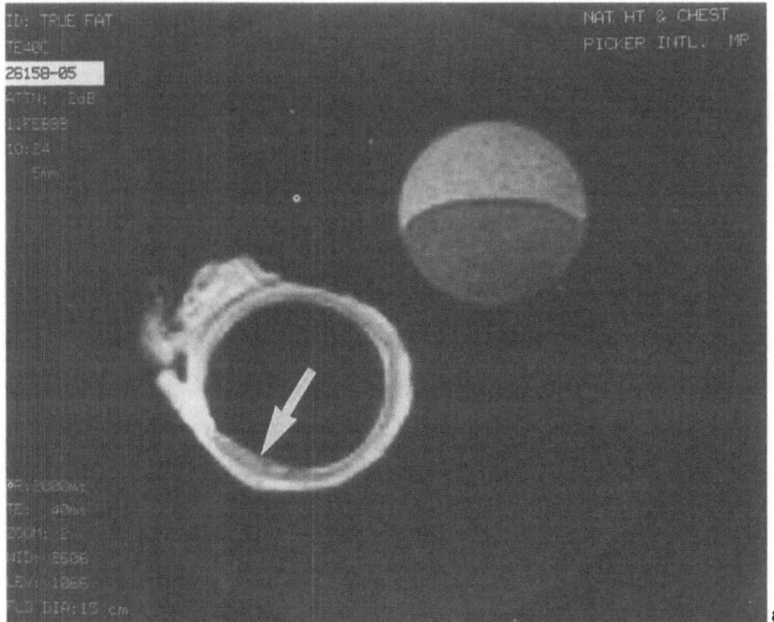

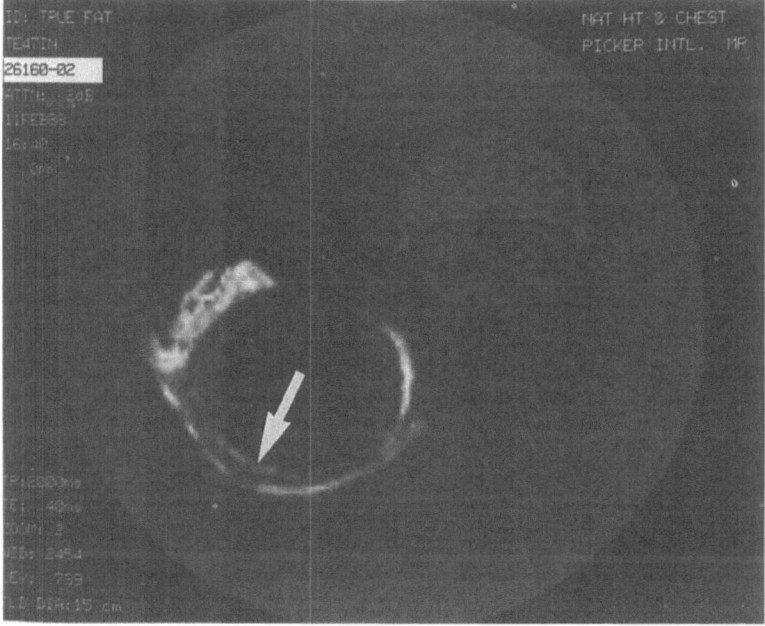

Fig. 5.1. a Post-mortem aortic specimen imaged using a conventional spin echo sequence together with a phantom containing oil floating on water. Atheroma involves approximately half of the circumference of the vessel and the largest accumulation is arrowed. b The subtraction image shows high signal in extravascular fat and in the lesion. The ratio of signal intensity compared with extravascular fat is 45%. (Reproduced with permission from Br Heart J (1989), 62:81–90.)

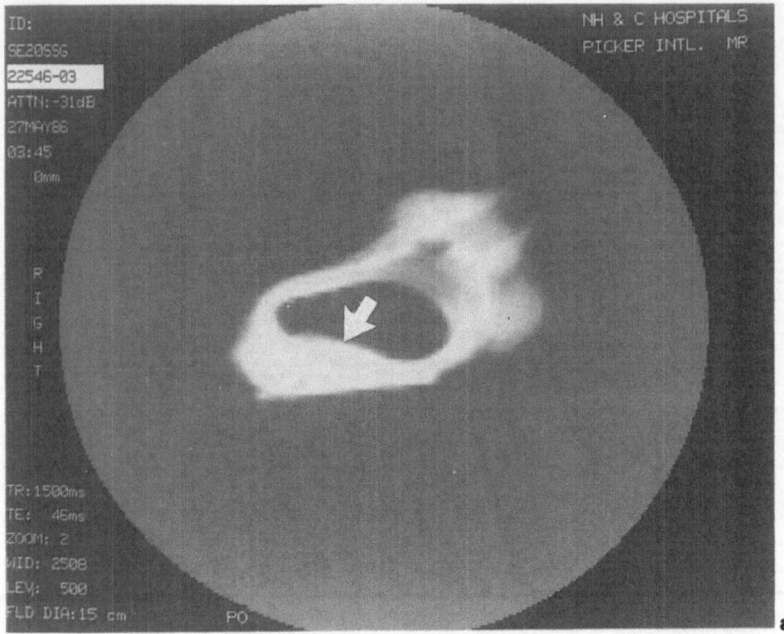

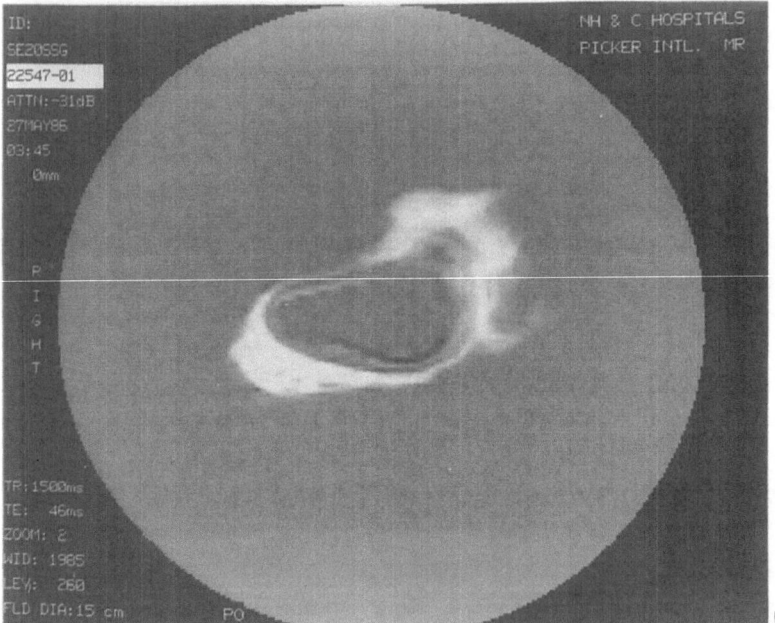

Fig. 5.2. a Conventional image of a lipid deficient plaque in a post-mortem aortic specimen (arrowed).
b The subtraction image shows very low signal (8%) compared with that from the extravascular fat.
(Reproduced with permission from Br Heart·J (1989), 62:81–90.)

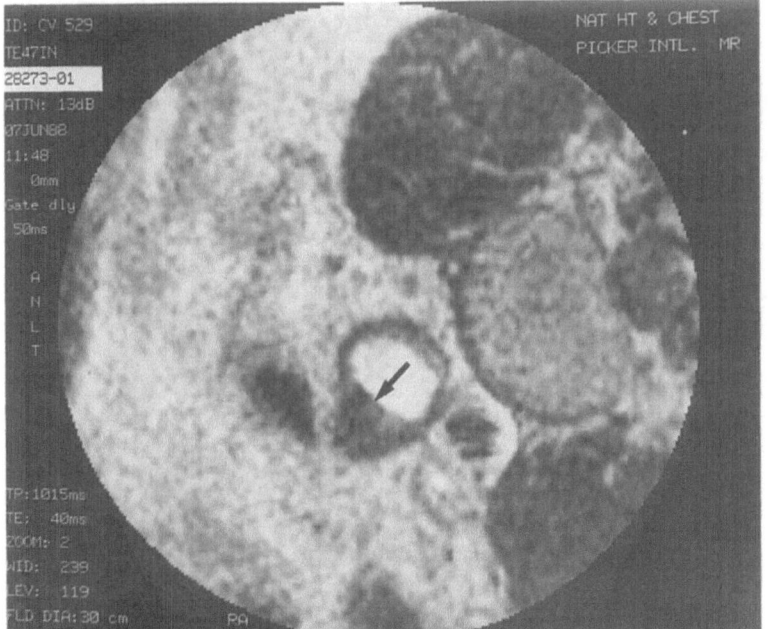

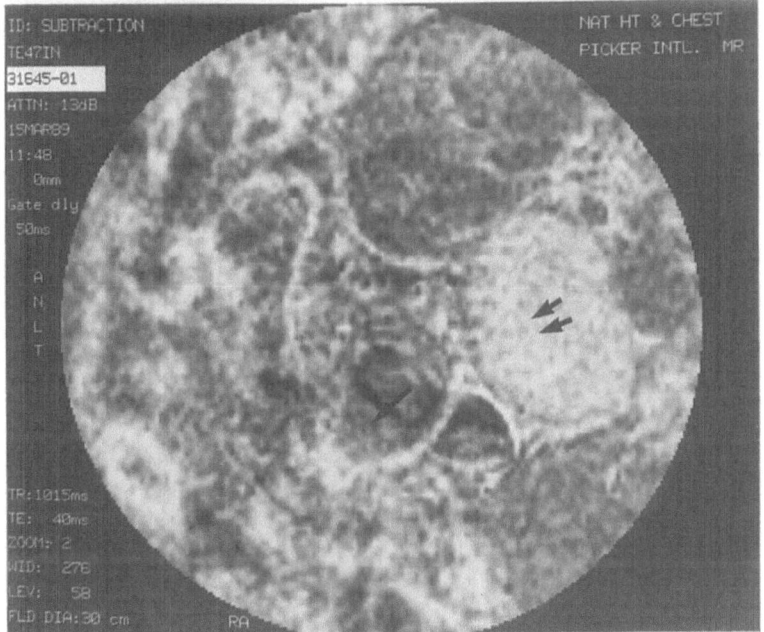

Fig. 5.3. a Spin echo image perpendicular to the descending aorta of a 60-year-old man with peripheral vascular disease, showing a large atheromatous plaque expanding the vessel wall (arrowed). **b** The subtraction image shows low signal in the lesion (arrow) indicating low lipid content. There is high signal from extravascular fat and from the bone marrow in the body of the lumbar vertebra (double arrow). (Reproduced with permission from Br Heart J (1989), 62:81–90.)

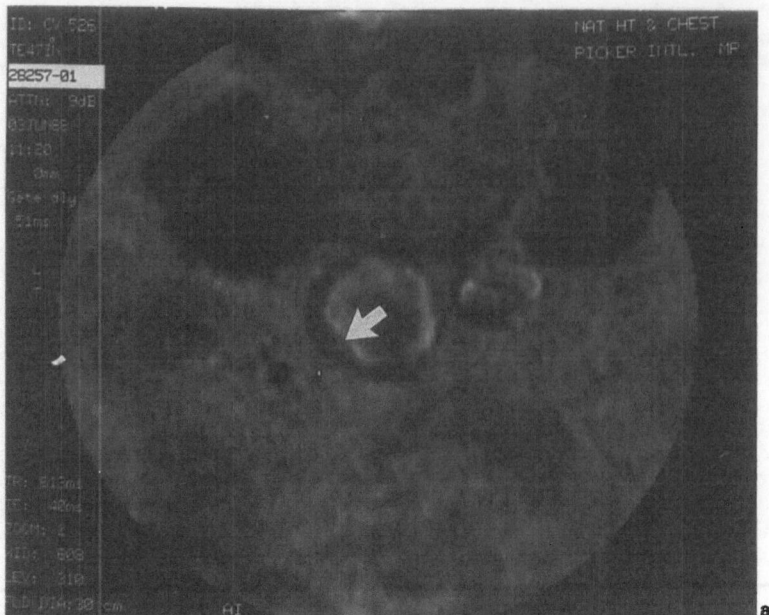

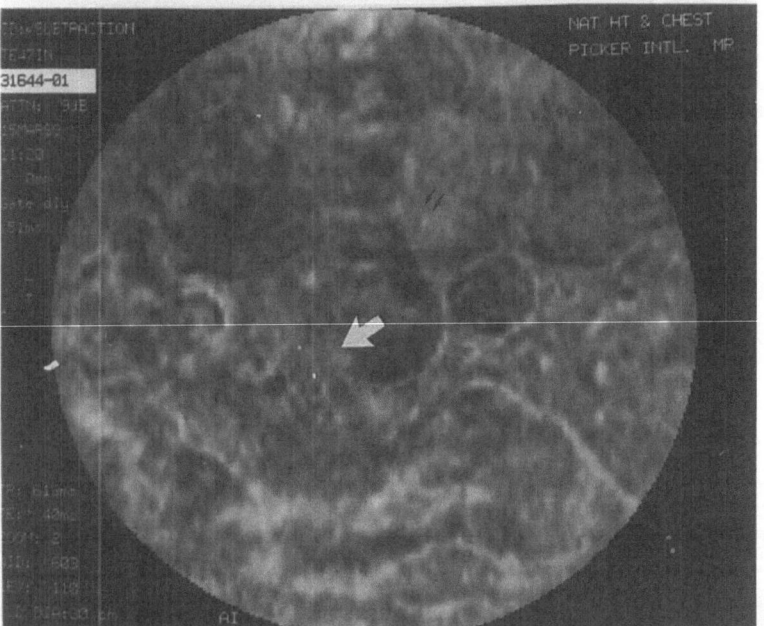

Fig. 5.4. a Spin echo image perpendicular to the descending aorta of a 57-year-old man with peripheral vascular disease. The wall is irregularly thickened with atheroma and the main accumulation is arrowed. The subtraction image **b** shows high signal within the lesion indicating high lipid content (similar to the marrow in the lumbar vertebral body [double arrow]). (Reproduced with permission from Br Heart J (1989), 62:81–90.)

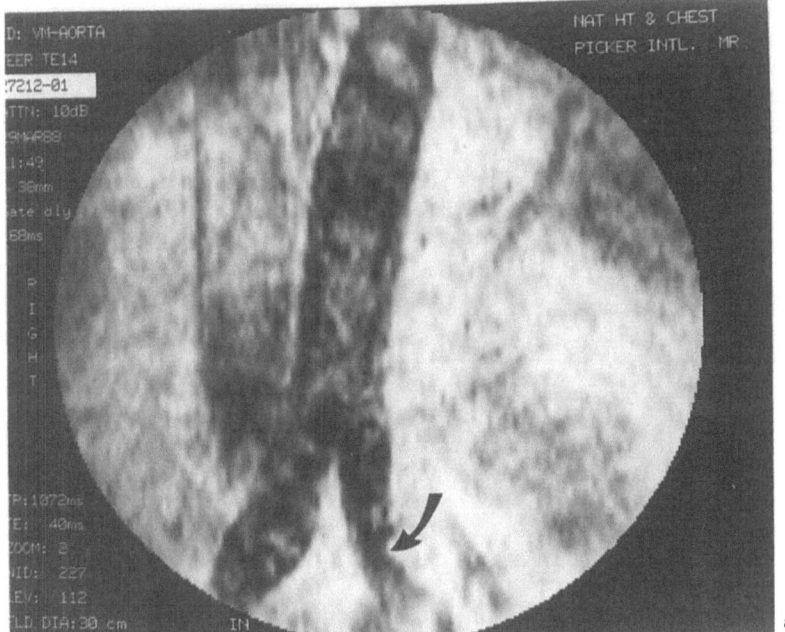

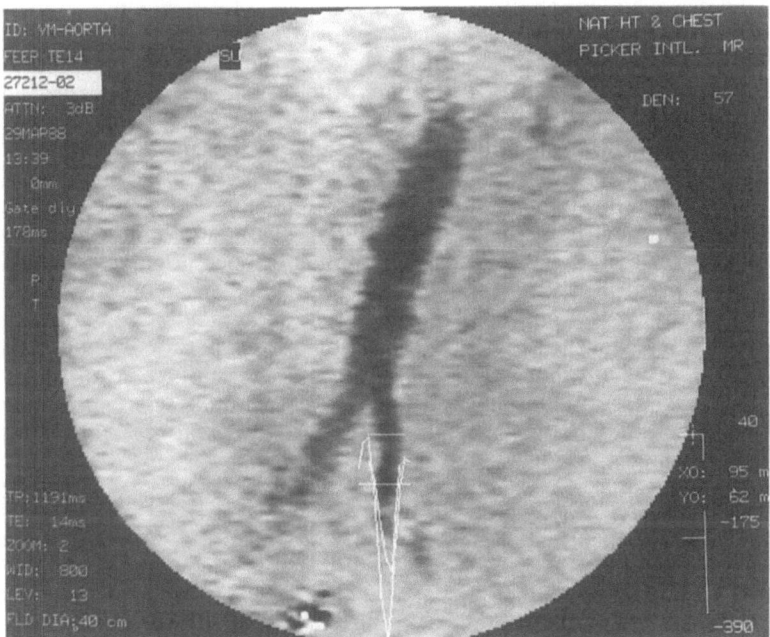

Fig. 5.5. a A 57-year-old male with atheromatous disease. The left iliac artery is narrowed (arrow) on the anatomic spin echo image. b The velocity map shows high velocities past the obstruction. (Reproduced with permission from Br Med Bull (1989), 45:968–990.)

which shows a patient with peripheral vascular disease causing a stenosis of the left iliac artery corresponding with high velocity.

Magnetic resonance vascular imaging shares with other tomographic techniques the disadvantage of interrupted display of the vessel. This problem has been solved in magnetic resonance angiography by three-dimensional acquisition and display. An alternative approach for velocity mapping is to acquire information in multiple slices and then to combine these in an image in which only the maximum of each pixel is displayed, irrespective of the slice in which it is seen.

The Assessment of Sclerosis

In 1904 Marchand recognised the association between fatty degeneration of the arteries and stiffening of the arterial wall and coined the term atherosclerosis (Aschoff 1933). Since then the fatty component of the disease (atherosis) has been extensively investigated, but less attention has been paid to the sclerosis. The elasticity or compliance of an artery is determined by the ratio of elastin, smooth muscle and collagen in the wall, and sclerosis arises from structural changes which include loss of elastin and an increase in collagen (Roach and Burton 1957). The changes are widespread and they are not necessarily associated with atherosis. Both components of the disease are closely linked, however, and both are adversely affected by hypercholesterolaemia, hypertension, and diabetes.

That atherosis is reversible has been shown in several studies (Blankenhorn and Kramsch 1989). The reversal of coronary atherosis has been confirmed in animals (Armstrong et al. 1970; Armstrong and Megan 1972) but reversal in humans is more difficult to demonstrate. Certainly, lesions in the peripheral circulation (Duffield et al. 1983) and the coronary arteries (Brensike et al. 1984) progress less rapidly following treatment of hyperlipidaemia, but regression has also been documented (Blankenhorn et al. 1987). Regression is more readily achieved in early lesions (Stary 1972), hence the importance of early detection and treatment. Atherosis will, however, usually be well advanced before it has a significant effect upon arterial flow or flow reserve, and many patients present well before this stage with sudden death from thrombosis superimposed upon an active and fissured plaque that is not necessarily flow limiting (Hackett et al. 1988). For this reason, strategies that rely upon the detection of sclerosis may be better suited to detect patients at a stage of the disease when reversal of risk factors may be beneficial.

Aortic Compliance

Compliance is the change in volume per unit change in pressure and it is a measure of stiffness and distensibility. The combination of elastic arteries and resistant arterioles constitutes a hydraulic filter enabling the intermittent cardiac output to be converted to a steady capillary flow. Part of the energy of left

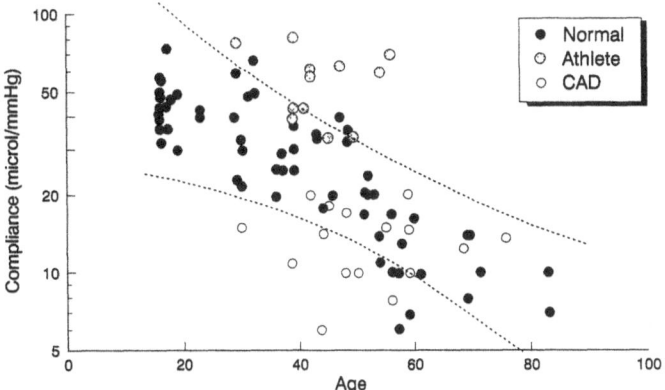

Fig. 5.6. Ascending aortic compliance displayed using a logarithmic scale and plotted against age. The 95% confidence intervals in a group of normal volunteers are shown, and compliance can be seen to fall with age. In a group of athletes, compliance is abnormally high and in patients with coronary artery disease (CAD) it is low. (Reproduced with permission from Br Heart J (1989), 62:90–96.)

ventricular contraction produces forward flow during systole, but the remainder is stored as potential energy in the distended arteries. During diastole, elastic recoil converts this potential energy into forward flow. A fall in aortic compliance increases the impedance to ventricular ejection and decreases capillary blood flow (Berne and Levy 1981; Wilcken et al. 1964). Aortic compliance, therefore, is a determinant of left ventricular afterload and it is important in patients with ventricular disease (Urschel et al. 1968).

Magnetic resonance imaging provides a direct non-invasive way of studying regional aortic compliance (Mohiaddin et al. 1989c). Images can be acquired at end diastole and end systole through the thoracic aorta and the change in volume between diastole and systole can be measured. The pulse pressure is measured using a conventional sphygmomanometer. In normal subjects, the ascending aorta is the most compliant region and compliance falls more distally. In athletes, regional aortic compliance is higher than normal, and in patients with coronary artery disease under the age of 50 years it is lower (Fig. 5.6). These changes are likely to be the result of structural changes in the aorta and they are a potential screening tool for the detection of vascular disease. Because there is overlap between normal compliance and compliance in patients with coronary artery disease, the test cannot have perfect sensitivity and specificity. Below the age of 50 years however there is much less overlap and the test is more sensitive.

Aortic Flow Wave Velocity

The repeated ejection of the ventricles generates pressure and flow waves in the aorta and pulmonary artery, and these pulsations are transmitted through the arterial tree (Milner 1982). The velocity of such waves depends principally on the distensibility and compliance of the vessel wall. If the vascular tree were rigid and blood incompressible, the motion of blood in the root of aorta would be transmitted peripherally instantaneously. Blood is indeed incompressible, but the vessels are distensible in varying degrees, and the pulse-wave velocity is therefore finite.

The flow-wave velocity can be measured by Doppler ultrasound (Caro et al. 1985) or by magnetic resonance velocity mapping (Mohiaddin et al. 1988). Although measurements cannot be made in the coronary arteries, the aorta is an ideal target because it is easy to image and because fatty streaks in the aorta and coronary arteries progress to raised lesions in parallel. In the magnetic resonance technique, high temporal resolution (10 ms) velocity maps are acquired in a plane perpendicular to the ascending and descending aorta and the time taken for the flow wave to travel between the two points is measured. Flow-wave velocity increases with age and with blood pressure and there is an inverse relationship between flow-wave velocity and regional compliance. In the same way that aortic compliance may be used to detect vascular disease, flow-wave velocity may also be a valuable parameter of the state of the arterial system.

Clinical Importance of Sclerosis

The clinical effects of sclerosis are subtle and it is not clear that it affects tissue perfusion in the same way as flow atheroma (Pickering 1963). Certainly, variations in compliance along an artery can lead to connective tissue overgrowth (Hasson et al. 1985) and so local compliance changes could influence disease progression. In addition, aortic compliance is an important component of left ventricular afterload and reduced compliance will not only increase myocardial oxygen demand, but it may also reduce oxygen supply through its affect on the normal aortic diastolic backflow which aids coronary perfusion (Klipstein et al. 1987; Bogren et al. 1989a,b). Similarly, reduced pulmonary arterial compliance may be a factor in the onset of right heart failure in patients with chronic lung disease (Bogren et al. 1989c). Another area of clinical importance may be following surgical repair of aortic coarctation, where persistent systolic hypertension may be the result of more widespread structural changes in the aorta (Rees et al. 1989).

Acknowledgement. This chapter is based with permission on: Mohiaddin RH (1990) Atherosclerotic vascular disease. In: Underwood SR, Firmin DN (eds) Magnetic resonance of the cardiovascular system. Blackwell Scientific Publications, Oxford, pp 131–142.

References

Armstrong ML, Megan MB (1972) Lipid depletion in atheromatous coronary arteries in rhesus monkeys after regression diets. Circ Res 30:675–80

Armstrong ML, Warner ED, Connor WE (1970) Regression of coronary atheromatosis in rhesus monkeys. Circ Res 27:59–67

Aschoff L (1933) In: Cowdry EV (ed) Arteriosclerosis: a survey of the problem. Macmillan, New York, pp 1–8

Band W, Goedhard WJ, Knoop AA (1973) Comparison of effects of high cholesterol intake on viscoelastic properties of the thoracic aorta in rats and rabbits. Atherosclerosis 18:163–72

Banga I, Balo J (1961) Elasticity of the vascular wall. I. The elastic tensibility of the human carotid as a function of age and arteriosclerosis. Acta Physiol Acad Sci Hung 20–21:237–47

Berne RM, Levy MN (1981) Cardiovascular physiology. CV Mosby, St Louis, pp 94–7

Blankenhorn DH, Kramsch DM (1989) Reversal of atherosis and sclerosis. Circulation 79:1–7

Blankenhorn DH, Nessim SA, Johnson RL, Sanmarco ME, Azen SP, Cashin-Hemphill L (1987) Beneficial effects of combined colestipol-niacin therapy on coronary atherosclerosis and coronary venous bypass grafts. JAMA 257:3233–40

Bogren HG, Klipstein RH, Firmin DN et al (1989a) Quantitation of antegrade and retrograde blood flow in the human aorta by magnetic resonance. Am Heart J 117:1214–22

Bogren HG, Mohiaddin RH, Klipstein RH et al. (1989b) The function of aorta in ischemic heart disease: A magnetic resonance and angiographic study of aortic compliance and blood flow patterns. Am Heart J 118:234–47

Bogren HG, Klipstein RH, Mohiaddin RH et al. (1989c) Pulmonary artery distensibility and blood flow patterns: a magnetic resonance study of normal subjects and of patients with pulmonary arterial hypertension. Am Heart J 118:990–9

Brateman L (1986) Chemical shift imaging: a review. AJR 146:971–80

Brensike JF, Levy RI, Kelsey SF et al. (1984) Effects of therapy with cholestyramine on progression of coronary arteriosclerosis: results of the NHLBI Type II Coronary Intervention Study. Circulation 69:313–24

Caro CG, Fish PJ, Goss DE (1985) Effect of isosorbide dinitrate on arterial haemodynamics in man. J Physiol [Lond] 365:93P

Dixon T (1984) Simple proton spectroscopic imaging. Radiology 153:189–94

Duffield RG, Lewis B, Miller NE, Jamieson CW, Brunt JN, Colchester AC (1983) Treatment of hyperlipidaemia retards progression of symptomatic femoral atherosclerosis: a randomised controlled trial. Lancet ii:639–42

Farrar DJ, Green HD, Wanger WD, Bond MG (1980) Reduction in pulse wave velocity and improvement of aortic distensibility accompanying regression of atherosclerosis in Rhesus monkey. Circ Res 47:425–32

Gotto AM (1983) Regression of atherosclerosis. Am J Med 70:989–91

Hackett D, Davies G, Maseri A (1988) Pre-existing coronary stenoses in patients with first myocardial infarction are not necessarily severe. Eur Heart J 9:1317–23

Hasson JE, Megerman J, Abbott WM (1985) Increased compliance near vascular anastomoses. J Vasc Surg 2:419–23

Herfkens R, Higgins C, Hricak H et al. (1983) Nuclear magnetic resonance imaging of atherosclerotic disease. Radiology 148:161–6

Hinks RS, Quencer RM (1988) Multislice chemical shift imaging by slice-selection gradient reversal. Magn Reson Imaging 6:22 (abstract)

Klipstein RH, Firmin DN, Underwood SR, Rees RSO, Longmore DB (1987) Blood flow patterns in the human aorta studied by magnetic resonance. Br Heart J 58:316–23

Milner WR (1982) Hemodynamics. Williams & Wilkins, Baltimore, pp 211–43

Mohiaddin RH, Firmin DN, Underwood SR et al. (1988) Magnetic resonance measurements of aortic flow wave velocity: the effect of age and disease. Magn Res Med 7:180 (abstract)

Mohiaddin RH, Longmore DB (1989a) MRI studies in atherosclerotic vascular disease: structural evaluation and physiological measurements. Br Med Bull 45:968–90

Mohiaddin RH, Underwood SR, Bogren HG et al. (1989b) Regional aortic compliance studied by magnetic resonance imaging: the effect of age, training and coronary artery disease. Br Heart J 62:90–6

Mohiaddin RH, Firmin DN, Underwood SR et al. (1989c) Magnetic resonance chemical shift imaging of human atheroma. Br Heart J 62:81–90

Mohiaddin RH, Sampson C, Firmin DN et al. (1989d) MR anatomic and flow imaging in peripheral vascular disease. Radiology 173:321 (abstract)

Nayler GL, Firmin DN, Longmore DB (1986) Blood flow imaging by cine magnetic resonance. J Comput Assist Tomogr 10:715–22

Pickering G (1963) Arteriosclerosis and atherosclerosis: the need for clear thinking. Am J Med 34:7–18

Potkin BN, Roberts WC (1988) Effects of percutaneous transluminal coronary angioplasty on atherosclerotic plaques and relation of plaque composition and arterial size to outcome. Am J Cardiol 62:41–50

Pykett I, Rosen L (1983) Nuclear magnetic resonance: in vivo proton chemical shift imaging. Radiology 149:197–201

Rees RSO, Somerville J, Ward C et al. (1989) Magnetic resonance imaging in the late postoperative assessment of coarctation of the aorta. Radiology 173:499–502

Roach MR, Burton AC (1957) The reason for the shape of the distensibility curves of arteries. Can J Biochem Physiol 35:681–90

Ross R (1986) The pathogenesis of atherosclerosis – an update. N Engl J Med 314:488–500

Smith E (1986) Development of atherosclerotic plaque. In: Shillingford J, Birdwood G (eds) Impact of research on the practice of cardiology. British Heart Foundation, London, pp 1–5

Stary HC (1972) Progression and regression of experimental atherosclerosis in rhesus monkeys. In: Goldsmith EF, Morr-Hankowsky J (eds) Medical primatology. S Karger, Basel, pp 356–67

Underwood SR, Firmin DN, Klipstein RH, Rees RSO, Longmore DB (1987) Magnetic resonance velocity mapping: clinical application of a new technique. Br Heart J 57:404–12

Urschel CW, Covel JW, Sonnenblick EH, Ross J, Braunwald E (1968) The effect of decreased aortic compliance on performance of the left ventricle. Am J Physiol 214:298–304

Virchow R (1856) Phlogose und thrombose im gefassystem, Gesammelte Abhandlungen zur wissenschaftlichen Medicin. Meidinger Sohn, Frankfurt-am-Main 458

Wesby GE, Higgins CB, Amparo EG, Hale JD, Kaufman L, Pogany AC (1985) Peripheral vascular disease: correlation of MR imaging and angiography. Radiology 156:733–9

Wilcken DE, Charlier AA, Hoffman JI, Guz A (1964) Effect of alterations in aortic impedance on the performance of the ventricles. Circ Res 14:283–93

Wissler RW, Vesselinovitch D, Davis HR, Lambert PH, Bekermeier M (1985) A new way to look at atherosclerotic involvement of artery wall and the functional effects. Ann NY Acad Sci 454:9–22

SECTION II

Descriptive Epidemiology

6 Historical Perspectives and the Basle Study

L.K. WIDMER and A. DA SILVA

Discoveries made by medical archeologists in Egyptian mummies of the 18th Dynasty have enabled the history of atherosclerosis to be traced back to between 1320 and 1580 AD (Leca 1971). Accounts by Thucydides indicated that an acute form of the disease was responsible for an epidemic of gangrene in young men in 450 AD (Ratschow 1939). The path of the disease can then be followed through the medieval universities of Salerno and Montpellier to the second half of the 16th century and the Renaissance with its outstanding illustrations by Leonardo da Vinci and Andreas Vesalius "de humano corpore fabrica" (Lyons and Petrucelli 1978). Internationally the disease was mentioned in communications in the 17th century by Fallopius Morgagni, vo Hilden, and in the discovery of the circulation by Harvey (da Silva and Widmer 1980, Lyons and Petrucelli 1978).

The first textbooks dealing with non-degenerative arteriopathy were Maurice Raynaud's "Asphyxie locale et gangréne symétrique des extrémités" (1852) and Leo Bürger's "Thrombangitis obliterans" (1908, 1924). However, peripheral vascular disease, which is of atherosclerotic origin, remained below the clinical cardiovascular horizon until the 1960s – in spite of the publications by Brown, Allen, Barker, Hines at the Mayo Clinic (Allen et al. 1946), Lewis (1936), and Ratschow (1939). Peripheral vascular disease remained a minor entity, first in the shadow of rheumatic diseases, and then during decades of atherosclerotic heart disease, in spite of textbooks issued by Klotz (1911), Ratschow (1939), Long (1933), Wright (1952), Hasse (1959), Hess et al. (1959), Wollheim and Zissler (1960), and Mitchell and Schwarz (1965).

The Basle Study

In 1959 L.K. Widmer, G. Hartmann and H. Staub of the Medical Clinic at the University of Basle – surprised that peripheral vascular disease was frequently detected in patients referred for other diagnoses – decided to investigate the epidemiology of peripheral vascular disease in a prospective field study of apparently healthy workers and employees of the Basle chemical industry (Marti 1961). The aim was to acquire information on peripheral vascular disease,

Fig. 6.1. The Railway Carriage made available by the Swiss Federal Railway Company and converted into an examination centre for history-taking, examinations of arteries (auscultation, oscillography), heart (ECG at rest and after maximum work capacity) and peripheral veins.

especially on early diagnosis, incidence, risk factors, and the course of the disease.

Apparently healthy people were called for an examination of the arteries and heart. The study sample comprised 4858 men aged 20–60 years (corresponding to the working population) and 1476 young women, (a smaller number because at that time few married women were working in the chemical industry). Each subject was examined three times, the examinations taking place in a railway car located in the factory grounds (Table 6.1; Fig. 6.1). The examination included assessments of peripheral arteries, coronary heart disease and peripheral vein disease (Widmer 1963; Widmer et al. 1967; da Silva and Widmer 1980; Nissen and Schweizer 1981). Persons with doubtful or pathological findings were invited for a check-up by a physician specialising in that particular field (Fig. 6.2). Data were analysed using punch-cards as 1965 was before the era of electronic data processing.

During the screening examination symptomatic cases with intermittent claudication were diagnosed on the basis of the questionnaire published by Rose (1962). Oligo- or asymptomatic cases were diagnosed by:

Auscultation (at rest and after three deep knee-bends) of arteries in the groin, the internal aspect of the thigh, the abdominal region, the supraclavicular space and – with breath held – in the neck.

Palpation of the pulse in the dorsalis pedis and tibialis posterior arteries; a grading

Table 6.1. Outline of the Basle Study

	Study I 1960–62	Study II 1965–68	Study III 1971–73	11-year follow-up 1982–83
Men	4858	3796	3528	239 peripheral vascular disease
Women	1476	829	711	239 controls

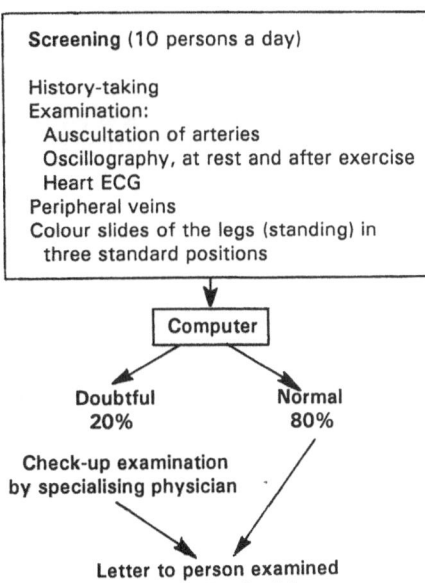

Fig. 6.2. Outline of screening and check-up examination. Screening was followed by a check-up examination performed by a specialised physician if the initial examination yielded doubtful or pathologic findings (as was the case in 20% of those screened).

system had originally been planned, but was abandoned because of poor reproducibility.

Electronic oscillography performed on the wrist, near the ankle and on the instep. The pressure of the semicircular cuffs was increased to 200 mmHg and slowly reduced whilst oscillations were recorded by differential manometers (da Silva and Widmer 1980). Oscillography was performed in the resting subject and after exercise (30 toe stands at a rate of one per second prior to recording of the ankle pulse).

For the check-up examination, subjects with intermittent claudication, oscillographic changes or arterial bruits – approximately 20% of those initially examined – were invited again to the railway carriage for examination by a specialising physician. The examination procedures which yielded doubtful findings were repeated. Subjects with pathological changes were invited for clinical examination.

Population Screening Methods

Intermittent claudication, the major symptom of peripheral vascular disease, was first observed in 1831 by the veterinary M. Boullay at the Académie Royale de Médicine in Paris (Boullay 1831). Peripheral vascular disease sufferers were described by Benjamin Brodie, Sergeant-surgeon to Queen Victoria, in 1846, as follows: "Such patients walk a short distance very well, but when they attempt more than this, the muscles seem to be unequal to the task and they can walk no further" (Brodie 1841, 1846). Characteristics of peripheral vascular disease – e.g. its appearance after a latent period, which in turn depends upon the load, and its rapid disappearance – were excellently discussed in 1887 by J. Charcot at the Salpêtrière in Paris (Charcot 1887). In 1936, Collens and Wilensky (in New York) and later, in 1949 Boyd, a vascular surgeon in Manchester, drew attention to the poor sensitivity of this symptom and proposed "walking machines" (Collens and Wilensky 1939, 1953; Boyd et al. 1949). According to Lindström, claudication appeared only on considerable exertion (Lindström 1958).

Artery auscultation is a method pioneered by Laënnec (Paris) who in 1826 concluded: "il me semble que les faites positifs et negatifs . . . tendent tous a prouver que le bruit de soufflet est le produit d'un simple spasme et ne suppose aucune lesion organique des arteres" (Laënnec 1826). Auscultation soon fell into oblivion, and found no mention in the first angiological textbooks. It was rediscovered, in 1952, by Edwards, a surgeon at Peter-Bent-Brigham Hospital in Boston: "it is but little appreciated that bruits may be heard over blood vessels characterized in a great variety of conditions, particularly those characterized by reduction in size of the vascular lumen" (Edwards and Levine 1952). A few years later in 1960, Strano (Rome) and Ratschow recommended auscultation of arteries as "indispensable for the early diagnosis of occlusive peripheral artery disease" (Ratschow 1959, Strano and Monaco 1953).

Pulse palpation – oscillography. In 1855, Vierordt, a physiologist at the University of Tübingen, judged the palpating finger method to be insufficient for making a complete judgement on the pulse and proposed using one of the first sphygmographs (Vierordt 1855). Sphygmography underwent a stormy development, but was not intended for detecting peripheral vascular disease, which was almost unknown, but for accurately measuring blood pressure (Marey 1878). Pachon's and von Recklinghausen's oscillometers, the first devices for detecting peripheral vascular disease, were replaced in the 1950s by oscillography thanks to the work of Gesenius, Ejrup (a Swedish internist) and Kappert, (the first Swiss specialist in cardiovascular medicine) (Kappert 1952; Laszt et al. 1956; da Silva and Widmer 1980). By recording pulses under varying cuff pressures, Ejrup achieved a sensitivity of 94% and a specificity of 67% for early stenotic lesions. Exercise doubles the sensitivity (Kocher and Widmer 1966). Fig. 6.3 compares oscillographic findings at rest and after exercise.

In the Basle Study sensitivity and specificity of screening procedures were determined by comparison with aortography (courtesy of Professors H. Ludin and H.E. Schmitt) in 147 men with a mean age ($\pm$ SD) of 57.6 ($\pm$ 8.5) years. Since peripheral vascular disease is, in general, not symmetrical we analysed individual extremities. There were 98 extremities with occlusions, 145 with stenosis and 35 without angiographic changes (da Silva and Widmer 1980) (Table 6.2).

Reported intermittent claudication and palpation of the pulse in the foot were poor indicators of early stage peripheral vascular disease, whereas combined oscillography and auscultation identified arterial changes with high sensitivity and specificity. For early diagnosis of stenotic lesions, the sensitivity of the combined procedure was 84%. The sensitivity and specificity of oscillography was similar to that of continuous-wave Doppler examination at rest and after exercise (da Silva et al. 1974).

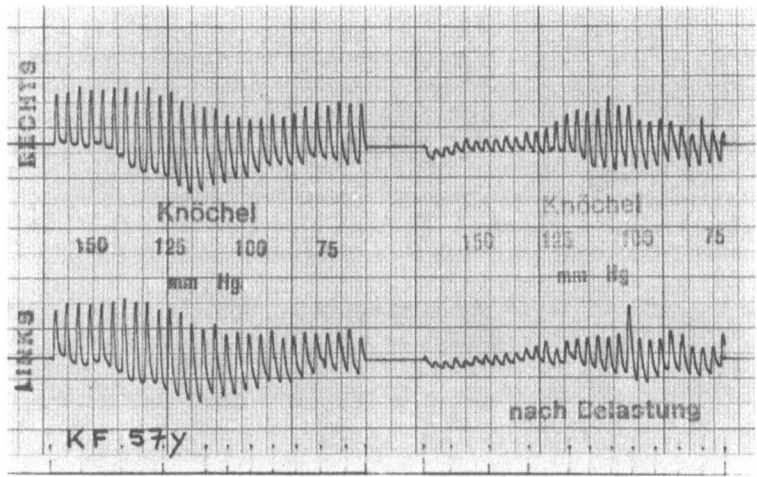

Fig. 6.3. Oscillography at rest and after exercise. The oscillogram of a 57-year-old man at rest shows apparently normal amplitudes (left). After exercise (30 toe-stands) there is a significant decrease due to bilateral stenosis of the femoral arteries (right) (confirmed by angiography).

Table 6.2. Validity of screening procedures in the Basle Study

Procedure	Sensitivity %	Specificity %
Intermittent claudication	13	100
Pulse palpation	18	94
Oscillography[a]	20 (38)	91 (71)
Auscultation	82	44
Combined oscillography and auscultation	88	93

[a] Amplitude differences above 25%; (after exercise).

Incidence and Prevalence

Peripheral vascular disease was recognised as an occlusive disease in 1575 by Fallopius: "Cessante ergo pulsu in aliquo membro indicat brevem vitae extinctionem in eadem parte" (da Silva and Widmer 1980; Lyons and Petrucelli 1978). In 1603, von Hilden, A Swiss Feldscherer (Feldscherer, a predecessor of actual surgeons), considered the medical and socioeconomic aspects of peripheral vascular disease "what help is it, that the head of a man is endowed with understanding and wisdom, when one of the limbs is diseased . . . because the spirits, especially those which the heart sends through the arteries to the limbs, are withheld" (Fabry von Hilden 1965). Two hundred years later Tiedemann, a pathologist in Heidelberg, recognised "that arteries are narrowed or closed by earthy concrements deposited in their walls, which are observed twice as much in men as in women" (Tiedemann 1843). An unprecedented increase in the information on peripheral vascular disease was generated in 1950 by prospective epidemiological field studies. Epstein and colleagues observed in 1957, that of the

30 000 workers of the amalgamated New York clothing industry, 4% of the men and 1% of the women suffered from intermittent claudication (Epstein et al. 1957). A few years later, comparative prevalence data were published by Mahmoud, Puchmayer on inhabitants of North Bohemia, Rose on 2000 London male civil servants, Richard on 3500 workers in Paris and Widmer on 4800 chemical workers in Basle (summarised by da Silva and Widmer 1980). The first 5-year incidence figures reported new occurrences of intermittent claudication in 5% of the men examined by Jouve in Marseille (Jouve et al. 1973), Kannel in Framingham and Puchmayer in North Bohemia (summarised by da Silva and Widmer 1980).

According to recent studies the prevalence of intermittent claudication is quite constant: 2% according to Criqui's 5-year follow-up of 612 men and women (average age 66 years), 2% according to Reunanen (10 962 inhabitants of Finland, 36–59 years old) and Hughson in Oxfordshire (population 40–69 years old) (Criqui et al. 1985a; Reunanen et al. 1982; Hughson et al. 1978). The prevalence of non-invasively diagnosed peripheral vascular disease is 11.7%, i.e. approximately five times higher than the symptomatic form.

Relationship to Coronary Heart Disease

In 1846 Sir Benjamin Brodie made the remark: "I say that this (angina pectoris) exactly corresponds to the sense of weakness and want of muscular power which exists in persons who have the arteries of the legs obstructed or ossified" (Brodie 1841, 1846). The idea that "L'angine de poitrine qui, dans l'espèce n'est que la claudication intermittente du coeur relève du même mécanisme" was generally accepted at the beginning of our century by pathologists and internists including Aschoff, Bier, Bumm and Otfried Muller (da Silva and Widmer 1980). Comparative prevalence data are provided by Epstein: intermittent claudication occurred 3.5 times more frequently in patients with coronary heart disease than in controls (Epstein et al. 1957). The Framingham study showed the 12-year incidence of intermittent claudication to be 10 times higher in coronary heart disease patients than in controls, and that the incidence of coronary heart disease is 5 times higher in claudicators than in controls (Kannel and McGee 1985). The correlation between the disease had already been proclaimed for clinical patients in 1953 by the London cardiologist McDonald: "Peripheral vascular disease in 24% of coronary cases and coronary heart disease in 39% of peripheral vascular patients, is now generally accepted, thanks to large series of patients examined both by coronary angiography and by recent epidemiological studies" (da Silva and Widmer 1980, Dormandy et al. 1989).

In the 1950s peripheral vascular disease was considered a rare occurrence which affected only the elderly and seldom led to amputation. This opinion was reinforced by the fact that peripheral vascular disease usually disappeared in health statistics under the heading "cardiovascular disease". The incidence is defined as the number of new cases of peripheral vascular disease occurring in a given period in a population "at risk", i.e. free of peripheral vascular disease at the beginning of the period. In the Basle Study II/III, 2630 men aged between 35 and 72 years were at risk. Symptomatic peripheral vascular disease was diagnosed on the basis of responses to Rose's questionnaire. Oligo- or asymptomatic peripheral vascular disease were diagnosed if one of three criteria was present:

1. Unilateral bruit in the distal thigh and/or in the pelvis as well as difference between the amplitude for the left and right extremities or between the pre- and post-exercise amplitude of more than 33%
2. Bilateral bruit in the distal thigh and/or in the pelvis as well as a difference between the amplitude for the left and right extremities or between the pre- and post-exercise amplitude of 25%–32%
3. Bilateral bruit in the distal thigh and/or pelvis.

Five-year statistics are only available for men because women were not representative of the working population. During a period of 5 years symptomatic peripheral vascular disease with intermittent claudication was observed in 51 men, the incidence being 19.4% per 1000 (Table 6.3); and was approximately three times higher in men aged 55–64 years than in men aged 35–44 years. Oligo- or asymptomatic peripheral vascular disease occurred in 57 men per 1000, and was approximately three times as frequent as the symptomatic form (da Silva and Widmer 1980).

Table 6.3. Five-year incidence of symptomatic and asymptomatic peripheral vascular disease in working men in Basle Studies II and III

Age (years)	At risk (*n*)	Symptomatic		Asymptomatic	
		n	per 1000	*n*	per 1000
35–44	1102	11	10	35	32
45–54	962	20	21	63	66
55–64	501	17	34	44	88
>65	65	3	47	9	138

The incidence rate of claudication was the same as those observed in Framingham and Marseille (27/1000 and 25/1000, respectively) (Jouve et al. 1973; Kannel and McGee 1985). Asymptomatic cases did not occur in the elderly!

Degenerative signs were observed in 97.6% of subjects with peripheral vascular disease and non-degenerative signs (acral lesions in Buerger's disease) in 2.4% of subjects. No aneurysms were found.

Risk Factors

The study of risk factors may be said to have gone through three generations (Feinstein 1985). Open studies gave rise to curiosity: "It is one of the penalties paid by those who enjoy the advantage of ease and affluence, and who live luxuriously. It is persons who eat too much, and drink too much fermented liquor, and do not take sufficient exercise that are especially liable to this disease, and not the labouring poor" (Brodie 1841, 1846). Comparative prevalence studies have analysed whether or not the given risk factor is found more frequently among persons with peripheral vascular disease than among those free of the disease. Those studies have enabled several factors (cold, alcohol, constitution) to be removed from the field of vision, but have not led to a "Unité de doctrine", mainly because of the heterogeneity of diseased and control groups. Prospective incidence studies have attempted to avoid these drawbacks. Risk factors investigated include the following:

1. *Sex*. In 1843, F. Tiedemann remarked that "men, by nature of a busy lifestyle associated with many bodily stresses, more often become very heated and also exposed to cold, in addition they enjoy alcoholic drinks, which are irritant to the circulatory system, more than do women" (Tiedemann 1843). Peripheral vascular disease is now known to occur three times more often in middle-aged men than in women. However, over the age of 70 years the ratio equals out.

2. *Diabetes mellitus*. In 1843, A. Faber, a Heidelberg pathologist, concluded on the basis of large clinical statistics than "an underlying process which scarcely ever spares the vessel walls is responsible for the predisposition of the diabetic to infection, gangrene, emaciation and cachexia" (Charcot 1858). In 1950 gangrene, besides coma, was the most frequent cause of diabetic death. According to Bell (1950), 13% of 50 000 diabetics died from gangrene. Current prospective studies indicate a rapid progression of peripheral vascular disease in diabetics with 40% developing critical limb ischaemia (controls 18%). The risk for amputation is up to 15 times higher among such patients than among controls (Stern 1988).

3. *Cigarette smoking*. During the Council of Basel in 1728 Pope Urban VIII pronounced that it was "a sin, to abuse the mouth as the chimney hood of Satan". At the beginning of this century smoking was only a doubtful risk factor. According to E. Romberg (1921) "the role played by the great abuse of alcohol, tobacco, coffee and tea is not worth mentioning. Of 49 patients with atherosclerosis indeed only 12 were alcoholics and only four indulged in the excessive use of tobacco". Ottfried Müller noted that "tobacco occasionally contributes to very severe vascular changes, whereas in other cases it could be enjoyed in abundance apparently without ill-effect" (da Silva and Widmer 1980). Speaking in 1911 about his first comparative studies, Erb, a Heidelberg neurologist, reported: "amongst my patients with claudication, smokers are three times, heavy smokers even six times, more frequent than in my general clientele" (Erb 1911). Half a decade later, large prospective studies demonstrated that the risk of peripheral vascular disease increases in proportion to cigarette consumption. Even for moderate smokers the risk is double that for non-smokers (Juergens et al. 1960; da Silva and Widmer 1980). In the Framingham study there was a four times higher claudication incidence in heavy smokers, clearly indicating that cigarette smoking heads the list of risk factors with a regression coefficient of 0.451 (Kannel and Shurtleff 1973; Kannel and McGee 1985). For the multifactorial actions of smoking on the pathogenesis of arteriopathy, see Greenhalgh (1981).

4. *Hypertension*. In 1832 J. Hope, a pathologist at St. Mary-le-Bone Parochial Infirmary in Edinburgh noted that "over-distension" of the arteries by the force of circulation is what, principally at least, produces arterial ossifications. They are found in stags long and often exercised in running and not in those which lead a tranquil life in the parks of the great" (da Silva and Widmer 1980). Marchand who was a pathologist at Strasbourg University and the father of the term atherosclerosis, reported in 1904 that "atheromatous degeneration and sclerosis is in the main, due to the action of a sustained increase in blood pressure (Marchand 1904). The role of essential hypertension, noted in 1933 by Sydenstricker in "Statistical study of arteriosclerosis", has been confirmed by numerous

prevalence studies which revealed that hypertension was present in 25% of peripheral vascular disease patients in comparison with 8% of the controls. Multifactorial analysis attributes a correlation coefficient of 0.178 for hypertension and intermittent claudication (Sydenstricker 1933).

5. *Hyperlipidaemia*. In 1852, Rokitansky, a Viennese pathologist, proposed that atherosclerosis was a result of "the endogenous secretion of a solidifying blastema from out of the blood onto the inner surface of arteries. Rudolf Virchow, on the other hand, spoke of "fat metamorphosis and imbibation" (Jouve et al. 1965; da Silva and Widmer 1980). Anitschokow's work in 1913 "Ueber die Veranderüng der kaninchen-aorta bei experimenteller Cholesterin-Steatose", Goffman's work on lipids and the Keys Seven Country study were convincing (Anitschkow and Chalatow 1913; da Silva and Widmer 1980; Keys 1980). The correlation with peripheral vascular disease was demonstrated in open studies by Bugar-Meszaris at the Istaan Hospital in Budapest and by Hasse in Darmstadt (Hasse 1959; da Silva and Widmer 1980). Comparative data collected by Schrade, Jouve, Juergens, Hartmann and Schwartz showed hypercholesterolaemia to be present in about 35% of peripheral vascular disease patients compared with controls (Juergens et al. 1960; Hartmann et al. 1962; Jouve et al. 1973; da Silva and Widmer 1980). In multifactorial analysis, cholesterol ranks third among six risk factors for intermittent claudication (Kannel and McGee 1985). "L'hypertriglycéridémie', se montre un stigme plus fidèle que l'hypercholestérinémie": this observation by Ponnoussamy in 1963 was confirmed by Greenhalgh and Sirtori, but refuted by the cautious observation of Leren and Ballentyne (summarised by da Silva and Widmer 1980). Actually high density lipoproteins are considered to be the most sensitive predictor.

6. *Combination of risk factors*. "In the majority of the cases the co-operation of several harmful factors is necessary to initiate the disease, intermittent claudication". In spite of this observation by Erb (1898) analysis was, for many years, directed mostly to individual risk factors and not to the risk profile. Jouve, however, observed in the 1960s that, several risk factors were present in more than half of peripheral vascular disease patients (Jouve et al. 1965). The "weight" of the combination was proved in Framingham, where intermittent claudication developed four times more frequently among risk-loaded people than among the risk-free. Similar figures were later recorded for asymptomatic peripheral vascular disease (Widmer et al. 1967, 1981; da Silva and Widmer 1980; Criqui et al. 1985a,b).

7. *Other factors*. The list of atherosclerotic risk factors has actually reached 261 parameters. Several psychological and environmental factors, such as hyperactive personality and lack of physical exercise, have not been sufficiently validated at the moment. The same is true for coagulation and fibrinolytic parameters in chronic peripheral vascular disease (Schlierf and Mörl 1987).

In the Basle study, symptomatic peripheral vascular disease occurred 1.7 times, and asymptomatic peripheral vascular disease two times more often among diabetics, cigarette smokers or men with initially elevated systolic blood pressure and lipid values than among risk-free men. On the contrary, no increase in the incidence of peripheral vascular disease was observed in men with initially elevated fasting blood sugar or post-load glucose, or in those who were initially overweight (da Silva and Widmer 1980) (Table 6.4). The combination of risk factors had a strong impact. The number of new occurrences of peripheral

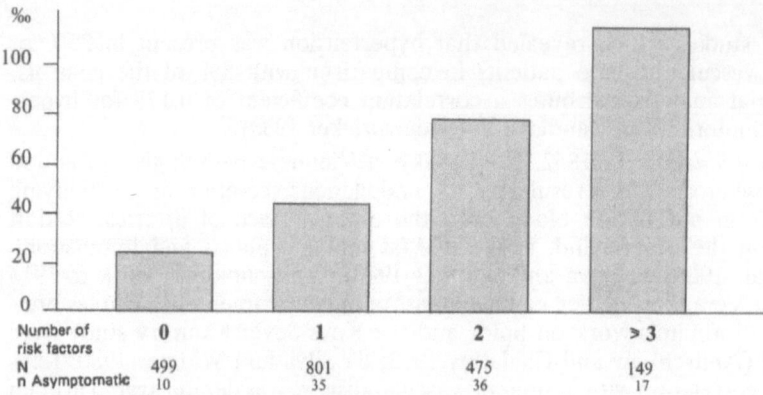

Number of risk factors	0	1	2	> 3
N	499	801	475	149
n Asymptomatic	10	35	36	17

Fig. 6.4. An impressive correlation between the number of risk factors present in a person at the beginning of the observation period and the 5-year incidence of peripheral vascular disease was noticed. Men with 3 risk factors at the beginning of observation developed peripheral vascular disease about 6 times more often than those initially free of risk factors.

Table 6.4. Five-year incidence of asymptomatic peripheral vascular disease per 1000 men classified by risk factors in Basle Studies II and III

Risk factor	Absent	Present
Elevated systolic blood pressure	48	85**
Elevated diastolic blood pressure	53	70
Elevated cholesterol	48	89*
Elevated triglycerides	43	2**
Elevated beta-lipoproteins	44	73*
Diabetes mellitus	52	17*
Cigarette smoking (>8/d)	39	80**
Elevated fasting blood sugar	52	53
Elevated post-load glucose	50	47
Overweight	60	49

*$p<0.05$; **$p<0.001$. Source: da Silva and Widmer (1980); Widmer et al. 1981.

vascular disease among men with three risk factors was six times the number among the risk-free (Fig. 6.4). In other words, 45% of all new peripheral vascular disease occurrences were in the small "high-risk" group of 8%.

Course of the Disease

A serious fate was ascribed to peripheral vascular disease in 1940 by Hines (Allen et al. 1946); half of all patients died within 3 years, so the author concluded that arterial surgery was superfluous. This gloomy picture was then considerably brightened in the 1950s by retrospective studies of Spaudling, Richards and Silberg. Not 55%, but only 5% of deaths, not 25% but 17% amputations occurred within 5 years (summarised by da Silva and Widmer 1980). A list of the wide range of mortality and amputation rates according to predictive parameters (Fontaine-stage, extension of the occlusion, type of treatment, pre-existing disease) has been published and discussed by da Silva and Widmer (1980).

The fate of the patients with peripheral vascular disease is mainly determined by coronary heart disease. In prospective population studies 5-year mortality for peripheral vascular disease subjects is 13% vs. 6% for controls, and 37% vs. 31%, for 11 years of follow-up. Cardiac death is responsible for 40% of deaths (Widmer et al. 1981; Reunanen et al. 1982; Kannel and McGee 1985). Cerebrovascular disease has already been discussed by the American Societies of Neurology and Cardiology in 1961 (Whisnant 1961). Studies limited to clinical history yielded 8%, Szillagy's angiographical study 14% and continuous wave Doppler examination up to 70% correlation between peripheral vascular disease and coronary heart disease. Interesting results were reported by Rose (1986). Data from population screened groups showed cerebrovascular death rates of approximately 9%.

In Basle an 11-year follow-up was performed on 239 peripheral vascular disease sufferers and 239 controls from the same population of the same age (drop-out rate was 8%). Some 121 subjects had died and 328 had survived. The mortality was analysed using data from the Swiss Federal Statistics Office. The survivors were examined and detailed histories taken. The death rate among peripheral vascular disease subjects (37%) was significantly higher than among the controls (13%) ($p<0.0001$). Death occurred at mean age ($\pm$ SD) of 66 ($\pm$ 7) years compared with 71 ($\pm$ 6) years in the controls. Further significant findings were observed: (a) for young peripheral vascular disease subjects (<55 years of age) the death rate was six times greater than among the age-matched controls; and (b) the death rate was correlated with the condition of the arteries at entry. Peripheral vascular disease subjects with several risk factors at the beginning of the study, had a three times higher death rate (37%) than those without or with one single risk factor (12%) ($p<0.0001$).

Analysis of the cause of death showed that cardiovascular disease – and predominantly coronary artery disease – was responsible for a significantly larger proportion of deaths among peripheral vascular disease subjects (57%) than among the controls (26%). Peripheral vascular disease itself was the cause of death in only 7%. Differences between the proportion of deaths due to tumours and the proportion due to cerebrovascular disease were not significant (Table 6.5).

Table 6.5. Cause of death in 30 controls and 80 cases of peripheral vascular disease during 11-year follow-up in Basle Study

	30 controls		80 cases	
	n	%	n	%
Degenerative cardiopathies	4	13	36	41*
Cerebrovascular diseases + hypertension	1	3	8	9
Arteriopathy	3	10	6	7
Cardiovascular disease (total)	8	26	50	57*
Tumours	12	40	28	32
Other diseases	10	34	10	11

*$p<0.001$.

Table 6.6. Concomitant cardiovascular disease and local complications in survivors with peripheral vascular disease and controls during 11-year follow-up in Basle Study

	199 controls (%)	129 survivors (%)
Angina pectoris	3.3	17.0**
Myocardial infarction	5.5	15.0*
Cerebrovascular diseases	4.0	12.4*
Acral lesions	3.5	7.7
Acute occlusions	1.5	8.0*
Amputations	0.5[a]	2.4

*$p<0.05$; **$p<0.001$; [a]due to a motorcycle accident.

Even in early peripheral vascular disease the fate of patients is burdened with non-fatal cardiac and cerebrovascular accidents (Table 6.6). Locally the disease is relatively benign: in 10 years 20% developed claudication, and 8% critical limb ischaemia. Amputation with its considerable personal and socioeconomic consequences was necessary in only 2.4% (Gautier 1978; Stirnemann et al. 1986; Zemp and Widmer 1988; Dormandy et al. 1989). Significant differences were observed for concomitant cardiovascular disease, carotid pathology (diagnosed by continuous wave Doppler), pronounced limitations in walking, and local sequelae.

Conclusion

This historical review reveals that interest in peripheral vascular disease has extended over several centuries, and that in more recent years several epidemiological studies have been conducted in this field. The Basle Study was one of the first epidemiological studies and shows that:

Early diagnosis of peripheral vascular disease is possible with simple methods within the range of a practitioner interested in cardiovascular disease

Early diagnosis seems to be useful as it frequently reveals concomitant cardiovascular disease which determines the fate of peripheral vascular disease patients

Early diagnosis allows prophylactic measures to be taken against risk factors and ischaemia, which is threatening to life and limbs.

The follow-up part of the study shows that:

Coronary heart disease, present in 41% who died and in 41% who survived, determines life expectancy

The local course of peripheral vascular disease is relatively benign

A complete cardiovascular check-up is necessary even for disease in its early stages

Prophylaxis is necessary against criticial limb ischemia and risk factors.

Acknowledgements. We are indebted to our colleagues Stähelin, Hartmann, Nissen and Schweizer, the factory physicians (Ciba, Geigy, Roche, Sandoz), our nursing staff Crista Haenel, Alice Metz, Gaby Regli and to F. Rösel, Head of the University Information. We also thank the participants and sponsors (Swiss Society of Angiology, Swiss Society of Phlebology, Swiss Foundation of Cardiology and the Basle Chemical Industry).

References

Albutt C (1915) Diseases of the arteries. Macmillan, New York

Allen EV, Barker NW, Hines EA (1946, 1955) Peripheral vascular diseases. Saunders, Philadelphia.

Anitschkow N, Chalatow SS (1913) Ueber experimentelle Cholesterin-Steatose und ihre Bedeutung für die Entstehen einiger pathologischer Prozesse. ZBL Allg Path u Path Anat 241

Aschoff L (1925) Vorträge über Pathologie, gehalten an den Universitäten und Akademien Japans im Jahr 1924. Gustav Fischer, Jena

Bell ET (1950) A post-mortem study of 1214 diabetic subjects with special reference to the vascular lesions. Proc Diab Assoc 10:62

Boullay M (1831) Oblitération des artères fémorales. Méd Vét. Arch Gén Méd 227:425

Boyd AM, Ratcliff AH, Jepson RP, James GWH (1949) Intermittent claudication (a clinical study). J Bone Joint Surg 31B:325

Brodie BC (1841) Lectures on mortification delivered at the medical theatre of St. George's Hospital. Lecture V. New series of London Med Gazette, p. 714

Brodie BC (1846) Lectures illustrative of various subjects in pathology and surgery, p. 360. Longman, Brown, Green and Longmans, London

Buerger L (1908) Thrombangiitis obliterans: a study of the vascular lesions leading to presenile spontaneous gangrene. Am J Med SCT 136:567

Buerger L (1924) The circulatory disturbances of the extremities. Saunders, Philadelphia

Charcot J (1858) Sur la claudication intermittente observée dans un cas d'oblitération complète de l'une des artères iliaques primitives, p. 225. rend hebd des séances et mémoires Soc Biologie, Paris

Charcot J (1887) Claudication intermittente et diabète. Leçons du Mardi à la Salpêtrière. Prog Méd 6:99

Collens WS, Wilensky ND (1936, 1953) Peripheral vascular diseases. Diagnosis and treatment. Thomas, Springfield

Criqui MH, Fronek A, Barrett-Connor E, Klauber MR, Gabriel S, Goodman D (1985a) The prevalence of peripheral arterial disease in a defined population. Circulation 71:516–21

Criqui MH, Coughlin SS, Fronek A (1985b) Noninvasively diagnosed peripheral arterial disease as a predictor of mortality: results from a prospective study. Circulation 71:768–73

da Silva A, Widmer LK (1980) Occlusive peripheral artery disease. The Basle Study. Huber, Bern

da Silva A, Mall Th, Widmer LK (1974) Zur Leistungsfähigkeit des Doppler-Verfahrens für den Nachweis der arteriellen Verschlusskrankheit. Vergleich mit der Oszillographie. Schweiz Med Wochenschr 104:1596

Dormandy J, Mahir M, Ascady G et al. (1989) Fate of the patient with chronic leg ischaemia. J Cardiovasc Surg 30:50–7

Edwards EA, Levine HD (1952) Peripheral vascular murmurs. Mechanism of production and diagnostic significance. Arch Intern Med 90:284

Epstein FH (1987) Coronary heart disease epidemiology – perspectives. In: Schlierf G, Mörl H (eds) Expanding horizons in atherosclerosis research. Springer-Verlag, Berlin

Epstein FH, Arbor A, Boas EP, Simpson R (1957) The epidemiology of atherosclerosis among a random sample of clothing workers of different ethnic origins in New York City. I. Prevalence of atherosclerosis and some associated characteristics. Associations between manifest atherosclerosis, serum lipid level, blood pressure, overweight and some other variables. J Chron Dis 5:300, 329

Erb W (1898) Ueber das "intermittierende Hinken" und andere nervöse Störungen infolge von Gefässerkrankungen. Dtsch Zschr Nervenheilk 13:1

Erb W (1911) Kliniche Beiträge zur Pathologie des intermittierenden Hinkens. Münch Med Wschr 2:2487

Fabry von Hilden W (1965) Vom heissen und kalten Brand: Nach der 1603 publizierten zweiten deutschen Ausgabe bearbeitet und herausgegeben von E. Hintzsche. Huber, Bern

Feinstein AR (1985) Clinical epidemiology – The architecture of clinical research. Saunders, Philadelphia

Gautier R (1978) Les ischémies gravissimes de jambe. Doc Méd Oberval

Greenhalgh RM (ed) (1981) Smoking and arterial disease. Pitman Medical, London

Hartmann G, Widmer LK, Creux G, Greensher A, Kaufmann L (1962) Basler Studie – Serumlipide bei 1900 berufstätigen Männern: Beziehung zu Körpergewicht, Konstitution und Blutdruck. Zeitschr Krforsch. Steinkopff, Darmstadt

Hasse HM (1959) Statistische Daten zur Prognose der arteriellen Verschluss-krankheiten. In: Ratschow M (ed) Angiologie. Thieme, Stuttgart, p. 609

Herrlinger R (1967) Geschichte der medizinischen Abbildung. Moos, München

Hertzer NR, Young JR, Beven EG et al. (1984) Late results of coronary bypass in patients with peripheral vascular disease: five year survival according to age and clinical cardiac status. Cleveland Clin Quart 53:133

Hess H, Kunlin J, Mittelmeier H, Schlicht L, Stampfl B (1959) Die obliterie-renden Gefässerkrankungen. Urban and Schwarzenberg, München

Hughson WG, Mann JI, Garrod A (1978) Intermittent claudication: prevalence and risk factors. Br Med J i:1379–81

Jelnes R, Gaardsting O, Hougaard IK (1988) Fate in intermittent claudication: outcome and risk factors. Br Med J 293:1137

Jonason T, Bergström R (1987) Cessation of smoking in patients with intermittent claudication. Effects on the risk of peripheral vascular complications, myocardial infarction and mortality. Acta Med Scand 221:253

Jouve A, Gérard R, Vague Ph, Gaymard PJ (1965) Les troubles du métabolisme des glucides dans l'artériosclérose (maladie coronarienne, artériopathie dégénérative des membres inférieurs). Bull Soc Méd Hôp (Paris) 116:1323

Jouve A, Sommer A, Avierinos Ch, Fondarai J (1973) Premiers résultats d'une enquête prospective sur les maladies cardiovasculaires dans une grande administration. Arch Mal Coeur 66:25

Juergens JL, Barker NW, Hines EA Jr (1960) Arteriosclerosis obliterans: review of 520 cases with special reference to pathogenic and prognostic factors. Circulation 21:188–95

Käilerö KS (1981) Mortality and morbidity in patients with intermittent claudication as defined by venous occlusion plethysmography. A ten year follow-up study. J Chronic Dis 34:455–62

Kannel WB, Shurtleff D (1973) Cigarettes and development of intermittent claudication. The Framingham Study. Geriatrics 28:61–8

Kannel WB, McGee D, Gordon T (1976) A general cardiovascular risk profile: the Framingham Study. Am J Cardiol 38:46–51

Kannel WB, McGee DL (1985) Update on some epidemiological features of intermittent claudication: the Framingham Study. J Am Geriatr Soc 22:13–18

Kappert A (1952) Die Diagnostik der peripheren Durchblutungsstörungen mit Hilfe des Ruhe- und Arbeits-Oszillogramms. Praxis 41:980

Keys A (1980) Seven Countries: a multivaried analysis of death and coronary heart disease. Harvard University Press, Cambridge, Mass

Klotz O (1911) Arteriosclerosis. University of Pittsburgh School of Med, Pittsburgh

Kocher R, Widmer LK (1966) Zur Leistungsfähigkeit der Oszillographie. Vergleich von Aortogramm und Oszillogramm. Cardiologia 49:166

Laënnec RTH (1826) Traité de l'ausculation médiate des maladies des poumons et du coeur (tome second). JS Chaudé, Paris

Lake M, Pratt GH, Wright IS (1942) Arteriosclerosis and varicose veins: occupational activities and other factors. JAMA 119:696

Lasila R, Lepantalo M, Lindfors O (1966) Peripheral arterial disease – natural outcome. Acta Med Scand 220:295–301

Laszt L, Meier R, Müller A (1956) Compt 2ème Congrès International d'Angéiologie. Ed. Universitaire, Fribourg

Leca AP (1971) La Médecine Egyptienne au Temps des Pharaons. Roger Dacosta, Paris

Lewis TH (1936) Vascular disorders of the limbs. Described for practitioners and students. Macmillan, London

Lindström BL (1958) Function test for peripheral arterial circulatory insufficiency in the lower extremities in obstructive arterial disease. Acta Chir Scand [Suppl] 242

Long (1933) In: Cawdry (ed) Arteriosclerosis: the survey of the problem. Macmillan, New York

Lyons AS, Petrucelli RJ (1978) Medicine. An illustrated history. Abrams, New York

Marchand (1904) Ueber Arteriosklerose (Athero-Sklerose). Vrhl Kongr Inn Med Leipzig XXI, 23

Marey EJ (1878) La méthode graphique dans les sciences expérimentales et principalement en physiologie et en médecine. Masson, Paris

Marti WK (1961) Häufigkeit klinisch nicht erkannter peripherer Arterienverschlüsse bei 461 Hospitalisierten. Schweiz Med Wochenschr 91:1424

Mitchell JRA, Schwartz CI (1965) Arterial disease. Blackwell, Oxford

Nissen C, Schweizer W (1981) Koronare Herzkrankheit – Basler Studie. In: Widmer LK, Stählin HB, Nissen C, da Silva A (eds): Die Venen-Arterien-Krankheiten, koronare Herzkrankheit bei Berufstätigen – Basler Studie 1959–1978. Huber, Bern

Ratschow M (1939) Die peripheren Durchblutungstörungen. Steinkopff, Dresden

Ratschow M (ed) (1959) Angiologie, Pathologie, Klinik und Therapie der peripheren Durchblutungstörungen. Thieme, Stuttgart

Raynaud AGN (1852) De l' asphyxie locale et de la gangréne symétrique des extrémitiés. Rignoux, Paris

Reunanen A, Takkunen H, Aromaa A (1982) Prevalence of intermittent claudication and its effects on mortality. Acta Med Scand 211:249–56

Romberg E (1921) Lehrbuch der Krankheiten des Herzens und der Blutgefässe. (3. ed). Enke, Stuttgart

Rose FC (ed) (1986) Stroke: epidemiological, therapeutic and socio-economic aspects. Royal Society of Medicine, London

Rose GA (1962) The diagnosis of ischaemic heart pain and intermittent claudication in field surveys. Bull Org Mond Santé 27:645–58

Schlierf G, Mörl H (1987) Expanding horizons in atherosclerosis research. Springer-Verlag, Berlin

Stähelin H, Spahr A, Schweizer W (ed) (1985) Cardiovaskuläre Risikofaktoren bei Kindern und Jugendlichen. Huber, Bern

Stern PH (1988) Occlusive vascular disease of the lower limbs: diagnosis, amputation surgery and rehabilitation. A review of the Burke experience. Am J Phys Med Rehabil 67:145

Stirnemann P, Nachbur B, Oesch A (1986) In: Zemp E, Widmer LK (eds). Angiologie 86. Huber, Bern

Strano A, Monaco R (1953) Phonoarteriographic registration of the autochtonic murmurs on peripheral arteries due to organic arteriopathies. Cardiologia 23:230

Sydenstricker E (1933) Statistical study of arteriosclerosis. In: Cawdry (ed) Arteriosclerosis, a survey of the problem. Macmillan, New York, p. 131

Tiedemann F (1843) Von der Verengung und Schliessung der Pulsadern in Krankheiten. Gross, Heidelberg

Vierordt K (1855) Die Lehre vom Arterienpuls in gesunden und kranken Zuständen, gegründet auf eine neue Methode der bildlichen Darstellung des menschlichen Pulses, Vieweg, Braunschweig

Virchow R (1856) Gesammelte Abhandlungen zur wissenschaftlichen Medicin. Meidingen Sohn, Frankfurt

Widmer LK (1963) Morbidität an Gliedmassenarterien-Verschluss bei 6400 Berufstätigen – Basler Studie. Bibl Cardiol 13:67

Widmer LK, Greensher A, Kannel WB (1964) Occlusion of peripheral arteries. A study of 6400 working subjects. Circulation 30:836–42

Widmer LK, Kaufmann L, Hartmann G et al. (1967) Organisation der Basler Studie über Arterien-, Venen- und Herzkrankheiten. Schweiz Med Wochenschr 97:99

Widmer LK, Stähelin HB, Nissen C, da Silva A (1981) Venen-, Arterien-Krankheiten, koronare Herzkrankheit bei Berufstätigen – Basler Studie 1959–1987. Huber, Bern

Whisnant JP (1961) Atherosclerosis and its prevalence. In: Millikan, Siekert, Whisnant (eds) Cerebral vascular diseases. Grune & Stratton, New York

Wollheim E, Zissler J (1960) Handbuch der Inneren Medizin (4th edn). Springer-Verlag, Berlin

Wright IS (1952) Vascular diseases in clinical practise. Year Book, Chicago

Zemp E, Widmer LK (eds) (1986) Angiologie 86. Huber, Bern

7 Large Vessel and Isolated Small Vessel Disease

M.H. CRIQUI, R.D. LANGER, A. FRONEK, H.S. FEIGELSON, M.R. KLAUBER

For more than a decade our group has been involved in the study of the epidemiology of peripheral vascular disease, using both traditional clinical assessments as well as highly accurate and reliable non-invasive measurements (Criqui 1985a,b,c, 1989). The non-invasive measures used are able to assess hemodynamics in the lower extremities from the largest (common femoral) to the smallest (arteries less than 2 mm in diameter) components of the peripheral arterial circulation. In evaluating circulation by these various non-invasive tests, we found that three groups of patients could be differentiated: a group with reduced flow in large vessels but normal small vessel (<2 mm diameter) flow; a group with normal large vessel flow but reduced small vessel flow; and a group with reduced flow in both large and small vessels. We labelled the first group large vessel peripheral arterial disease (LV-PAD), and the second group isolated small vessel (ISV-PAD). However, while we could characterise this third group as having LV-PAD, we were uncertain as to whether their reduced small vessel flow was due to the small vessels *per se*, or simply reduced flow due to proximal large vessel obstruction.

In a series of studies over the past decade, we have been able to compare and contrast patients with LV-PAD (the first and third group above), patients with ISV-PAD, and patients with normal peripheral circulation. In this chapter we will discuss what is known about the pathophysiology and epidemiology of these two conditions.

Pathophysiology

LV-PAD

The pathophysiology of this condition is well known. The lumen of a major artery or arteries in the lower extremities is progressively narrowed by a typical atherosclerotic plaque (Krupski and Effeney 1988). When the angiographic diameter of a large artery is reduced by 50%, which corresponds to a reduction in cross-sectional area of 75%, flow begins to decrease (Krupski and Effeney 1988). Reduced flow leads to the typical symptoms and physical findings of LV-PAD, and total occlusion can produce severe symptomatology and may lead to limb loss.

ISV-PAD

ISV-PAD is a functional diagnosis, since this condition cannot be reliably assessed angiographically. Thus, it is currently not known whether the reduced flow in this condition is a result of small vessel occlusion, vascular spasm or other non-occlusive dysfunction of small vessels, or some combination of anatomical and functional abnormalities.

Non-invasive Measurement Criteria

LV-PAD

Tests Employed. In our laboratory we use two tests:

1. Segmental systolic blood pressure measurements are taken at five levels of the lower extremity (upper thigh [UT], above knee [AK], below knee [BK], above ankle [AA], and toe [T]), and expressed as a proportion of the arm systolic pressure.

2. Flow velocity by Doppler ultrasound is measured in the femoral and posterior tibial arteries. This technique utilises the well-known Doppler effect produced by back-scattered ultrasound from red blood cells in motion.

Reproducibility. Reproducibility is high for segmental blood pressure (Carter 1968; Baker and Dix 1981; Fowkes 1988), and these estimates are similar in subjects with and without LV-PAD (Osmundsen et al. 1985). Flow velocity by Doppler ultrasound is also reproducible. Three consecutive measurements at 15-minute intervals, with reposition of the probe each time, yielded measurements within one standard deviation of the mean (Fronek et al. 1976).

Using both segmental blood pressure and flow velocity in a reproducibility study of 45 patients in our laboratory, no change sufficient to reclassify a patient (normal vs. LV-PAD) occurred.

Accuracy. Segmental blood pressure ratios of ≤ 0.85 have over 90% sensitivity and 99% specificity in detecting angiogram positive disease (Carter 1968; Winsor 1950; Yao 1973; Hummel et al. 1978; Bernstein and Fronek 1982; Laing and Greenhalgh 1983; Fowkes 1988; Malinow et al. 1989). Using the criteria noted below, flow velocity measurements produce sharp discrimination between normals and those with angiographic disease (Fronek et al. 1976; Bernstein and Fronek 1982).

Employing both segmental blood pressure and flow velocity produces 95% sensitivity and 99% specificity for hemodynamically significant peripheral arterial disease, which corresponds to angiographic luminal obstruction $\geq 50\%$. The degree of abnormality in the non-invasive tests reflects the severity of disease at angiography (Chamberlain et al. 1975; Bernstein and Fronek 1982).

Specific LV-PAD Criteria. LV-PAD is defined as *either* abnormal segmental pressures (LV_{BP+}) or an abnormal flow velocity (LV_{FV+}).

LV_{BP+} is defined as either:

1. An AA and BK ratio both ≤0.8

or

2. An AA or BK ratio ≤0.8
    and
    A UT ratio ≤0.8 *or* an AK ratio ≤0.8 *or* a T ratio ≤0.7

LV_{FV+} is defined as either:

1. A femoral peak forward flow of ≤20 cm/s
    and
    A femoral pulse decay of ≥260 ms *or* a femoral deceleration of ≤100 cm/s^2

or

2. A posterior tibial peak forward flow of ≤10 cm/s
    and
    A posterior tibial pulse decay of ≥220 ms *or* a posterior tibial deceleration of ≤70 cm/s^2

ISV-PAD

Tests Employed. Two tests are employed in our laboratory. The first is the toe pressure ratio, as described for LV-PAD. The second is the pulse reappearance half-time, (PRT half) following 4 minutes of flow occlusion, which is the time in seconds it takes for pulse amplitude at the toe to reach one-half the baseline value. A third test, post-occlusive reactive hyperemia (PORH), is often abnormal as well in peripheral arterial diseases. However, in our studies the PORH test did not add additional information beyond the toe pressure ratio and the PRT half in the diagnosis of ISV-PAD.

Reproducibility. The toe pressure has high reproducibility (Gunderson 1972; Nielsen 1972), as does the PRT half.

Accuracy. As noted, ISV-PAD cannot be reliably assessed angiographically and, while indicating decreased perfusion, may not indicate anatomical obstruction. Thus, the cutpoints chosen for toe pressure and PRT half represent extreme values from previous studies (Fronek et al. 1977; Fronek and Bernstein 1985).

Specific ISV-PAD criteria. As discussed earlier, ISV-PAD can be determined only in the absence of LV-PAD. The criteria (given the absence of LV-PAD) are:

A T ratio ≤0.7, with the AA ratio and the BK ratio normal

or

a PRT half ≤20 seconds

Symptomatology

LV-PAD

The classic symptom of this condition is intermittent claudication, defined as exercise calf pain not present at rest, and which is relieved only by rest within 10 minutes. The standard instrument in epidemiologic surveys for assessing claudication is the Rose questionnaire (Rose 1962). However, in our population study we found that including atypical findings ("possible" claudication) in subjects with exercise calf pain not present at rest, such as the ability to walk through the pain, actually doubled the sensitivity of such symptoms for detecting non-invasively diagnosed peripheral arterial disease from 10% to 20%, with little loss of specificity (Criqui et al. 1985b). This finding indicates many patients with true LV-PAD will not have classic claudication. However, additional analyses revealed that in subjects with bilateral, and in particular, subjects with severe LV-PAD, the sensitivity of claudication was sharply increased. Thirty-six per cent of subjects with bilateral LV-PAD and 65% of subjects with severe LV-PAD (LV_{BP+} and LV_{FV+} in the same limb) had Rose or possible claudication (Criqui et al. 1985b).

The overall sensitivity of 20% seems low, but this reflects the finding in our population study that about half of the cases had LV-PAD confined to one or both posterior tibial arteries, as determined by abnormal posterior tibial flow velocities. These patients had normal femoral flow velocities and normal segmental pressures, and were all asymptomatic. Removal of these patients from analysis increased the sensitivity of Rose and possible claudication to more than 40% (Criqui et al. 1985b).

The positive predictive value of claudication, or the proportion of subjects with claudication who had LV-PAD, ranged from 40% to 55% depending on the claudication definition. This surprising finding indicates that as many as half of patients, even with classic Rose claudication, will not have LV-PAD (Criqui et al. 1985b).

ISV-PAD

Subjects with ISV-PAD in our population study were no more likely to report Rose (1.1%) or possible (4.3%) claudication than subjects free of peripheral arterial disease (1.0% for Rose and 3.0% for possible claudication) (Criqui et al. 1985b). This was an expected finding since the flow restriction in ISV-PAD is distal to the calf muscles.

Physical Findings

LV-PAD

The classic findings in LV-PAD are reduced or absent peripheral pulses. In addition, the affected extremity may be pale and/or cold. In severe disease, frank ischaemia and rest pain may be present.

In our population-based study, we explored the association of anormal pulse findings, defined as a femoral, posterior tibial or dorsalis pedis pulse lower than grade 3 on a subjective 0–4 scale, or a femoral bruit (Criqui et al. 1985b). Similar to claudication, an abnormal femoral pulse (reduced and/or bruit) showed high specificity and reasonably predictive values, but low sensitivity (13%–20%). An abnormal dorsalis pedis pulse showed a sensitivity of 50%, but poor specificity (73%), due to the frequent congenital absence of this artery (Ison 1968, Barnhorst and Barner 1968). The best single indicator of LV-PAD was an abnormal posterior tibial pulse, which had a sensitivity of 71%, a specificity of 91%, a positive predictive value of 49%, and a negative predictive value of 97%. Surprisingly, selecting various combinations of claudication and pulse abnormalities did not improve on the single criterion of an abnormal posterior tibial pulse. Improvements in sensitivity were invariably associated with reduced specificity and positive predictive values, and vice versa (Criqui et al. 1985b).

ISV-PAD

It seems possible that this condition could be associated with pain, pallor, and/or sensory changes in the foot. However, we did not specifically solicit this information in our study, since the distinct character of ISV-PAD did not become apparent until after these data had been collected. Thus, the symptomatology, if any, of this condition awaits further research.

Prevalence

LV-PAD

Since angiography evaluation would be inappropriate for a population-based sample, and symptoms and physical examination produce significant numbers of false positives and false negatives, the best estimates of prevalence come from using non-invasive testing in population studies. Our population-based study yielded an LV-PAD prevalence of 11.7% in an older population (mean age of 66 years with an age range of 38–82 years (Criqui et al. 1985a). Hyperlipidemic subjects were somewhat over-represented in our population, and after adjustment for this selection, the overall rate was 10.2%. Fig. 7.1 indicates that this rate was slightly higher in men than in women, and increased sharply with age (Criqui et al. 1985a).

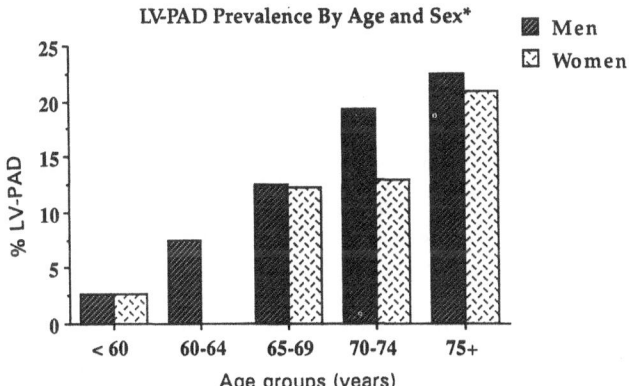

Fig. 7.1. Large vessel peripheral arterial disease (LV-PAD) prevalence by age and sex. Adapted from Criqui et al. (1985a) with permission.

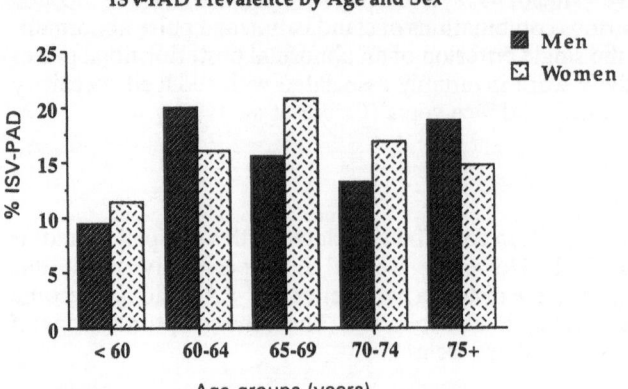

ISV-PAD Prevalence by Age and Sex*

Fig. 7.2. Isolated small vessel peripheral arterial disease (ISV-PAD) prevalence by age and sex. Adapted from Criqui et al. (1985a) with permission.

Of the subjects with LV-PAD, 44% also had reduced small vessel flow. However, as noted earlier, we could not determine whether this was due to small vessel disease or proximal LV-PAD.

ISV-PAD

The overall prevalence of ISV-PAD in our population was 16.0%, and this estimate was not biased by oversampling of hyperlipidemic subjects since hyperlipidemia was unassociated with ISV-PAD (RR = 1.0, p = 0.82) (Criqui et al. 1985a). Unlike LV-PAD, Fig. 7.2 shows that rates of ISV-PAD were equivalent in men and women, and that there was no overall association with age, although ISV-PAD appeared somewhat less common before age 60 years (Criqui et al. 1985a).

Risk Factors

LV-PAD

Risk factors for LV-PAD have been evaluated in both cross-sectional and prospective analyses, using various endpoints including claudication (Keen et al. 1965; Bothig et al. 1976; Reunanen et al. 1982; Kannel and McGee 1985), non-invasive assessment (Isacsson 1972; Olsson and Eklund 1975; Hughson et al. 1978a; Schroll and Munck 1981; Criqui et al. 1989), and angiography in clinical studies (Greenhalgh et al. 1971; Sirtori et al. 1974; Davignon et al. 1977; Bradby et al. 1978; Cavallo-Perin et al. 1984). Major risk factors for coronary heart disease, including cigarette smoking, dyslipidemia, hypertension, and diabetes, are also associated with LV-PAD. However, cigarette smoking and diabetes may be particularly important risk factors for LV-PAD (Criqui et al. 1989). In addition, the dyslipidemia characteristic of LV-PAD patients may differ somewhat from that characteristic of coronary disease patients, despite the frequent coexistence of coronary disease and LV-PAD (Dormandy et al. 1989). Elevated triglyceride and VLDL cholesterol may be somewhat more characteristic for LV-PAD than for coronary disease (Criqui et al. 1989), and reduced HDL is also prominent in LV-PAD (Bradby et al. 1978). This constellation of lipids may reflect insulin resistance, which is consistent with the strong association between diabetes and LV-PAD. Table 7.1 shows the risk factor differences in our population-based study for subjects free of peripheral arterial disease, subjects with ISV-PAD, and subjects with LV-PAD, including the sub-categories of moderate and severe LV-PAD.

Table 7.1. Age and sex-adjusted mean levels of risk factors by peripheral arterial disease status

	Peripheral arterial disease				
	Normal	ISV-PAD	All LV-PAD	Moderate LV-PAD[a]	Severe LV-PAD[a]
n	408	90	67	49	18
Pack-yrs cigs.	18.7	17.5	29.5*	25.8	39.2*
Triglyceride (mg/dl)	134.5	140.9	159.9*	153.7	176.2*
HDL chol (mg/dl)	55.5	53.4	53.9	55.6	49.5
LDL chol (mg/dl)	158.5	154.5	155.2	153.6	159.5
Systolic BP (mmHg)	129.6	130.2	132.8	131.0	137.7*
Diastolic BP (mmHg)	76.8	75.3	79.8*	78.7	82.9*
BMI (kg/m²)	24.5	24.8	26.8*	27.1*	25.9
Fasting glucose (mg/dl)[b]	94.9	97.4	Men 104.8*	95.2	122.5*
			Women 96.3	96.4	95.7

* p≤0.05 compared with normals. [a] Subset of all LV-PAD. [b] Sex-specific values given for LV-PAD because·of an interaction by gender. Adapted from Criqui et al. (1989).

Cigarette smoking was a potent risk factor for LV-PAD. Fasting glucose was a powerful predictor in men, but not in women. Blood pressure was also significantly associated with LV-PAD. The dyslipidemia characteristic of LV-PAD was prominent, with significantly elevated triglyceride, and lower (albeit non-significantly) HDL cholesterol. LDL cholesterol levels were similar to normals. Severe LV-PAD was associated with greater risk factor differences than moderate LV-PAD for each risk factor, except for BMI, where only the result for moderate disease was significant. Overall, these findings for non-invasively diagnosed peripheral arterial disease were concordant with earlier studies.

ISV-PAD

Examination of potential risk factors for ISV-PAD in Table 7.1 revealed no significant associations. Even age was unrelated to ISV-PAD, after age 60 years (Fig. 7.2). These results strongly suggest ISV-PAD may be a non-atherosclerotic phenomenon (Criqui et al. 1989). Subjects with ISV-PAD did have slightly higher levels of glucose and triglyceride, and slightly lower levels of HDL cholesterol, as well as a slightly higher BMI. Although these findings may hint that ISV-PAD may have some association with the dyslipidemia of LV-PAD, none of these results approached statistical significance.

In our population, the association of both fasting plasma glucose and glycosylated hemoglobin with ISV-PAD was non-significant. These findings challenge the concept of ISV-PAD being closely associated with diabetes. Consistent with these data is the finding in a review that diabetics do not have an excess of microvascular disease of the lower extremities (LoGerfo and Coffman 1984).

Natural History

LV-PAD

A recent review concludes that, for patients with LV-PAD surviving several years after diagnosis, two-thirds to three-quarters will have no worse than stable symptoms, while the remaining quarter to one-third will show deterioration (Dormandy et al. 1989). Amputation will eventually be required in 1%–5%. Angioplasty or surgery can favourably influence the natural history of LV-PAD in selected patients.

The above figures are from data in survivors. However, as discussed below, survival is sharply decreased in patients with LV-PAD, primarily due to cardiovascular disease. Intuitively, LV-PAD might progress faster in those with early cardiovascular mortality.

Progression of LV-PAD is more rapid in patients who smoke cigarettes (Jonason and Ringquist 1985) or who have diabetes (Strandness 1987). Risk factor intervention such as smoking cessation (Jonason and Ringquist 1985) and therapy for dyslipidemia (Barndt et al. 1977; Blankenhorn et al. 1991) can slow progression and, in some cases, cause regression of LV-PAD (Barndt et al. 1977; Blankenhorn et al. 1991).

ISV-PAD

The natural history of ISV-PAD is unknown. Unanswered questions include the stability of this diagnosis over time, the relationship of progression to risk factors, and whether ISV-PAD patients are more likely to develop LV-PAD than normal subjects. We are currently engaged in a study addressing these questions.

Morbidity and Mortality

LV-PAD

Earlier studies have reported a greatly increased prevalence of coronary disease and cerebrovascular occlusive disease in patients with LV-PAD (Dormandy et al. 1989). The simultaneous occurrence of clinical atherosclerotic disease in more than one arterial bed is to be expected, given the similarity of risk factors for these conditions.

Earlier studies have also reported a doubling of mortality, mostly due to cardiovascular disease, in patients with claudication (Peabody and Kannel 1974; Hughson et al. 1978b; Reunanen et al. 1982). In some studies, much of this risk was due to coronary heart disease risk factors or existing baseline cardiovascular disease (Reunanen et al. 1982). We speculated that the increased accuracy of non-invasive testing for LV-PAD should provide more precise and higher estimates of the risk of LV-PAD for cardiovascular and all-cause mortality. In 1985 we published the follow-up results from our population study and showed, after four years, a four- to five-fold increased risk of LV-PAD for all-cause mortality, independent of cardiovascular risk factors or other cardiovascular conditions (angina, myocardial infarction, or stroke) at baseline (Criqui et al. 1985c). We have recently reported the 10-year follow-up results for all-cause and cardiovascular-cause specific mortality, for both LV-PAD (Criqui et al. unpublished observations) and ISV-PAD (Langer et al. unpublished observations). Table 7.2 shows our results.

Table 7.2. Relative risks[a] of large vessel peripheral arterial disease (LV-PAD) and isolated small vessel disease (ISV-PAD) for coronary, stroke, all-cardiovascular, and all-cause mortality. 10-year follow-up

Mortality	LV-PAD ($n=67$)	ISV-PAD ($n=90$)
Coronary	5.8 (2.5–13.0)	2.6 (1.1–6.4)
Stroke	1.0 (0.1–11.0)	2.3 (0.4–14.0)
All-cardiovascular	5.1 (2.7–9.9)	2.5 (1.2–5.1)
All-cause	3.0 (1.8–4.8)	1.2 (0.7–2.1)

[a] Compared to subjects free of peripheral arterial disease; adjusted for age, sex, blood pressure, pack-years of cigarettes, HDL cholesterol, LDL cholesterol, log triglyceride, fasting plasma glucose, and body mass index; 95% confidence interval in parentheses. Adapted from Criqui et al. (unpublished observations); Langer et al. (unpublished observations).

LV-PAD after 10 years remained strong and significantly associated with all-cause mortality. As expected, the associations for coronary and all-cardiovascular mortality were even higher, with relative risks in the 5–6 range. The estimate for stroke mortality was unity, but the 95% confidence interval was wide, reflecting the fact that there have been only 4 stroke deaths in our population.

ISV-PAD

In our initial analyses at 4 years, there was a small and statistically insignificant increase in all-cause mortality in subjects with ISV-PAD (Criqui et al. 1985c).

After 10 years, the results for all-cause mortality were similar, but the results for cause-specific deaths indicate a two-and-a-half times greater risk of cardiovascular death (Table 7.2). This finding is especially interesting in light of the absence of the traditional cardiovascular risk factor associations with ISV-PAD. We have speculated that ISV-PAD may reflect either a variant of atherosclerosis with different risk factor associations (Reed et al. 1988; Reed 1990), or a vascular hyper-reactivity syndrome similar to that described for microvascular angina (Cannon et al. 1985; Maseri et al. 1990). It has been reported that patients with microvascular angina have PORH abnormalities in a forearm occlusion test (Sax et al. 1987) and the vast majority of our ISV-PAD patients had PRT half abnormalities, which reflect abnormal hyperemic response. Whatever the mechanism, persons with ISV-PAD appear to be at elevated risk of fatal cardiovascular disease.

These provocative results for both LV-PAD and ISV-PAD, independent of cardiovascular risk factors and existing disease at baseline, indicate the potential for identifying patients at high risk of early mortality by non-invasive testing for LV-PAD and ISV-PAD. At least for patients with LV-PAD, risk factor modification, and when appropriate, surgical correction for associated cardiovascular disease, should improve survival. Risk factor modification and/or surgical intervention for associated cardiovascular disease may help patients with ISV-PAD as well, although other therapies may need to be devised if the underlying pathophysiology principally involves spasm.

Future Research

LV-PAD

The epidemiology of LV-PAD is increasingly being better defined, although important questions remain. The progression of this disease, as well as the effects of risk factor modification and various medications on progression, need further clarification. The degree of overlap between LV-PAD and carotid, coronary, and renal atherosclerosis, and the important determinants of multi-system disease, need better evaluation. Our ongoing LV-PAD research is focusing on these questions.

ISV-PAD

The recent demonstration of the apparently non-atherosclerotic nature of this condition, despite its prognostic significance for associated cardiovascular disease, demands further investigation. Important questions include the stability of this condition over time, and whether ISV-PAD can evolve to LV-PAD. We are currently focusing on these questions in our research. The lack of clear risk factor associations with ISV-PAD makes identification of those at risk problematic at this time. Further evaluation of any symptoms associated with ISV-PAD, and of associated abnormalities of vascular physiology and possibly anatomical pathology, should be the focus of future research.

References

Baker JD, Dix D (1981) Variability of Doppler ankle pressures with arterial occlusive disease: an evaluation of ankle index and brachial-ankle pressure gradient. Surgery 89:134–7

Barndt R Jr., Blankenhorn DH, Crawford DW et al. (1977) Regression and progression of early femoral atherosclerosis in treated hyperlipoproteinemic patients, Ann Intern Med 86:139–46

Barnhorst DA, Barner HB (1968) Prevalence of congenitally absent pedal pulses. N Engl J Med 278:264–5

Bernstein EF, Fronek A (1982) Current status of non-invasive tests in the diagnosis of peripheral arterial disease. Surg Clin-North Am 62:473–87

Blankenhorn DH, Azen SP, Crawford DW et al. (1991) Effects of colestipol-niacin therapy on human femoral atherosclerosis. Circulation 83:438–47

Bothig S, Metelitsa VI, Barth W et al. (1976) Prevalence of ischaemic heart disease, arterial hypertension and intermittent claudication, and distribution of risk factors among middle-aged men in Moscow and Berlin. Cor Vasa 18:104–18

Bradby GVH, Valente AJ, Walton KW (1978) Serum high density lipoproteins in peripheral vascular disease. Lancet ii:1271–4

Cannon RO III, Leon MB, Watson RM, Rosing DR, Epstein SE (1985) Chest pain and "normal" coronary arteries-role of small coronary arteries. Am J Cardiol 55:50B–60B

Carter SA (1968) Indirect systolic pressures and pulse waves in arterial occlusive disease of the lower extremities. Circulation 37:624–38

Cavallo-Perin P, Barile C, Ozzello A et al. (1984) Peripheral vascular disease and risk factors of atherosclerosis: an epidemiologic study. Panminerva Med 26:139–43

Chamberlain J, Housley E, Macpherson AIS (1975) The relationship between ultrasound assessment and angiography in occlusive arterial disease of the lower limb. Br J Surg 62:64–7

Criqui HM, Fronek A, Barrett-Connor E, Klauber MR, Gabriel S, Goodman D (1985a) The prevalence of peripheral arterial disease in a defined population. Circulation 71:510–15

Criqui MH, Fronek A, Klauber MR, Barrett-Connor E, Gabriel S (1985b) The sensitivity, specificity, and predictive value of traditional clinical evaluation of peripheral arterial disease: results from a non-invasive testing in a defined population. Circulation 71:516–21

Criqui MH, Coughlin SS, Fronek A (1985c) Non-invasively diagnosed peripheral arterial disease as a predictor of mortality. Results from a prospective population-based study. Circulation 72:768–73

Criqui MH, Browner D, Fronek A et al. (1989) Peripheral arterial disease in large vessels is epidemiologically distinct from small vessel disease: an analysis of risk factors. Am J Epidemiol 129:1110–19.

Davignon J, Lussier-Cacan S, Ortin-George M et al. (1977) Plasma lipids and lipoprotein patterns in angiographically graded atherosclerosis of the legs and in coronary heart disease. Can Med Assoc J 116:1245–50

Dormandy J, Mahir M, Arcady G et al. (1989) Fate of the patient with chronic leg ischaemia. J Cardiovasc Surg 30:50–57

Fowkes FGR (1988) The measurement of atherosclerotic peripheral arterial disease in epidemiologic surveys. Int J Epidemiol 17:248–54

Fronek A, Bernstein EF (1985) Post-occlusive reactive hyperemia in the testing of the peripheral arterial system: pressure, velocity, and pulse reappearance time. In: Bernstein EF (ed) Non-invasive diagnostic techniques in vascular disease. Mosby, St. Louis, p. 584

Fronek A, Coel M, Bernstein EF (1976) Quantitative ultrasonographic studies of lower extremity flow velocities in health and diseases. Circulation 53; 957–60

Fronek A, Coel M, Bernstein EF (1977) The pulse reappearance time – an index of overall blood flow impairment in the ischemic extremity. Surgery 81:376–81

Greenhalgh RM, Lewis B, Rosengarten DS et al. (1971) Serum lipids and lipoproteins in peripheral vascular disease. Lancet ii:947–50

Gunderson J (1972) Multisegmental measurements of blood pressure. Reproducibility in healthy subjects and in patients with arterial insufficiency. Acta Chir Scand 426 (Suppl):51

Hughson WG, Mann JI, Garrod A (1978a) Intermittent claudication prevalence and risk factors. Br Med J 1:1379–81

Hughson WG, Mann JI, Tibbs DJ et al. (1978b) Intermittent claudication: factors determining outcome. Br Med J i:1377–79

Hummel BW, Hummel BA, Mowbry A et al. (1978) Reactive hyperemia vs. treadmill exercise testing in arterial disease. Arch Surg 113:95–8

Isacsson S (1972) Venous occlusion plethysmography in 55-year-old men: a population study in Malmo, Sweden. Acta Med Scand 537 (Suppl):1–62

Ison JW (1968) Palpation of dorsalis pedis pulse. JAMA 206:2745

Jonason T, Ringquist I (1985) Factors of prognostic importance for subsequent rest pain in patients with intermittent claudication. Acta Med Scand 218:27–33

Kannel WB, McGee DL (1985) Update on some epidemiologic features of intermittent claudication: The Framingham Study. J Am Geriatr Soc 33:13–18

Keen H, Rose G, Pyke DA et al. (1965) Blood-sugar and arterial disease. Lancet ii:505–8

Krupski WC, Effeney DJ (1988) Arteries. In: Way L (ed) Current surgical diagnosis and treatment, 8th edn. Appleton and Lange, Norwalk, CT

Laing S, Greenhalgh RM (1983) The detection and progression of asymptomatic peripheral arterial disease. Br J Surg 70:628–30

Langer RD, Criqui MH, Fronek A, Feigelson HS, Klauber MR (1991) Isolated small vessel peripheral arterial disease predicts cardiovascular events. (Submitted for publication.)

LoGerfo FW, Coffman JD (1984) Vascular and microvascular disease of the foot in diabetes: implications for foot care. N Engl J Med 311:1615–19

Malinow MR, Kang SS, Tayler LM et al. (1989) Prevalence of hyperhomocyst(e)inemia in patients with peripheral arterial occlusive disease. Circulation 79:1180–8.

Maseri A, Davies G, Hackett D, Kaski JC (1990) Coronary artery spasm and vasoconstriction, the case for a distinction. Circulation 81:1983–91

Nielsen PE (1972) The measurement of digital systolic pressure by strain guage technique. Scand J Clin Lab Invest 29:371

Olsson AG, Eklund B (1975) Studies in asymptomatic primary hyperlipidemia. V. Peripheral circulation. Acta Med Scand 198:197–206

Osmundson PJ, O'Fallon WM, Clements IP, Kazmier FJ, Zimmerman BR, Palumbo PJ (1985) Reproducibility of non-invasive tests of peripheral occlusive arterial disease. J Vasc Surg 2:678–83

Peabody CN, Kannel WB (1974) Intermittent claudication: surgical significance. Arch Surg 109:693–97

Reed DM (1990) The paradox of high risk of stroke in populations with low risk of coronary heart disease. Am J Epidemiol 131:579–88

Reed DM, Resch JA, Hayashi T, MacLean C, Yano K (1988) A prospective study of cerebral artery atherosclerosis. Stroke 19:820–5

Reunanen A, Takkunen H, Aroma A (1982) Prevalence of intermittent claudication and its effect on mortality. Acta Med Scand 211:249–56

Rose GA (1962) The diagnosis of ischemic heart pain and intermittent claudication in field surveys. Bull WHO 27:645–658

Sax FL, Cannon RO, Hanson C, Epstein SE (1987) Impaired forearm vasodilator reserve in patients with microvascular angina. Evidence of a generalized disorder of vascular function? N Engl J Med 317:1366–70

Schroll M, Munck O (1981) Estimation of peripheral arteriosclerotic disease by ankle blood pressure: measurements in a population study of 60-year-old men and women. J Chron Dis 34:261–9

Sirtori CR, Biasi G, Vercellio G et al. (1974) Diet, lipids, and lipoproteins in patients with peripheral vascular disease. Am J Med Sci 268:325–32

Strandness DE (1987) Vascular disease of the extremities. In: Harrison's Principles of Internal Medicine, 11th edn, McGraw-Hill, New York, pp. 1040–6

Winsor T (1950) Influence of arterial disease on the systolic blood pressure gradients of the extremity. Am J Med Sci 220:117

Yao ST (1973) New techniques of objective arterial evaluation. Arch Surg 106:600–4

8 Symptomatic and Asymptomatic Disease

C.V. RUCKLEY

As interest grows in the possibilities for earlier intervention in occlusive vascular disease, whether it be by health education, pharmacological means or the newer minimally invasive radiological techniques there is a need to look afresh at the natural history of the condition, its epidemiology and the tools for early detection. Although a considerable amount of research has centred on occult disease of the coronary arteries (Margolis et al. 1973) little is known about the distribution of "subclinical" occlusive disease of the lower limbs.

Difficulties in Assessing Severity of Disease

In measuring asymptomatic disease in a population the "gold standard" of arteriography cannot be employed. Recourse is therefore made to indirect measures of lesions which interfere with blood flow (Fowkes 1988a). The ankle-brachial systolic pressure ratio has been found to be up to 95% sensitive in detecting angiogram positive disease (Bernstein and Fronek 1982) and additional cases may be identified by using the test after a standard exercise (Carter 1972; Laing and Greenhalgh 1980) or a period of tourniquet-induced hyperaemia (Baker 1978). The recently available duplex scanning technique has the potential to provide more direct non-invasive assessment of atheromatous disease (Jager et al. 1985). Until recently studies of peripheral vascular disease in the UK (Reid et al. 1966, 1974; Hughson et al. 1978) have reported only the prevalence of symptomatic disease using the WHO/Rose questionnaire on intermittent claudication (Rose 1962).

Intermittent claudication, i.e. exercise-related discomfort in specific muscle groups, is the salient symptom. Indeed claudication is virtually the only symptom in early occlusive disease of the lower limbs. This is not an easy topic to explore, nor to document accurately, since there are many factors other than simply the severity of the occlusive process which determine whether a patient will experience or report symptoms. These factors include the availability of collateral circulation; the location of the atheroma; anatomical arterial variations; the patient's occupation, personality and levels of activity; the limiting effects of other diseases, drug therapy and possible psychological or sensory impairment. Space does not allow a full analysis of these factors. However, two of them, collateral circulation and the anatomical location of the lesion, are sufficiently important in the present context that some comment is called for.

Although atherosclerosis is pathologically a continuously progressive disease – assuming that the promoting factors such as smoking and hypercholesterolaemia continue unabated – the clinical pattern consists of episodes of deterioration followed by resolution with long periods of stability or apparent improvement due to compensatory mechanisms, particularly the development of collateral circulatory pathways around obstructed segments. This stepwise progression of symptoms, which is illustrated diagramatically in Fig. 8.1, is of fundamental importance in understanding the relationship between pathology and symptoms, the detection of disease and the interpretation of apparent therapeutic responses. In everyday clinical practice most therapies are prescribed after a patient has

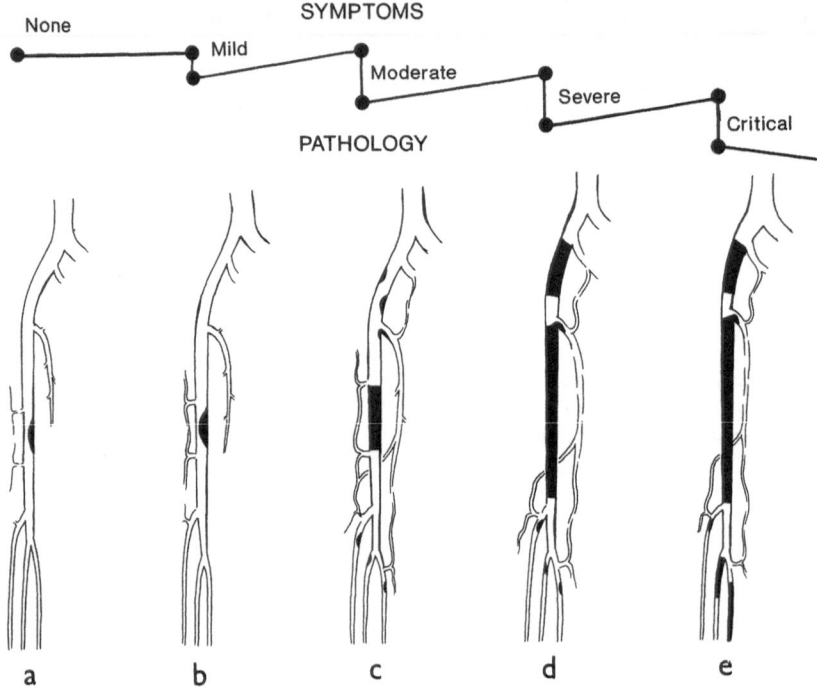

Fig. 8.1. Diagram relating progression of the pathological changes in obliterative arterial disease of the lower limb to the symptoms. Each deterioration is followed by a period of spontaneous symptomatic improvement due to collateral circulation and other factors. Note that the time sequence may be very different. There is commonly only a relatively short interval between a stenosis becoming symptomatic and its progression to total occlusion, whereas a patient with femoral artery occlusion may remain relatively stable for many years. Once rest pain has occurred the time interval before gangrene supervenes is usually short.

experienced an exacerbation of symptoms, because that is the event which stimulates the patient to seek medical advice. It follows that the therapy will be applied during the phase of resolution, hence the probability that symptomatic improvement will be inappropriately ascribed to the treatment. This has to be distinguished from the placebo response which is also a very commonly observed phenomenon in vascular disease.

The location of the occlusive lesion determines its capacity to give rise to symptoms. Fig. 8.2 illustrates two short occlusions of similar size within a short distance of each other in the arterial tree, one certain to give rise to severe symptoms and the other to slight or no symptoms. The differences are determined by the presence or absence of collateral channels. Major arteries whose occlusion may be relatively "silent" include the internal carotid, the first part of the subclavian, the inferior mesenteric, the internal iliac and the superficial femoral. Moderate symptoms will usually accompany occlusion of the axillary, the brachial, the external iliac and the popliteal. Arteries whose occlusion is usually catastrophic include the middle cerebral, the ophthalmic, the superior mesenteric and the common femoral. Thus the severity of the symptoms in no way equates to the severity of the pathological process.

For these reasons the need for objective measures of arterial disease were well recognised in a study which my colleagues and I conducted recently in Edinburgh (Fowkes et al. 1991). The aims of the Edinburgh Artery Study were:–

1. To determine the prevalence of symptomatic and pre-symptomatic disease in the general population
2. To study the aetiology and natural history of peripheral vascular disease
3. To evaluate arterial disease in the lower limbs as a predictor of later cardiovascular events

This chapter is concerned with the first of these objectives and much is based on the results of the Edinburgh Artery Study – the first large scale cross-sectional study in the UK of symptomatic and asymptomatic peripheral vascular disease.

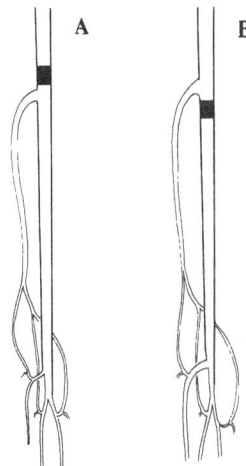

Fig. 8.2. Diagram to show how two occlusions of similar size may differ greatly in the potential to produce symptoms, depending on anatomical location and the availability of collateral channels.

Edinburgh Artery Study: Methods

Study Population

Inhabitants of Edinburgh aged 55–74 years comprised the target population. An age-stratified random sample was selected from the age–sex registers of 10 general practices with catchment populations spread geographically and socio-economically throughout the city. The sample size of 1500 participants was estimated on the basis of the number required to conduct a subsequent follow-up study with adequate power to detect differences in the incidence of subsequent vascular episodes, taking into account the baseline characteristics. From each practice 272 subjects were selected: 34 males and 34 females from each 5-year age band. All the general practitioners reviewed lists of their patients selected for the study and excluded those unfit to participate due, for example, to severe mental illness or terminal disease. These exclusions were replaced by other randomly selected patients.

The project was assisted by publicity in the local media and followed by letters of invitation, signed jointly by a partner in each general practice plus the study director, which were sent to the subjects inviting them to attend a clinic for medical examination. For those having difficulty attending the clinic, transport or a home visit was arranged; travelling expenses were offered. Letters returned by the Post Office were replaced by letters to other randomly sampled patients. On receipt of a positive reply an appointment, map and details of the examination were sent. Non-responders were sent a second letter. Non-attenders were offered another appointment, usually by telephone.

The response rate of those receiving an invitation was 65%. The responders were reasonably typical of the whole target population. The crude response rate did not differ substantially by age nor sex, although there was slight underrepresentation of women aged 70–74 years and males aged 55–59 years who comprised 21.3% and 22.5% instead of 25.0% of women and men respectively. The social class distribution of responders was similar to that of Edinburgh adult residents in the 1981 census except that the responders contained fewer social class IV and V (13% compared with 19%). The crude response rate varied between the 10 general practices from 47% to 71% with the lower response rates occurring in practices serving deprived areas, thus also suggesting a slight underrepresentation of lower social classes.

Clinical Examination

The main study was preceded by a pilot study of all clinical and laboratory procedures on 50 volunteers from the general public. Before and periodically during the study the quality of the clinical measurements was checked by repeat measurements. Individual observer measurements were observed for drift. Studies of the variability of the laboratory measurements have been reported previously (Fowkes et al. 1988a,b).

The patients were seen on weekday mornings from August 1987 to September 1988. Occasional out-of-hours sessions were also held. Subjects were asked to fast from 11 p.m. the previous evening (if not diabetic) and to refrain from smoking for 2 hours prior to the examination. Each subject had two sets of clinical procedures performed by one of two teams, each comprising a nurse and a technician. A self-administered questionnaire was completed by the subject and then checked by a member of the survey team.

The measurements comprised: questionnaire, blood tests and laboratory measurements. The majority of the questions had been previously validated; they dealt with personal characteristics, social class (Office of Population Censuses and Surveys 1980), intermittent claudication and angina (WHO/Rose questionnaire) (Rose 1962), medical history, smoking, exercise, personality, diet and alcohol consumption.

In the first set of procedures a 20 ml blood sample was taken for estimation of biochemical, haemostatic and rheological factors. Standard height was measured to the nearest 5 mm without shoes. Weight was measured, without shoes and outer clothing to the nearest 100 g on a digital Soehnle scale. A 12-lead electrocardiogram (ECG) and rhythm strip were taken using a Hewlett Packard *Pagewriter* electrocardiograph. ECGs were coded later independently by two trained staff using the Minnesota code (Prineas et al. 1982). A third member of staff checked the results and if there was any disparity between the two codings coded a third time. If the third code did not agree with either of the previous two the ECG was read by a consultant cardiologist and final coding agreed following discussion between the coders.

In the second set of procedures systolic and diastolic (Phase V) blood pressures were taken in the right arm using a Hawksley random zero sphygmomanometer, after the patient had rested for 10 minutes in the supine position. Ankle Doppler ultrasound systolic blood pressures were recorded and the ankle-brachial pressure index (ABPI) calculated. The Doppler pressure measurement was repeated 15 seconds after release of a cuff occluding the arterial circulation for 4 minutes at 50 mmHg above systolic pressure (reactive hyperaemia test). The timing was standardised using an electronic timer.

Twenty per cent of subjects in each practice who did not respond were randomly selected for follow-up. Each was sent a letter enclosing a short questionnaire. Subjects who did not return the questionnaire were telephoned or visited at home on up to three occasions at different times of the day and evening.

Data Analysis

Information on the questionnaire and recording forms was checked by the clinic staff and entered into a DBase III ® (Ashton Tate Ltd) database by the research secretary. Error rates in the data entry were assessing by checking intermittent random samples of subjects and by carrying out logic checks. The data files were analysed on the Edinburgh University mainframe computer using BMDP statistical package (Dixon et al. 1988). The lesser ABPI and the greater reduction during reactive hyperaemia in either leg were used as end points.

The prevalence rates for the city of Edinburgh were estimated by weighting the age-specific rate derived from 1981 census data. Multiple logistic regression was used to investigate the relationship between peripheral vascular disease and age, sex and height, and to estimate adjusted odds ratios and confidence intervals.

Multiple linear regression was used to analyse the results of ABPI and reactive hyperaemia and to estimate the adjusted differences between the sexes. Analyses of cardiovascular risk factors including social status, smoking, personality, diet, biochemical, haemostatic and rheological factors were not included at this stage but preliminary results are reported in subsequent chapters. Confidence intervals for relative risks were calculated using a recommended method (Morris and Gardner 1989).

Edinburgh Artery Study: Symptomatic and Asymptomatic Disease

Prevalence

Seventy-three subjects (4.6%) had intermittent claudication according to responses to the WHO questionnaire: 32 (2%) had Grade 2 (pain walking at an ordinary pace on the level); 1.5% had Grade 1 (pain only walking uphill or hurrying) and 17 had "probable" claudication (calf pain but one WHO criterion not fulfilled). Adjusting for differences in the age and sex structures of the study and target populations, the overall prevalence of claudication in Edinburgh City among 55 to 74 year olds was 4.5% (95% CI 3.5%–5.5%). Among subjects with intermittent claudication, more than one-third recalled a doctor stating that they had "hardening of the arteries in the legs".

The distribution of the ABPI was slightly negatively skewed with a mean of 1.03 (SD 0.18). On average, pressures on the left leg were lower than on the right, with large differences tending to occur when the mean ABPI of both legs was low. The percentage drop in ankle systolic pressure during the reactive hyperaemia test was positively skewed with a mean reduction of 10.0% (SD 12.1%). Among the asymptomatic patients, 111 had grossly abnormal results, namely an ABPI ≤ 0.7 or hyperaemic drop of $>35\%$ or both APBI ≤ 0.9 and hyperaemic drop of $>20\%$ (Table 8.1). These subjects comprised 8.0% (95% CI 6.4%–9.4%) of the

Table 8.1. Ankle-brachial pressure index (ABPI) and reduction in systolic pressure during reactive hyperaemia in subjects with and without intermittent claudication

Reduction in systolic pressure with reactive hyperaemia	Ankle-brachial pressure index (ABPI)					
	Claudication present			Claudication absent		
	≤ 0.7	>0.7–0.9	>0.9	≤ 0.7	>0.7–0.9	>0.9
$\leq 20\%$	4	9	22	34	130	1061
>20–35%	5	2	2	15	20	103
$>35\%$	11	4	0	10	10	18
No result	7	1	5	4	23	82
Total	27	16	29	63	183	1264

Note: ABPI result missing in 1 subject with claudication and 9 subjects without claudication. Reproduced from Fowkes et al. (1991) by courtesy of the *International Journal of Epidemiology*.

Edinburgh City population aged 55–74 years and were classified as having major asymptomatic disease. A further 233, 16.6% (95% CI 14.6%–18.5%) were asymptomatic and had moderately abnormal results which would be used commonly in clinical practice as indicators of disease: 130 (9.0%) had ABPI ≤0.9 and 103 (7.6%) had a hyperaemic pressure reduction of >20%. Table 8.1 also shows some inconsistencies in the results of these clinical measurements. For example in subjects with intermittent claudication 22 (37%) had an ABPI >0.9% and reactive hyperaemia of ≤20%. In comparing claudicants with non-claudicants, however, 60% vs. 16% had ABPI ≤0.9, and 44% vs. 13% had a hyperaemic drop of >20%.

Age and Sex Distribution

The increasing prevalence of intermittent claudication with age is shown in Fig. 8.3. Men aged 65–69 years were an exception but this may be accounted for by the small numbers in each age category. After adjusting for age the odds ratio of intermittent claudication in women compared to men was 1.11 (95% CI 0.70–1.79) suggesting that prevalence was similar in men and women.

Mean ABPIs decreased with age in both sexes (Fig. 8.3) and after adjusting for age were found to be higher in men than women, the difference being 0.053 (95% CI 0.036–0.071). However the ABPI correlated with height (males: $r = 0.14$, females: $r = 0.11$) and after adjusting for age and height, the sex difference was reduced to 0.020 (95% CI 0.004–0.044). Despite a higher overall mean, men had relatively more ABPIs below 0.8. Taking this level as an arbitrary cut-off for disease, the odds ratio of disease in men compared with women was 1.39 (95% CI 0.86–2.2).

The reactive hyperaemia test showed a less consistent reduction with age in both sexes (Fig. 8.3). In particular the drop in pressure in 70–74-year-old women was less than expected. After adjusting for age the mean reduction in pressure was higher in men than in women, the difference being 3.0% (95% CI 1.8%–4.3%).

Relationship with Ischaemic Heart Disease

Evidence of ischaemic heart disease was noted in 614 (38.6%) of the study population: 232 (14.6%) recalled a doctor stating they had angina or a heart attack; 296 (18.6%) had angina or possible myocardial infarction according to the WHO/Rose questionnaire and 248 (15.6%) had none of the above but had ECG evidence of ischaemia or previous myocardial infarction. In those with intermittent claudication 52 (71%) had evidence of ischaemic heart disease (relative risk 1.9; 95% CI 1.6 to 2.3) (Table 8.2). In subjects with major asymptomatic peripheral vascular disease 57 (54%) had evidence of ischaemic heart disease (relative risk 1.6; 95% CI 1.3–1.9). Conversely in subjects with evidence of ischaemic heart disease a smaller proportion, 52 (8%) had intermittent claudi-

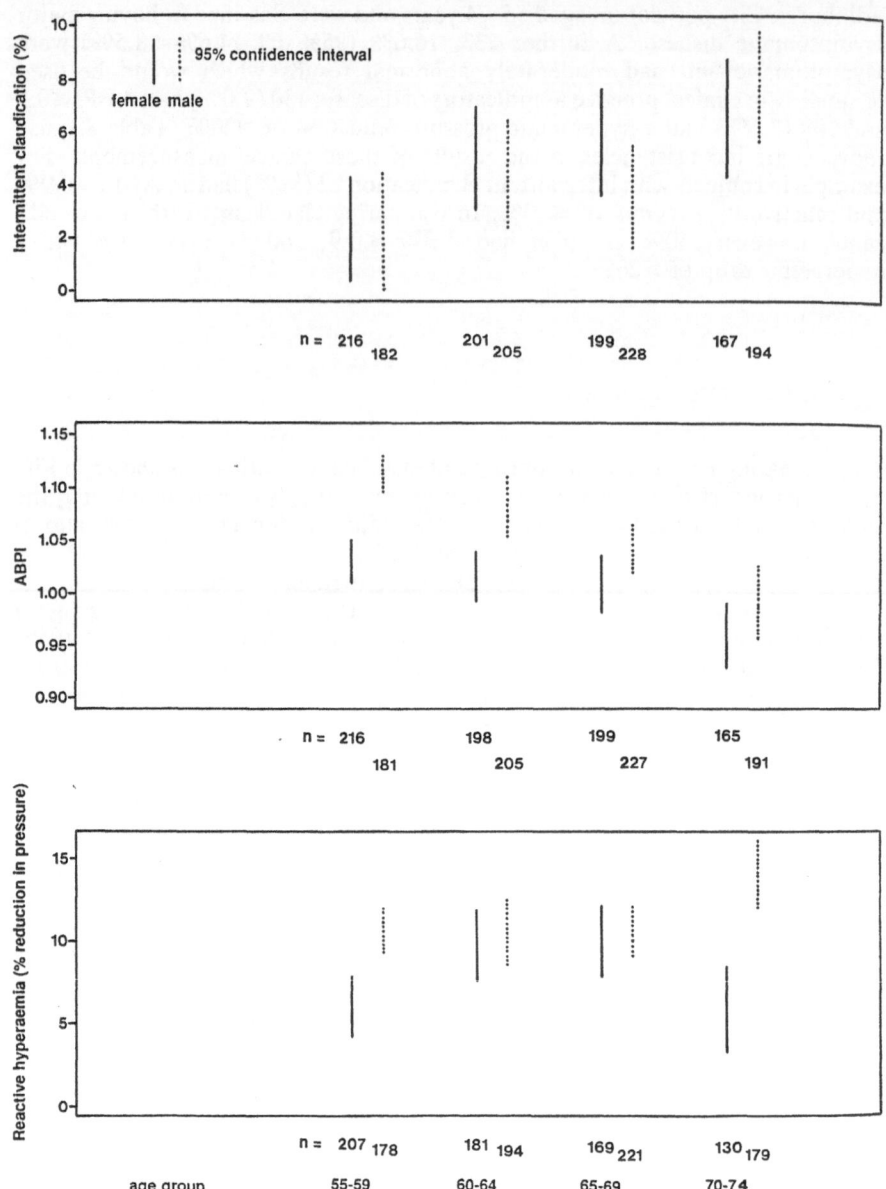

Fig. 8.3. Intermittent claudication, ABPI and reactive hyperaemia related to age. Reproduced from Fowkes et al. (1991) by courtesy of the *International Journal of Epidemiology*.

cation but they had a higher relative risk namely 3.9 (95% CI 2.4–6.5). The relative risk of asymptomatic peripheral vascular disease in subjects with ischaemic heart disease was 2.3 (95% CI 1.6–3.4). The numbers were insufficient for separate analyses to be carried out for males and females.

Table 8.2. Relative risks of ischaemic heart disease in subjects with intermittent claudication and major asymptomatic peripheral vascular disease

Ischaemic heart disease (IHD)	Intermittent Claudication			Major asymptomatic peripheral vascular disease		
	Present (n=73)	Absent (n=1519)	Rel. risk of IHD (95% CI)	Present (n=101)	Absent (n=1299)	Rel. risk of IHD (95% CI)
Any IHD (below)	52	562	1.9 (1.6–2.3)	57	445	1.6 (1.3–1.9)
Recall of doctor diagnosis	28	204	2.9 (2.1–3.9)	19	164	1.4 (0.8–2.1)
WHO questionnaire positive	29	267	2.3 (1.7–3.1)	24	218	1.4 (0.9–2.0)
	(n=35)	(n=1191)		(n=73)	(n=1030)	
ECG codes[a] and asymptomatic	14	234	2.0 (1.3–3.1)	26	178	2.1 (1.5–2.9)

[a] ECG Minnesota codes: 1.1–1.2 (probable myocardial infarction); 1.3, 4.1–4.4, 5.1–5.3, 7.1 (possible ischaemia). Reproduced from Fowkes et al. (1991) by courtesy of the *International Journal of Epidemiology*.

Possible Bias in Estimating Prevalence of Disease

The progression of the atheromatous plaque from haemodynamically insignificant stenosis to haemodynamically significant stenosis and then to eventual occlusion (Fig. 8.1) is well understood by the pathologist and the vascular specialist alike. So also is the potential of occlusive lesions at different sites to produce symptoms of different severity or even no symptoms at all. However, there has hitherto been little information about the relative frequencies of these phenomena and virtually no information about time-scales and the speed of transition from one category to another. We hope eventually to throw light on the latter aspect with further longitudinal studies of the Edinburgh cohort; although the difficulties of accurately measuring symptomatic and asymptomatic disease should not be underestimated.

The Edinburgh Artery Study has shown that in addition to 4.5% of the population aged between 55 and 74 having intermittent claudication almost twice that number (8.0%) had non-invasive investigations which suggested significant asymptomatic disease. It is vital to know the validity of these non-invasive tests and the nature of the lesions they are detecting. These investigations had been previously evaluated in pilot studies for their reproducibility and inter-observer variability (Fowkes et al. 1988a,b). To validate these measurements further a subsample of 46 of the asymptomatic patients and 86 healthy controls (selected randomly from subjects with ABPI >0.9 and reactive hyperaemia <20%) were examined "blind" with a duplex scanner by an experienced radiologist. This study (as yet unpublished) showed gross differences in disease between the two groups: one-third of cases having total occlusion of either a femoral or popliteal artery and 28 out of 31 of the remainder had evidence of moderate or severe stenosis. In contrast none of the controls had occlusion and only three had evidence of >50% narrowing of the femoral or popliteal artery.

Lesser degrees of abnormality on non-invasive tests (ABPI ≤0.9 or a >20%

reduction in systolic pressure during reactive hyperaemia) were shown in a further 16.6% of the population. Although these cut-off points have been shown to be valid in identifying angiogram positive disease in patients in the clinical setting (Baker 1978; Bernstein and Fronek 1982; Fowkes 1988a) their validity in the general population is not clearly known and further studies against gold standards such as duplex scanning are required. However, since the trends with age and other parameters were as expected and the ABPIs reflected severity of disease as confirmed by duplex it seems reasonable to assume a spectrum of disease with the 16.6% representing a further substantial sector of less extensive asymptomatic disease. This assumption would be supported by the fact that the distribution of the ABPI had a negative skew and the distribution of the reactive hyperaemia test had a positive skew. One-third of claudicants had an ABPI of >0.9 and reactive hyperaemia pressure reduction of <20% but this was not unexpected since the "claudicants" identified by the WHO/Rose questionnaire would be expected to include a number of false positives and some true claudicants would be expected to have "normal" results for any continuous biological variable.

Some other potential sources of bias have to be acknowledged, as in any epidemiological study. First, in planning the study a deliberate attempt was made to include individuals from socially deprived areas and this may have caused a lower response and an underestimate of total prevalence. Second, the WHO/ Rose questionnaire is known to lack sensitivity so that the claudication rates may have been an underestimate. Third, fewer tests of reactive hyperaemia were carried out on older subjects giving rise to the possibility that major asymptomatic disease might have been undetected. Thus, overall, the bias may have been towards an underestimate. However our findings on intermittent claudication were comparable with other studies (Fowkes 1988b) suggesting that substantial bias did not occur. A prevalence of 2.2% was reported by Hugheson in 1978 but this was in the south of England, generally regarded as a more "healthy" area and comprised a younger age group and excluded cases with a normal ABPI. By contrast a survey in rural Finland, also in 1978, found a prevalence of 7.7% in men aged 55–74 (Heliovaara et al. 1978).

Surveys of Symptomatic and Asymptomatic Disease

In general, cardiovascular surveys in most countries have not included laboratory tests of peripheral arteries. In two population studies in Belgium and Israel among asymptomatic subjects between 40 and 60 years the prevalence of ABPI <0.9 was about 4% (De Backer et al. 1979; Gofin et al. (1987) whereas in a Danish study of 60-year-olds the prevalence was nearly 12% (Schroll and Munck 1981). These rates were lower than the 18% observed in our study, probably due to the differences in age groups. Stress tests were not employed in these studies. But in a Californian Lipid Research Clinics study, partly based on a hyperlipid-aemic population a combination of leg pressures, waveform analysis and a reactive hyperaemia test was used (Criqui et al. 1985). Adopting conventional clinical cut-off points for disease the prevalence of asymptomatic disease increased from 3% at less than 60 years to over 20% at 75 years and older. In the

Basle study of a selected population of factory workers the prevalence of arterial occlusion in asymptomatic men aged 60–64 years was 6% (Widmer et al. 1964). Although these various studies are not directly comparable due to differences in population selection and in tests employed, our figure of 8% of the population aged 55 to 74 years having major asymptomatic arterial disease is of a similar order of magnitude and has the advantage of being based on a combination of tests applied to a random sample of the general population.

As expected the prevalence of intermittent claudication and of asymptomatic disease increased with age in the Edinburgh Artery Study. However, overall, the disease rates did not differ between the sexes, a finding at variance with others who have found a higher frequency of intermittent claudication in males (Hughson et al. 1978; Schroll and Munck 1981; Criqui et al. 1985). This was unlikely to be due to an age effect because in our youngest age group (55–59 years) claudication rates were no higher in men (Fig. 8.3).

Another unexpected finding in the Edinburgh Artery Study was higher mean ABPI in males, only partly explained by height. A population study in Israel found no sex difference in ABPI (Gofin et al. 1987). We conclude that in the normal "healthy" population ABPIs are on average higher in men than in women, although the distribution of ABPI and reactive hyperaemia tests indicate more arterial disease in men.

Obliterative arterial disease being a multifocal process, an association between asymptomatic arterial disease and coronary artery disease was expected in the Edinburgh Artery Study. Definitive tests such as coronary angiography or exercise ECGs were not feasible in a screening study but the group with asymptomatic disease did show increased reports of angina, recall of doctor diagnosis of heart disease, and ischaemia or myocardial infarction on resting ECG. The prevalence of heart disease has been shown, as in the Edinburgh Artery Study, to be two to four times higher than in normal subjects (Reid et al. 1966; Hughson et al. 1978) and to be associated with increased risk of future heart disease (Kannel et al. 1970, Reunanen et al. 1982). However the association between asymptomatic peripheral vascular disease and risk of cardiac disease is not clear. Improvements in methods of identifying asymptomatic disease by non-invasive tests together with advances in intervention radiology and the therapeutic prevention of cardiovascular events suggest that further studies of the risks in subjects with asymptomatic peripheral disease are indicated. This group also offers the opportunity to study and to influence risk factors at a stage when life style has not been altered by the impact of symptoms.

Acknowledgements. My colleagues and I are grateful to the British Heart Foundation for financial support for the Edinburgh Artery Study, and to the many general practitioners, nursing and technical staff involved in the project.

References

Baker JD (1978) Post-stress Doppler ankle pressures. Arch Surg 113:1171–3
Bernstein EF, Fronek A (1982) Current status of noninvasive tests in the diagnosis of peripheral arterial disease. Surg Clin North Am 62:473–87

Carter SA (1972) Response of ankle systolic pressure to leg exercise in mild or questionable arterial disease. N Engl J Med 287:578–82

Criqui MH, Fronek A, Barrett-Connor E (1985) The prevalence of peripheral arterial disease in a defined population. Circulation 71:510–15

De Backer IG, Kornitzer M, Sobolski J, Denolin H (1979) Intermittent claudication – epidemiology and natural history. Acta Cardiol 34:115–24

Dixon WJ, Brown MB, Engelman L et al. (1988) BMDP Statistical Software Manual. University of California Press, Berkeley.

Fowkes FGR (1988a) The measurement of atherosclerotic peripheral arterial disease in epidemiological surveys. Int J Epidemiol 17:201–7

Fowkes FGR (1988b) Epidemiology of atherosclerotic arterial disease in the lower limbs. Eur J Vasc Surg 2:283–91

Fowkes FGR, Housley E, Macintyre CCA, Prescott RJ, Ruckley CV (1988a) Variability of ankle and brachial systolic pressures in the measurement of atherosclerotic peripheral arterial disease. J Epidemiol Comm Health 42:128–33

Fowkes FGR, Housley E, Macintyre CCA, Prescott RJ, Ruckley CV (1988b) Reproducibility of reactive hyperaemia test in the measurement of peripheral arterial disease. Br J Surg 75:743–6

Fowkes FGR, Housley E, Cawood EHH, Macintyre CCA, Ruckley CV, Prescott RJ (1991) Edinburgh Artery Study; prevalence of asymptomatic and symptomatic peripheral disease arterial in the general population. Int J Epidemiol 20:384–92

Gofin R, Kark JD, Friedlander Y, Lewis BS, Witt H et al. (1987) Peripheral vascular disease in a middle-aged population sample. Isr J Med Sci 23:157–67

Heliovaara M, Karvonen MJ, Vilhunen R, Punsar S (1978) Smoking, carbon monoxide and atherosclerotic diseases. Br Med J i:268–70

Hughson WG, Mann JI, Garrod A (1978) Intermittent claudication: prevalence and risk factors. Br Med J i:1379–81

Jager KA, Phillips DJ, Martin RL et al. (1985) Noninvasive mapping of lower limb arterial lesions. Ultrasound Med Biol 11:515–21

Kannel WB, Skinner JJ, Schwartz MJ, Shurtleff D (1970) Intermittent claudication: incidence in the Framingham Study. Circulation 41:875–83

Laing SP, Greenhalgh RM (1980) Standard exercise test to assess peripheral arterial disease. Br Med J 280:13–6

Margolis JR, Kannel WB, Feinleib M, Dawber TR, McNamara PM (1973) Clinical features of unrecognized myocardial infarction – silent and symptomatic. Am J Cardiol 32:1–7

Morris JA, Gardner MJ (1989) Calculating intervals for relative risks (odds ratios) and standardised ratios and rates. Br Med J 296:1313–16

Office of Population Censuses and Surveys (1980) Classification of occupations 1980. Her Majesty's Stationery Office, London

Prineas RJ, Crow RS, Blackburn H (1982) The Minnesotta code manual of electrocardiographic findings. John Wright/PSG, Boston

Reid DD, Brett GZ, Hamilton PJS, Jarrett RJ, Keen H, Rose G (1974) Cardiorespiratory disease and diabetes among middle-aged male civil servants. Lancet i:469–73

Reid DD, Holland WW, Humerfelt S, Rose G (1966) A cardiovascular survey of British postal workers. Lancet i:614–18

Reunanen A, Takkunen H, Aromaa A (1982) Prevalence of intermittent claudication and its effect on mortality. Acta Med Scand 211:249–56

Richard JL, Ducimetiere P, Elgrishi I et al. (1972) Dépistage par questionnaire de l'insuffisance coronarienne et de la claudication intermittente. Rev Epidemiol Med Soc Sante Publ 20:735–55

Rose GA (1962) The diagnosis of ischaemic heart pain and intermittent claudication in field surveys. Bull WHO 27:645–658

Schroll M, Munck O (1981) Estimation of peripheral arteriosclerotic disease by ankle blood pressure measurements in a population study of 60-year-old men and women. J Chron Dis 34:261–9

Widmer LK, Greensher A, Kannel WB (1964) Occlusion of peripheral arteries: a study of 6400 working subjects. Circulation 30:836–42

9 Prevalence in General Practice

H.E.J.H. STOFFERS, V. KAISER, J.A. KNOTTNERUS

Importance of Peripheral Vascular Disease in General Practice

Peripheral vascular disease is a term often used loosely to describe the manifestation of atherosclerosis below the bifurcation of the abdominal aorta. It is the most frequent disease affecting peripheral arteries in the elderly. The population is ageing in western societies, therefore the prevalence of atherosclerotic disease will rise and the diagnostic and therapeutic management of this chronic disease will become more important to general practitioners.

Although coronary artery disease and cerebrovascular disease get more attention in popular and medical publications, peripheral vascular disease is a relevant subject to general practice. From results published in the literature it can be concluded that intermittent claudication has a relatively benign course in 75% of cases (Dormandy et al. 1989). Only in a minority of cases will serious complications occur: reduced mobility, severe pain at rest or gangrene, for which percutaneous transluminal angioplasty, vascular surgery or amputation might be indicated (Dormandy et al. 1989). The presence of intermittent claudication is one of the major predictors of disability in later life (Lammi et al. 1989).

A progressive course of the disease cannot be predicted in individual cases at the time of diagnosis (Jonason and Ringqvist 1985; Rosenbloom et al. 1988) and, therefore, it is necessary to examine the patient at regular intervals, especially in the first year after diagnosis (Jelnes et al. 1986; Naschitz et al. 1988). Conservative therapy of intermittent claudication – "start walking and stop smoking" (Housley 1988) – has been proven effective and should be the task of the general practitioner (Van Vroonhoven 1986; Blomberry 1987). Thus, referral to a vascular surgeon can be restricted to progressive cases.

Furthermore a high prevalence of vascular co-morbidity and mortality has been reported among patients with intermittent claudication (Reunanen et al. 1982; Dormandy and Mahir 1986; Hughson et al. 1987; Müller-Bühl et al. 1987; Allardice et al. 1988; Bengtsson et al. 1989; Collin et al. 1988; Dormandy et al. 1989). Recognising and treating intermittent claudication – being an indicator of general cardiovascular risk – will reduce vascular risk and thus will contribute to the prevention of atherosclerotic complications.

All these tasks – adequate diagnosis, regular follow-up, conservative therapy, preventive measures and selection for referral – belong to the domain of the general practitioner. Knowledge of the prevalence ("prior probability") is important in the diagnostic process of the general practitioner. Prevalence figures also depict the workload which a chronic disease entails. In this chapter the prevalence of peripheral vascular disease is described. Results from the literature and first results from our own research will be presented and discussed.

Methods of Estimating Prevalence

Figures on the prevalence of peripheral vascular disease differ depending on the study population and the diagnostic criteria that were used. Three types of study can be distinguished.

In *morbidity studies by general practitioners* symptomatic patients, i.e. patients suffering from intermittent claudication, who visit their general practitioner are considered for inclusion (Lamberts 1984; Hoogen et al. 1985; Lamberts 1987). A positive diagnosis is established by the general practitioner in dialogue with the patient. In many cases the diagnosis probably will not meet the exact definition of intermittent claudication according to the "Rose questionnaire" (Rose et al. 1982). In these studies the prevalence of intermittent claudication varied from 0.2% to 0.7% of the total practice population.

In *population studies on intermittent claudication*, generally questionnaires or interviews based on the Rose questionnaire are used. Relevant to general practice are those studies in which the study population is not highly selected by age, sex, risk factors or stage of the disease (Hughson et al. 1987; Reunanen et al. 1982; Criqui et al. 1985; Hale et al. 1988; Evans 1988). In these studies the prevalence of intermittent claudication, whether known or not known to a (family) doctor varied between 1.7% and 4.8% of the study population.

In a number of population studies *non-invasive diagnostic techniques* were used (Schroll and Munck 1981; Criqui et al. 1985; Cammer Paris et al. 1988; Hiatt et al. 1990; Newman et al. 1991). In these studies symptomatic and asymptomatic cases were observed. The prevalence of peripheral vascular disease in the study of Criqui et al. (1985), being the only study without a (highly) selected study population, was 11.7%.

The Limburg Peripheral Arterial Occlusive Disease (PAOD) Study

Since many clinical features of peripheral vascular disease relevant to general practice, such as prevalence, diagnostic features, prognostic indicators, and effects of conservative therapy are not known in sufficient detail (Stoffers et al. 1988), a study was set up by the University of Limburg to investigate some of these aspects.

The first part of the "Limburg PAOD Study" is a cross-sectional survey among 3654 subjects aged 40–79 years. Main research questions are "What is the prevalence of peripheral vascular disease?" and "What is the diagnostic value of typical signs and symptoms?". The patients were selected and sampled from a study base of approximately 30 000 patients (18 general practices). All patients in the study population were examined by their general practitioner. The general practitioners based their clinical diagnosis on medical history (an extensive questionnaire filled in by the patient), physical examination and data abstracted from the patient's record. Independently of the general practitioner the practice assistants assessed the ankle-brachial systolic pressure ratio ("AB-Ratio") using a pocket Doppler device and a mercury sphygmomanometer. A patient with an AB-Ratio lower than 0.95, measured twice within a week, was considered to have peripheral vascular disease. Results will be published in detail in 1992.

The second part of the study is a randomised controlled trial which is designed to answer the question whether aspirin has a therapeutic effect in peripheral vascular disease (Kaiser and Stoffers 1988). Studies on the reproducibility (Stoffers et al. 1991) and the validity of the AB-Ratio using a pocket-Doppler device have also been carried out. A study on the long-term prognosis of peripheral vascular disease is currently in preparation.

Comparison of Prevalences in Different Studies

The prevalence figures from different studies cannot be compared directly. In order to make an adequate comparison we estimated minimum and maximum age-standardised prevalence figures by adding data from different studies for different age categories. The results of this "educated guess" is presented in Table 9.1. In Table 9.2 a similar comparison is presented based on empirical data from the "Limburg Peripheral Arterial Occlusive Disease (PAOD) Study".

Table 9.3 shows the prevalence of peripheral vascular disease in the elderly according to the different types of study mentioned above. In Table 9.4 results from our own research are presented.

Table 9.1 shows a difference between the number of symptomatic patients in the population and the number of symptomatic patients who visited their general practitioner ("tip of the iceberg"). This suggests that there is a group of patients suffering from intermittent claudication which the general practitioner is not aware of. This is in accordance with data from an English population study which was conducted in two general practices: only half of the people with intermittent claudication had consulted a doctor for their complaints (Hughson et al. 1987). In a review by Dormandy et al. (1989) it was suggested that probably only half of the patients with intermittent claudication are known as such to a doctor. Table 9.2 shows that in our study, in which patients did not come spontaneously to the office of their general practitioner but were invited by him to take part in the study, approximately one-third of all patients who had been diagnosed by their general practitioner as a case of intermittent claudication, were "new" cases (2–3/8).

Table 9.1. Estimated prevalence of peripheral vascular disease, standardised for age, in different types of study

Type of study	Description of cases	Prevalence (per thousand) min–max[a]
Morbidity registration in general practice	Intermittent claudication not following strict Rose criteria presented to the general practitioner	2–8
Population studies using the Rose questionnaire	Intermittent claudication following	
	strict Rose criteria	4–12
	less strict criteria	19–25
Population study using non-invasive techniques	Symptomatic and asymptomatic peripheral vascular disease	27

[a] Standardised to the age distribution of the Dutch population 1 January 1989. Prevalence figures were estimated by comparing prevalence figures from different studies for different age categories (see text).

Table 9.2. Estimated prevalence of peripheral vascular disease, standardised for age, using different criteria in the Limburg peripheral arterial occlusive disease (PAOD) study

Variable	Inclusion criterion	Prevalence (per thousand)[a]
Intermittent claudication "Rose"	Strict criteria of Rose questionnaire	5
Intermittent claudication according to general practitioner	Diagnosed during research project	
	all cases	8
	new cases	2–3
Peripheral vascular disease, non-invasive	Ankle-brachial ratio <0.95	26

[a] Standardised to the age distribution of the Dutch population 1 January 1989. The prevalence for the age category "<45 years" was considered to be zero and for the age category ">75 years" was extrapolated linearly from previous age categories.

Table 9.3. Prevalence of peripheral vascular disease among the elderly in different types of study

Reference	Method	Sex	Prevalence (%)		
			65–69 yrs	70–74 yrs	75+ yrs
Hoogen et al. 1985	General practice morbidity registration	Male		5.4	9.6
		Female		1.5	3.6
		Total		3.2	5.9
Evans 1988	Population study, Rose criteria	Male	4.9	5.6	9.6
		Female	3.6	4.7	2.8
		Total	4.2	5.1	5.5
Criqui et al. 1985	Population study, non-invasive tests	Male	12	19	23
		Female	12	12.5	21.5

Fig. 9.4. Prevalence of peripheral vascular disease in different age categories, using different criteria in the Limburg peripheral arterial occlusive disease (PAOD) study[a]

Criterion	Sex	Prevalence (%)			
		45≤55 yrs.	55≤65 yrs.	65≤75 yrs.	All
Intermittent claudication	Male	0.6	2.5	8.8	3.0
according to	Female	0.6	1.6	—	0.8
general practitioner	Total	0.6	2.0	3.4	1.8[c]
Intermittent claudication	Male	0.7	3.0	5.0	2.5
"Rose"	Female	0.7	0.5	—	0.4
	Total	0.7	1.6	2.0	1.4
AB-Ratio <0.95[b]	Male	0.6	6.2	18.8	6.7
	Female	4.5	5.6	6.9	5.6
	Total	2.7	6.0	11.6	6.1

[a] Calculated for the study base population: age 45≤75 yrs; males = 8926, females = 9958.
[b] AB-Ratio: Ankle-brachial systolic pressure ratio using a pocket Doppler device.
[c] New cases: 0.5%.

Tables 9.1 and 9.2 also show a large difference between the number of cases with symptomatic peripheral vascular disease, i.e. intermittent claudication diagnosed by general practitioners or intermittent claudication according to the Rose questionnaire, and the number of cases with "objective" peripheral vascular disease, i.e. based on the results of noninvasive tests. The same difference can be observed in Tables 9.3 and 9.4. Many patients with "objective" peripheral vascular disease are not known as such to the general practitioner. Among these hidden patients there will be symptomatic and asymptomatic (or atypical) cases, and Table 9.4 suggests that the difference between "objective" and symptomatic patients is larger among females than among males.

Questions for Further Study

Some questions arise. What is the relevance of asymptomatic disease? If longitudinal studies confirm the suggestion that asymptomatic peripheral vascular disease is a risk factor for coronary artery disease or cerebrovascular disease (Widmer et al. 1983), or if it is demonstrated which asymptomatic patients eventually become symptomatic, it will be important to identify these patients in order to take preventive measures. Then a pocket Doppler device could become an important instrument in general practice.

Another question is whether the Rose questionnaire for intermittent claudication adequately describes the spectrum of peripheral vascular disease relevant to general practitioners. Maybe the "asymptomatic" stage does have certain symptoms after all and maybe there are less typical complaints which can be associated with peripheral vascular disease.

Data on the prevalence of peripheral vascular disease among females do not yield uniform conclusions. Does peripheral vascular disease present in a less typical way in females? Is the course of the disease in females more benign?

Cross-sectional and longitudinal research is necessary to answer these and other questions regarding diagnosis, prognosis and treatment of peripheral vascular disease.

Acknowledgements. The "Limburg PAOD Study" is funded by the 'Nederlandse Organisatie voor Wetenschappelijk Onderzoek (900–715.154) and the "Praeventiefonds" (28–1323). Doppler devices were financed by DAGRA-Pharma B.V.

References

Allardice JT, Allwright GJ, Wafula JMC, Wyatt AP (1988) High prevalence of abdominal aortic aneurysm in men with peripheral vascular disease. Screening by ultrasonography. Br J Surg 75: 140–2

Bengtsson H, Ekberg O, Aspelin P, Källerö S, Bergqvist D (1989) Ultrasound screening of the abdominal aorta in patients with intermittent claudication. Eur J Vasc Surg 3:497–502

Blomberry PA (1987) Intermittent claudication. An update on management. Drugs 34:404–10

Cammer Paris BE, Libow LS, Halperin JL, Mulvihill MN (1988) The prevalence and one-year outcome of limb arterial obstructive disease in a nursing home population. J Am Geriatr Soc 36:607–12

Collin J, Walton J, Araujo L, Lindsell D (1988) Oxford screening programme for abdominal aortic aneurysm in men aged 65–74 years. Lancet ii:613–15

Criqui MH, Fronek A, Barret-Connor E, Klauber MR, Gabriel S, Goodman D (1985) The prevalence of peripheral arterial disease in a defined population. Circulation 71:510–15

Dormandy JA, Mahir MS (1986) The natural history of peripheral atheromatous disease of legs [congress-paper]. Department of vascular surgery, St James' and St George's Hospitals, London

Dormandy J, Mahir M, Ascady G et al. (1989) Fate of the patient with chronic leg ischaemia. J Cardiovasc Surg 30:50–7

Evans JG (1988) Intermittent claudication. Age Ageing 17:139

Hale WE, Marks RG, May FE, Moore MT, Stewart RB (1988) Epidemiology of intermittent claudication: evaluation of risk factors. Age Ageing 17:57–60

Hiatt WR, Marshall WA, Baxter J et al. (1990) Diagnostic methods for peripheral arterial disease in the San Luis Valley Diabetes Study. J Clin Epidemiol 43:597–606

Hoogen HJM van den, Huygen FJA, Schellekens JWG, Straat JM, Van der Velden HGM (1985) Morbidity figures from general practice. Data from four general practices 1978–1982. Dept. of General Practice, University of Nijmegen, Nijmegen

Housley E (1988) Treating claudication in five words. Br Med J 296:1483–4

Hughson WH, Mann JI, Garrod A (1987) Intermittent claudication: prevalence and risk factors. Br Med J i:1379–81

Jelnes R, Gaardsting O, Hougaard Jensen K, Baekgaard N, Tønnessen KH, Schroeder T (1986) Fate in intermittent claudication: outcome and risk factors. Br Med J 293:1137–40

Jonason T, Ringqvist I (1985) Factors of prognostic importance for subsequent rest pain in patients with intermittent claudication. Acta Med Scand 218:27–33

Kaiser V, Stoffers HEJH (1988) Peripheral arterial obstructive disease in general practice: diagnosis, natural history and intervention. Allgemein Medizin; 17: Heft 1, S3 (Abstract)

Lamberts H (1984) Morbidity in general practice. Diagnosis-related information from the Monitoring Project. Huisartsenpers, Utrecht

Lamberts H (1987) Transition Project. Dept. of General Practice, University of Amsterdam, Amsterdam

Lammi UK, Kivelä SL, Nissinen A, Punsar S, Puska P, Karvonen M (1989) Predictors of disability in elderly Finnish men – a longitudinal study. J Clin Epidemiol 42:1215–25

Müller-Bühl U, Diehm C, Sieben U et al. (1987). Prävalenz und Risokofaktoren von peripher-arterieller Verschlusskrankheit und koronarer Herzkrankheit. Vasa 21 (Suppl):1–46

Naschitz JE, Ambrosio DA, Chang JB (1988) Intermittent claudication: predictors and outcome. Angiology 39:16–21

Newman AB, Sutton-Tyrell K, Rutan GH, Locher J, Kuller LH (1991) Lower extremity arterial disease in elderly subjects with systolic hypertension. J Clin Epidemiol 44:15–20

Reunanen A, Takkunen H, Aromaa A (1982) Prevalence of intermittent claudication and its effect on mortality. Acta Med Scand 211:249–56

Rose GA, Blackburn H, Gillum RF, Prineas RJ (1982) Cardiovascular survey methods, 2nd edn. World Health Organization, Geneva

Rosenbloom MS, Flanigan DP, Schuler JJ et al. (1988) Risk factors affecting the natural history of intermittent claudication. Arch Surg 123:867–70

Schroll M, Munck O (1981) Estimation of peripheral arteriosclerotic disease by ankle blood pressure measurements in a population of 60-year-old men and women. J Chron Dis 34:261–9

Stoffers HEJH, Kaiser V, Lemmens ThGJ, Knottnerus JA (1988) Perifeer arterieel obstructief vaatlijden in de huisartspraktijk: een verborgen ziektebeeld? [Peripheral arterial occlusive vasculopathy in general practice: a hidden disease?] Huisarts Wet 31:202–6 [summary in English]

Stoffers HEJH, Kaiser V, Kester ADM, Schouten HJA, Knottnerus JA (1991) Peripheral arterial occlusive disease in general practice: the reproducibility of the ankle-arm systolic pressure ratio. Scand J Prim Health Care 8 (in press)

Van Vroonhoven TJ (1986) Juist ook voor patienten met claudicatio intermittens geldt: de een is de ander niet. Ned Tijdschr Geneeskd 130:1345–8

Widmer LK, Biland L, Delley A, Da Silva A (1983) Zum Stellenwert der peripherer-arteriellen Verschlusskrankheit in der Praxis. Schweiz Med Wochenschr 113:1924–7

10 Intermittent Claudication in Scotland

W.C.S. SMITH, M. WOODWARD, H. TUNSTALL-PEDOE

Scotland has one of the highest national mortality rates from coronary heart disease both in men and in women (Tunstall-Pedoe et al. 1986). Mortality rates from cerebrovascular disease in Scotland are also high but not as high as those found in Eastern Europe (Uemura and Pisa 1988). However, the occurrence of the other major manifestation of circulatory disease in Western countries, peripheral vascular disease, is less well described. This is partly due to the fact that it is not as frequently identified as the underlying cause of death on a death certificate, and partly because it often occurs in people who already have either coronary or cerebrovascular disease.

Studies of the occurrence of peripheral vascular disease are based on detailed inspection of death certificates or on *ad hoc* surveys. Such surveys are often based in hospital clinics (Juergens et al. 1960) rather than in the community (Fowkes 1988a). Epidemiological tools are required to identify individuals who have signs and symptoms of peripheral vascular disease and ideally these should be validated by diagnostic tests (Fowkes 1988b).

The questionnaire for intermittent claudication developed by Rose at the London School of Hygiene and Tropical Medicine (Rose et al. 1977) has been used in an epidemiological survey of the Scottish adult population, the Scottish Heart Health Study (Smith et al. 1987). Data from this study have been used to determine the prevalence of intermittent claudication in the community, and to examine risk factors for intermittent claudication using a case–control methodology.

The Scottish Heart Health Study

The Scottish Heart Health Study was designed to establish the levels of coronary risk factors in a cross-sectional sample of the Scottish population and the extent to which the geographical variation in coronary heart disease could be explained in terms of these risk factors (Smith et al. 1987). The study was based in 22 Scottish districts and 10 359 men and women aged 40–59 years participated. The field work was conducted between 1984–1986 and the overall findings for coronary heart disease risk factors and life style have been published (Smith et al. 1989; Tunstall-Pedoe et al. 1989). Each participant completed a questionnaire which contained the self-administered version of the Rose Chest Pain Questionnaire with the section on leg pain (Rose et al. 1977). Each subject also underwent a clinical examination which included the measurement of height, weight, and blood pressure, the recording of a 12-lead electrocardiogram, and the collection of a non-fasting venous blood sample.

The presence of intermittent claudication was determined using the standard criteria (Rose et al. 1977), with two levels of severity (Grade 1 and 2) depending on whether or not the pain occurred when walking at an ordinary pace on the level. The prevalence rates of intermittent claudication were calculated by age and sex, and were compared with other studies which had used the same methods. For the case–control analysis, cases of intermittent claudication were identified using the described criteria, and each was matched for year of birth, sex and Scottish district with two control subjects who did not have intermittent claudication. At the point of data collection, the observers were blind to the case–control status of the individual subject. Odds ratios were estimated by conditional logistic regression (Breslow and Day 1980) using an iterative fitting method in GLIM (Adena and Wilson 1982). The base for the odds ratios for categorical variables is stated in the relevant tables, and the continuous variables have been divided into tertiles with the first tertile as the base. An alternative method of analysis is to compare the 92 cases with all the participants in the Scottish Heart Health Study without symptoms of intermittent claudication after adjusting for age and sex.

Prevalence of Intermittent Claudication

Analysis of the questionnaire data from the 10 042 subjects who adequately completed the questions on leg pain was carried out (317 participants did not complete the questions). A total of 1755 men and women reported pain in either leg on walking but only 92 met the criteria for intermittent claudication (41 with Grade 1 and 51 with Grade 2). The frequency of intermittent claudication by age and sex is shown in Table 10.1. The prevalence of intermittent claudication rose steeply with age and was higher in men than in women in the older age groups. Table 10.2 compares the rates of intermittent claudication found in the Scottish population with those found in other studies which had used the same question-naire method to determine intermittent claudication.

Table 10.1. Prevalence of intermittent claudication by age and sex in the Scottish heart health study

Symptoms	Men (Age in years)					Women (Age in years)				
	40–44	45–49	50–54	55–59	40–59	40–44	45–49	50–54	55–59	40–59
Intermittent claudication										
Rate/1000	4.0	10.0	8.1	21.6	10.9	4.5	7.1	3.7	12.4	6.9
No. of cases	5	12	11	28	56	6	9	5	16	36
Grade 1										
Rate/1000	1.6	4.2	4.4	10.0	5.1	1.5	3.2	1.5	5.4	2.9
No. of cases	2	5	6	13	26	2	4	2	7	15
Grade 2										
Rate/1000	2.4	5.8	3.7	11.6	5.9	3.0	4.0	2.2	7.0	4.0
No. of cases	3	7	5	15	30	4	5	3	9	21

Table 10.2. Rates of intermittent claudication in epidemiological studies using a questionnaire

Study (reference)	Men	Women	Age group (years)	Study size
Present study	11/1000	7/1000	40–59	10359
Whitehall (Rose et al. 1977)	8/1000		40–64	18403
Oxford (Hughson et al. 1978)	22/1000		45–69	1716
		12/1000	50–69	1535
Finland (Reunanen et al. 1982)	21/1000	18/1000	30–59	10962
France (Richard et al. 1972)	5/1000		22–59	7996
Belgium (Van Ganse et al. 1972)	13/1000		20–65	397
Denmark (Gyntelberg 1973)	20/1000		40–59	5249
Sweden (Isacsson 1972)	28/1000		55	703

The intermittent claudication cases were compared with their matched controls for the frequency of occurrence of other manifestations of cardiovascular disease (Table 10.3). This table shows that those men and women with intermittent claudication were more likely to have a past medical history of stroke, angina, and myocardial infarction than the controls. They were also more likely to be positive for angina using the Rose chest pain questionnaire and to have an ischaemic electrocardiogram.

Table 10.3. Prevalence of other forms of cardiovascular disease in cases and controls in the Scottish Heart Health Study

	Men				Women			
	Intermittent claudication group (n=56)		Control group (n=112)		Intermittent claudication group (n=36)		Control group (n=72)	
	n	%	n	%	n	%	n	%
Past medical history								
Angina	9	16.4	3	2.7	11	31.4	9	12.7
Myocardial infarction	7	12.7	2	1.8	4	11.8	3	4.3
Stroke	3	5.5	1	0.9	1	2.9	1	1.4
Rose questionnaire								
Angina	19	33.9	11	10.0	10	29.4	11	15.9
Electrocardiogram								
Ischaemia	11	19.6	14	12.8	9	25.0	7	9.7

Risk Factors For Intermittent Claudication

The case–control analysis is presented in Tables 10.4 and 10.5, with odds ratios for the risk factors studied and their 95% confidence limits. The odds ratios for male and female current smokers, compared with those who never smoked, were 2.67 and 2.75 respectively, which are significant at the 5% level. There is no significant difference between male ex-smokers and never-smokers, but there is for women. The results for expired air carbon monoxide, which is a biochemical marker of tobacco inhalation, also show a significant effect. Plasma fibrinogen was a significant risk factor only in men. Diabetes was not found to be a significant risk factor but the level of non-fasting serum glucose was in women. Moderate

Table 10.4. Odds ratios for intermittent claudication in men and women for smoking, alcohol and diabetes in the Scottish Heart Health Study

Risk factor	Men		Women	
	Odds ratio	(95% confidence interval)	Odds ratio	(95% confidence interval)
Cigarette smoking (base=never smoked)				
Current	2.67	(1.23,5.77)	2.75	(1.18,6.40)
Ex-smoking	1.23	(0.52,2.90)	2.70	(1.11,6.57)
Alcohol intake (base=non-drinker)				
1–9 Units/week	2.10	(0.93,4.75)	0.53	(0.27,1.04)
>9 Units/week	1.68	(0.80,3.52)	0.11	(0.01,0.78)
Diabetes (base=no)	1.06	(0.26,4.35)	0.67	(0.09,4.84)

Table 10.5. Odds ratios for intermittent claudication for continuous risk factors where second and third tertiles are compared with the first tertile as the base, in the Scottish Heart Health Study

Variable	Tertile	Men		Women	
		Odds ratio	(95% confidence interval)	Odds ratio	(95% confidence interval)
Expired air					
Carbon monoxide	2nd	2.54	(1.30,4.99)	1.77	(0.70,4.48)
	3rd	2.32	(1.18,4.56)	3.30	(1.38,7.93)
Plasma fibrinogen	2nd	2.33	(1.05,5.17)	1.06	(0.42,2.68)
	3rd	2.73	(1.24,6.00)	1.20	(0.49,2.94)
Blood pressure					
Systolic	2nd	1.75	(0.93,3.32)	1.56	(0.67,3.65)
	3rd	1.15	(0.58,2.28)	1.64	(0.71,3.78)
Diastolic	2nd	0.93	(0.50,1.74)	1.00	(0.42,2.35)
	3rd	0.66	(0.35,1.26)	1.67	(0.77,3.63)
Body Mass Index	2nd	0.63	(0.33,1.23)	2.59	(1.09,6.15)
	3rd	0.87	(0.47,1.60)	2.23	(0.94,5.30)
Serum glucose	2nd	0.67	(0.34,1.33)	3.29	(1.15,9.36)
	3rd	0.61	(0.30,1.22)	3.98	(1.42,11.18)
Serum total	2nd	2.01	(0.95,4.25)	2.25	(0.88,5.79)
cholesterol	3rd	2.15	(1.02,4.55)	1.99	(0.76,5.22)
Serum	2nd	0.81	(0.41,1.60)	0.84	(0.37,1.91)
HDL-cholesterol	3rd	0.57	(0.27,1.17)	0.34	(0.12,0.96)
Serum	2nd	0.73	(0.33,1.63)	3.41	(1.20,9.71)
triglycerides	3rd	2.23	(1.13,4.40)	4.46	(1.59,12.52)

alcohol consumption was associated with an odds ratio of 2.10 in men when compared with non-drinkers, but alcohol was a significant protective factor in women. An odds ratio of less than one in men and greater than one in women was observed for body mass index, although this was only moderately significant in women and not at all significant in men. The odds ratios for both systolic and diastolic blood pressure was not significantly different from unity. Serum triglyceride was a significant risk factor in both men and women. Serum total cholesterol was a significant risk factor in men and HDL-cholesterol a significant protective factor in women. After adjustment, taking all other risk factors as covariates for each individual risk factor in turn by multivariate analysis, significant effects remain for total cholesterol, fibrinogen and cigarette smoking for men, and expiratory carbon monoxide, HDL-cholesterol, serum glucose, triglycerides, ex-smoking and alcohol for women.

Intermittent Claudication in Epidemiological Studies

In the Scottish Heart Health Study, the prevalence of symptoms of intermittent claudication based on the Rose self-administered questionnaire was 10.9 per 1000 in men, and 6.9 per 1000 in women. These rates rose with age and were greatest in the 55–59 year age group in men at 21.6 per 1000. These rates are similar to those found in other British studies (Rose et al. 1977; Hughson et al. 1978.) if the age differences are taken into account. In contrast, results from Finland, Sweden and Denmark have shown rates of intermittent claudication higher than those found in our study (Isacsson 1972; Gyntelberg 1973; Reunanen et al. 1982). The Rose questionnaire has been the traditional epidemiological tool for investigating peripheral vascular disease, but while this method is highly specific, it may not be highly sensitive (Richard et al. 1972). More recently other non-invasive tests have been reviewed and their place in epidemiological research discussed (Fowkes 1988b).

Those subjects found to be positive for intermittent claudication using the Rose method in the Scottish Heart Health Study, were also more likely than their matched controls to have other manifestations of cardiovascular disease. Analysis of Whitehall Civil Servants (Reid et al. 1974) showed only a little overlap in individuals between the manifestations of cardiovascular disease elicited by the Rose questionnaire, and only a few of those with angina or a history of myocardial infarction had intermittent claudication. In our study a considerable proportion of those with intermittent claudication had evidence of coronary heart disease.

In the case–control analysis within the cross-sectional study of the Scottish Heart Health Study, the case or control status of the participants was determined after the clinical examination and collection of information, thus potentially reducing observer bias, but not excluding bias introduced by the responders. The analysis confirms the major role of cigarette smoking in peripheral vascular disease (Fowkes 1989). Smoking is associated with an increase in fibrinogen and blood viscosity which may also have a role in the pathogenesis of peripheral vascular disease (Lowe 1987). The importance of blood pressure, observed previously (Reunanen et al. 1982), is not seen in our study, and the effects of diabetes mellitus and of random blood glucose confirm findings from the Framingham study (Lowe 1987) but are significant only in women. The consistent finding of a significant effect of triglycerides (Kannel and McGee 1985) in other studies is seen in our results also, but while total cholesterol is noted to be a risk factor and HDL-cholesterol a protective factor, these reach statistical significance only in women. The different associations of alcohol with intermittent claudication between men and women are of interest, and may represent different biological mechanisms, as well as different patterns of alcohol consumption between men and women in Scotland (Smith et al. 1989). The results observed using all the study participants without intermittent claudication as controls show a broadly similar pattern of risk factors and some differences from the results of this analysis based on matched controls (Smith et al. unpublished observations).

In the Scottish Heart Health Study we have shown that the prevalence of intermittent claudication in a Scottish population assessed by the Rose questionnaire, is similar to that of other British studies but is lower than that found in the Nordic countries. The absolute prevalence rates of intermittent claudication are much lower than the prevalence rates of coronary heart disease in Scotland. This case–control study has identified the risk factors for peripheral vascular disease in Scotland although these are often statistically significant in women only.

Summary: Scottish Heart Health Study

The prevalence of intermittent claudication in 10 042 Scottish men and women aged 40–59 years was investigated using the Rose questionnaire. Ninety-two cases were found yielding a prevalence rate of 11/1000 for men and 7/1000 for women. Each case identified was compared with two controls, matched by age, sex and geographical location, for the frequency of other forms of cardiovascular disease and risk factors. Those with intermittent claudication were more likely to have a history of stroke, angina and myocardial infarction than the controls. Self-reported smoking was a statistically significant risk factor for intermittent claudication as were expired air carbon monoxide and fibrinogen. Serum glucose and body mass index were significant risk factors and alcohol consumption a protective factor in women. Serum triglyserides were risk factors in both men and women but cholesterol was only significant in men and HDL-cholesterol was a significant protective factor in women. The odds ratios for blood pressure was not significantly different from unity. Total cholesterol, fibrinogen and cigarette smoking remained significant for men in multivariate analysis (adjusting each risk factor for all the others) while for women, carbon monoxide, HDL-cholesterol, serum glucose, triglycerides, ex-smoking and alcohol all had independent effects.

Acknowledgements. We thank Dr G.D.O. Lowe, Senior Lecturer, University Department of Medicine, Glasgow Royal Infirmary who was responsible for the fibrinogen assays in this study. The Cardiovascular Epidemiology Unit is funded by the Chief Scientist Office of the Scottish Home & Health Department; however, the conclusions in this chapter are those of the authors, not of the Scottish Home & Health Department.

References

Adena MA, Wilson SR (1982) Generalised linear models in epidemiological research: case–control studies. Instat Foundation, Sydney

Breslow NE, Day NE (1980) Statistical methods in cancer research, vol. 1. The analysis of case–control studies. IARC, Lyon

Fowkes FGR (1988a) Epidemiology of atherosclerotic arterial disease in the lower limbs. Eur J Vasc Surg 2:283–91

Fowkes FGR (1988b) The measurement of atherosclerotic peripheral arterial disease in epidemiological surveys. Int J Epidemiol 17:248–54

Fowkes FGR (1989) Aetiology of peripheral atherosclerosis. Br Med J 298:405–6

Gyntelberg F (1973) Physical fitness and coronary heart disease – male residents in Copenhagen aged 40–59. Dan Med Bull 20:1–4

Hughson WG, Mann JI, Garrod A (1978) Intermittent Claudication: prevalence and risk factors. Br Med J i:1379–81

Isacsson S-O (1972) Validation of the WHO questionnaire and reports of clinical and electrocardiographic findings related to venous occlusion plethysmography. Acta Med Scand (Suppl) 537:1–62

Juergens JL, Barlar NW, Hines EA (1960) Arteriosclerosis obliterans: review of 520 cases with special reference to pathogenic and prognostic factors. Circulation 21:188

Kannel WB, McGee DL (1985) Update on some epidemiological features of intermittent claudication: The Framingham Study. J. Am Geriatr Soc 33:13–18

Lowe GDO (1987) Blood rheology in general medicine and surgery. Clin Haematol 1:827–61

Reid DD, Brett GZ, Hamilton PJS, Jarrett RJ, Keen H, Rose H (1974) Cardiorespiratory disease and diabetes among middle-aged male civil servants. Lancet 1:469–73

Reunanen A, Takkunen H, Aromaa A (1982) Prevalence of intermittent claudication and its effect on mortality. Acta Med Scand 211:249–56

Richard JL, Ducimetiere P, Elgrishi I (1972) Depistage par questionnaire de l'insuffisance coronarienne et de la claudication intermittente. Rev Epidemiol Med Soc Sante Publ 20:735–55

Rose G, McCartney P, Reid DD (1977) Self-administration of a questionnaire on chest pain and intermittent claudication. Br J Prev Soc Med 31:42–8

Smith WCS, Crombie IK, Tavendale R, Irving M, Kenicer MB, Tunstall-Pedoe H (1987) The Scottish Heart Health Study: objectives and development of methods. Health Bull (Edinb) 45:211–17

Smith WCS, Tunstall-Pedoe H, Crombie IK, Tavendale R (1989) Concomitants of excess coronary deaths: major risk factor and lifestyle findings from 10 359 men and women in the Scottish Heart Health Study. Scott Med J 34:550–5

Tunstall-Pedoe H, Smith WCS, Crombie IK (1986) Levels and trends of coronary heart disease mortality in Scotland compared with some other countries. Heath Bull (Edinb) 44:153–61

Tunstall-Pedoe H, Smith WCS, Crombie IK, Tavendale R (1989) Coronary risk factor and lifestyle variation across Scotland: results from the Scottish Heart Health Study. Scott Med J 43:556–60

Uemura K, Pisa Z (1988) Trends in cardiovascular disease mortality in industrialised countries since 1950. World Health Stat Q 41:155–78

Van Ganse W, Van Hoorne N, De Backer G (1972) L'interview et le questionnaire auto-administré de Rose dans une étude pilote d'atherosclerose. Rev Epidomiol Med Soc Sante Publ 20:7

SECTION III

Vascular Risk Factors

11 Risk Factors and Cardiovascular Outcome

Calf Pain on Walking, Risk Factors and Cardiovascular Outcome in Middle-Aged British Men

A.G. SHAPER, S.G. WANNAMETHEE and M.K. WALKER

It is widely accepted that peripheral vascular disease, ischaemic heart disease and cerebrovascular disease are all manifestations of atherosclerosis and that they share many risk factors associated with their development. Much less is known about the aetiology, pathogenesis and development of peripheral vascular disease than about the other two conditions, possibly because its major manifestation, intermittent claudication (calf pain on exercise), has been regarded as a relatively benign, certainly not life-threatening, condition. The prevalence of intermittent claudication is uncertain, in large measure due to the variability in the methods of ascertainment, and little attention has been paid to the outcome in terms of cardiovascular events or death. Most studies show an increased death rate from ischaemic heart disease and other cardiovascular causes in subjects with intermittent claudication (Kannel et al. 1970; Jelnes et al. 1986; Reunanen et al. 1982; Davey Smith et al. 1990) and this is not unexpected considering the nature of the underlying disorder. To what extent this increased mortality is due to the associated presence of established ischaemic heart disease is uncertain. This chapter examines the risk factors associated with calf pain on walking in a large prospective study of middle-aged men and the outcome in terms of cardiovascular morbidity and mortality. The purpose is to determine the importance of this symptom in the development and risk of cardiovascular morbidity and death, its dependence on the presence of established cardiovascular disease and risk factors, and its usefulness in clinical and epidemiological studies.

British Regional Heart Study: Methods

The British Regional Heart Study (BRHS) is a prospective study of cardiovascular disease involving 7735 men aged 40–59 years selected from the age–sex registers of a single group general practice in each of 24 towns in England, Wales and Scotland. The criteria for selecting the town, the general practice and the subjects as well as the methods of data collection have been reported (Shaper et al. 1981). Research nurses administered to each man a standard questionnaire which included questions on smoking habits, alcohol intake and medical history. Several physical measurements were made, and blood samples were taken for measurement of biochemical and haematological variables. Details of the measurement of serum lipid concentrations have been described (Thelle et al. 1983). The men were classified according to their current smoking status: those who had never smoked, ex-cigarette smokers and current smokers. Those who had only ever smoked pipe/cigars are grouped as "never smoked". Ex-cigarette smokers who are currently pipe/cigar smokers are classified as ex-cigarette smokers. Alcohol consumption was recorded using questions on frequency, quantity and type, similar to those used in the 1978 General Household Survey. Heavy drinkers were defined as those regularly drinking more than six drinks daily. The longest-held occupation of each man was recorded and then coded in accordance with the Registrar General's occupational classification. Body mass index calculated as weight/height^2 was used as an index of relative weight. The men were asked to indicate their usual pattern of physical activity, which included regular walking or cycling, recreational activity and sporting activity. A physical activity score was derived for each man based on frequency and type of leisure activity (Shaper and Wannemethee 1991). The men were grouped into 6 broad categories based on their total score: inactive, occasional, light, moderate, moderately vigorous and vigorous. Active men were those whose physical activity was moderate or greater.

Breathlessness. A modified version of the Medical Research Council Question-naire on Respiratory Symptoms (1966 version) was administered at the initial examination and the men were classified as having no breathlessness, mild, moderate or severe breathlessness (Cook and Shaper 1988). Forced expiratory volume (FEV1) was measured using a Vitalograph J49–B2 spirometer with the subject seated; the maximum of two readings was used, height standardised to 1.73 metres.

Pre-existing Disease. The men were asked to recall a doctor's diagnosis of angina, myocardial infarction, diabetes and a number of other disorders listed on the questionnaire. The WHO (Rose) chest pain questionnaire was administered to all men at the initial examination and a 3-orthogonal lead electrocardiogram (ECG) was recorded at rest (Shaper et al. 1984).

Ischaemic Heart Disease The men were separated into three groups according to the evidence of ischaemic heart disease present at screening:

1. No evidence of ischaemic heart disease on WHO chest pain questionnaire, electrocardiogram or recall of a doctor diagnosis of ischaemic heart disease.

2. Men with evidence suggesting ischaemic heart disease short of a definite myocardial infarction. This group contains those with electocardiographic evidence of possible or definite myocardial ischaemia or possible myocardial infarction, those with angina or a possible myocardial infarction on WHO chest pain questionnaire, or with recall of a doctor diagnosis of angina.

3. Men with a previous definite myocardial infarction on electrocardiogram or who recalled a doctor diagnosis of a heart attack.

In the analyses men with pre-existing evidence of ischaemic heart disease consist of men in Groups 2 and 3.

Measurement of Calf Pain on Walking

Each man was asked: (1) "Do you get pain in your calf muscles when you walk uphill or hurry?" and (2) "Do you ever get pain in your calf muscles on walking at an ordinary pace, on the level?".

The questions on leg pain were not administered satisfactorily in the first three towns and all 915 men from these towns have been excluded. In addition, 13 men did not provide full information on both questions and have been omitted from the analysis. The term calf pain on walking is used in this chapter rather than intermittent claudication, as we did not use the full enquiry included in the questionnaire developed by the London School of Hygiene and Tropical Medicine (LSHTM) (Rose 1962; Rose et al. 1977).

Follow-up of Subjects

All men, whether or not they showed evidence of ischaemic heart disease at initial examination were followed up for all-cause mortality and cardiovascular morbidity (Walker and Shaper 1984). Information on death was collected through the established "tagging" procedures provided by the NHS registers in Southport (for England and Wales) and Edinburgh (for Scotland). Mortality and major ischaemic heart disease events (fatal and non-fatal) are based on 9.5 years of follow-up for each man. A non-fatal myocardial infarction was diagnosed according to WHO criteria. Fatal events were defined as death from ischaemic heart disease, International Classification of Diseases (ICD, 9th revision codes 410–414) as the underlying cause. Non-fatal stroke events were those which produced a neurological deficit that was present for more than 24 hours. Fatal episodes were those coded on the death certificate to ICD 430–8.

British Regional Heart Study: Calf Pain on Walking

Calf pain on walking was present in 784 (11.5%) of the 6807 men who answered both questions. Those with calf pain only on going uphill or hurrying ($n=397$; 5.8%) were regarded as *moderate* and those with calf pain at an ordinary pace on the level as well as going uphill or hurrying ($n = 349$; 5.1%) were regarded as *severe*. A small group with calf pain at an ordinary pace but not going uphill or hurrying ($n = 38$; 0.6%) were also regarded as *severe*.

Cardiovascular Risk Factors and Calf Pain

Table 11.1 presents the level of cardiovascular risk factors in men with no calf pain, moderate and severe leg pain. The mean age increased significantly with increasing severity of calf pain. Mean systolic blood pressure rose significantly with increasing severity of calf pain; for diastolic blood pressure only those with severe calf pain had a higher mean than those who reported no calf pain. Men who reported calf pain were significantly heavier than men with no calf pain. There was no association with serum total cholesterol or triglycerides. HDL-cholesterol decreased significantly with increasing severity of calf pain. Non-fasting blood glucose levels increased progressively but not significantly in the three calf pain groups. However, those who reported calf pain, particularly of severe degree, were more likely to be diabetic. Those with calf pain were more likely to be manual workers, to be current smokers and to be heavy drinkers or non-drinkers. They were more likely to be inactive and less likely to be engaged in levels of moderate or greater physical activity. These prevalence rates increased with increasing severity of calf pain. Leg pain was strongly associated

Table 11.1. Mean levels and frequency (%) of risk factors in men with no calf pain, moderate or severe calf pain

	Calf pain on walking			
	None (n=6023)	Moderate (n=397)	Severe (n=387)	p-value for trend
Age (years)	50.2 (0.07)	51.4 (0.28)	52.1 (0.29)	***
SBP (mm Hg)	145.8 (0.27)	146.7 (1.08)	148.5 (1.14)	*
DBP (mm Hg)	82.7 (0.17)	81.8 (0.68)	84.1 (0.66)	NS
BMI (Kg/m^2)	25.42 (0.04)	26.04 (0.17)	26.00 (0.20)	***
Cholesterol (mmol/l)	6.27 (0.01)	6.44 (0.05)	6.24 (0.06)	NS
HDL-chol (mmol/l)	1.15 (0.0)	1.12 (0.01)	1.12 (0.02)	*
Triglyceride (mmol/l)	0.55 (0.008)	0.67 (0.03)	0.58 (0.03)	NS
Glucose (mmol/l)	5.64 (0.02)	5.69 (0.07)	5.79 (0.10)	NS
Diabetes	1.4	1.8	4.2	***
Manual class	57.3	68.8	73.8	***
Smoking				
Never	24.4	13.4	11.9	
Ex-smoker	35.2	32.7	30.2	
Light (<20/day)	15.0	16.5	25.3	
Mod/heavy (>20/day)	25.4	37.5	32.6	
All current	40.4	53.6	57.9	***
Alcohol				
Non-drinkers	5.8	7.3	10.3	**
Heavy drinkers	10.9	12.6	15.5	**
Physical activity				
Inactive	8.6	13.3	20.1	***
Active	38.0	24.8	15.6	***
Breathlessness				
Mild	8.2	22.4	14.0	
Moderate	3.5	9.8	18.3	
Severe	2.2	7.1	22.2	
All	13.9	39.3	54.5	***
FEV1	335.8 (0.96)	310.5 (3.49)	293.7 (4.39)	***

χ^2_1 test for trend: *** $p<0.001$, ** $p<0.01$, * $p<0.05$, NS = not significant.
SBP, systolic blood pressure; DBP, diastolic blood pressure; BMI, body mass index

with breathlessness, particularly of a severe degree. Over half the men who reported severe calf pain reported symptoms of breathlessness. FEV1, which is strongly correlated with breathlessness decreased significantly with increasing severity of calf pain.

Prevalence of Ischaemic Heart Disease and Calf Pain

Table 11.2 shows the percentage of men with varying evidence of ischaemic heart disease as measured by recall of doctor diagnosis, by WHO (Rose) questionnaire or by ECG, according to the presence or severity of calf pain. The proportion of men with evidence of ischaemic heart disease increased with increasing severity of leg pain. Nearly half of those who reported severe calf pain and 40% of those with moderate calf pain had evidence of ischaemic heart disease. Those with severe calf pain were four times more likely to have had definite myocardial infarction than those with no leg pain.

Table 11.2. Prevalence of ischaemic heart disease in men with no calf pain, moderate or severe calf pain

	Calf pain on walking			
	None ($n=6023$)	Moderate ($n=397$)	Severe ($n=387$)	p-value for trend
Recall (DR)	4.3	9.6	21.5	
IHD (Q)				
Angina only	4.5	11.3	11.1	
Possible MI only	6.0	8.1	10.6	
Both	1.6	8.6	15.8	
Total	12.1	28.0	37.5	p<0.0001
Electrocardiogram				
Possible myocardial ischaemia	7.2	6.6	9.0	
Definite ischaemia/possible MI	4.0	9.8	9.3	
Definite MI	2.9	4.8	7.2	
Total	14.1	21.0	25.6	p<0.0001
Pre-existing IHD				
II (myocardial ischaemia)	18.6	32.2	27.7	p<0.0001
III (definite MI)	4.8	8.3	19.4	p<0.0001

DR, recall of doctor diagnosis; Q, WHO (Rose) questionnaire; MI, myocardial infarction; IHD, ischaemic heart disease.

Cardiovascular Mortality and Morbidity and Calf Pain

Within the 9.5 year follow-up (after exclusion of the 928 men with incomplete data on leg pain), there were 548 major ischaemic heart disease events ("heart attacks"), and 121 stroke cases, fatal and non-fatal. There were 638 deaths from all causes of which 325 (51%) were attributed to cardiovascular disease and 313 to non-cardiovascular disease. Table 11.3 shows the age-adjusted rate for heart attacks and stroke (fatal and non-fatal) and cardiovascular mortality. The rate of heart attacks, stoke and cardiovascular mortality increased significantly with increasing severity of calf pain.

Table 11.3. Age-adjusted incidence rates (per 1000/year) of ischaemic heart disease and stroke events (fatal and non-fatal) and cardiovascular deaths in men with no calf pain, moderate or severe calf pain

Age-adjusted rates/1000/year (n)	Calf pain on walking		
	None (n=6023)	Moderate (n=397)	Severe (n=387)
	% (n)	% (n)	% (n)
IHD events (548)	7.4 (438)	12.1 (48)	15.2 (62)
Stroke events (121)	1.7 (95)	1.9 (8)	4.1 (18)
CVD mortality (325)	4.4 (241)	8.2 (33)	11.9 (51)

Overall test for difference: $p<0.0001$ for IHD events, for stroke and for CVD mortality; IHD, ischaemic heart disease; CVD, cardiovascular disease.

Adjustment for Cardiovascular Risk Factors. Because of the strong association between calf pain and cardiovascular risk factors we have examined the relationships adjusting first for the "major risk factors" i.e. age, smoking, social class, systolic blood pressure, serum total cholesterol, HDL-cholesterol, diabetes and physical activity. We have then examined the effects of adjusting in addition for the presence of ischaemic heart disease and breathlessness, and in the case of stroke, for left ventricular hypertrophy on ECG. After adjusting for the major risk factors, calf pain still exhibited significant associations with heart attack and stroke (fatal and non-fatal) and with cardiovascular mortality (Table 11.4). Moderate calf pain was associated with a small (42%) increase in risk of heart attacks, but the increase was not significant. Severe calf pain was associated with a nearly two-fold increase in risk ($p<0.05$) compared with those who reported no calf pain. For cardiovascular mortality, men with moderate calf pain showed a 70% increase in risk and in men with severe calf pain there was a greater than two-fold increase in cardiovascular mortality compared with men who reported no calf pain. There was a significant increase in risk of stroke in men with severe calf pain after adjusting for the major risk factors. Those who reported moderate leg pain showed lower risk than those who reported none, but the difference was not significant.

Since calf pain is strongly associated with evidence of ischaemic heart disease we have examined the relationship in each disease category, adjusting in addition for pre-existing ischaemic heart disease. This reduced the relative risks further in men with moderate and severe calf pain. For major heart attacks, the difference was now only marginally significant ($p = 0.06$). Moderate and severe calf pain were still significantly associated with an increase in risk of cardiovascular mortality. Adjusting for pre-existing ischaemic heart disease and left ventricular hypertrophy in addition to the major risk factors reduced the excess risk of stroke, but those with severe calf pain still showed nearly a two-fold increase in risk of stroke compared with those who reported no calf pain.

Adjustment for Breathlessness. Breathlessness appears to be an early indicator of ischaemic heart disease in the absence of angina or electrocardiographic evidence of myocardial ischaemia and is an independent predictor of both angina and heart attack (Cook and Shaper 1988; 1989a). Both angina and calf pain showed a strong association with breathlessness (Table 11.1). After adjusting for symptoms of breathlessness in addition to the major risk factors and the presence of ischaemic heart disease, the excess risk of heart attack associated with severe calf pain was

Table 11.4. Adjusted relative risks (95% confidence intervals) of ischaemic heart disease events, stroke events and cardiovascular deaths, in men with no calf pain, moderate or severe calf pain

	Calf pain on walking		
	None	Moderate	Severe
Heart attacks (+)	1.0	1.42 (1.01,1.99)	1.73 (1.26,2.39)
(+) and IHD	1.0	1.27 (0.89,1.82)	1.35 (0.96,1.90)
(+) IHD and breathlessness	1.0	1.22 (0.85,1.75)	1.17 (0.82,1.68)
CVD (+)	1.0	1.70 (1.14,2.53)	2.39 (1.66,3.42)
(+) and IHD	1.0	1.51 (1.01,2.25)	1.79 (1.22,2.61)
(+) IHD and breathlessness	1.0	1.42 (0.93,2.16)	1.54 (1.03,2.29)
Stroke (+)	1.0	0.87 (0.39,1.93)	1.92 (1.28,2.86)
(+) IHD, LVH	1.0	0.79 (0.35,1.77)	1.73 (1.00,3.00)
(+) IHD, LVH, and breathlessness	1.0	0.79 (0.35,1.77)	1.73 (0.95,3.16)

(+), Relative risk adjusted for age, smoking, cholesterol, HDL-Cholesterol, systolic blood pressure, physical activity, social class, diabetes.
(+) and IHD, Adjusted for the above and pre-existing ischaemic heart disease.
(+) IHD, LVH, Adjusted for major risk factors, ischaemic heart disease and left ventricular hypertrophy.

reduced to 17% and was no longer significant (Table 11.4). Severe leg pain was still associated with a 54% increase in risk of cardiovascular mortality and the relationship with stroke remained unchanged.

Pre-existing Ischaemic Heart Disease. In order to assess whether the presence or severity of calf pain was of similar prognostic significance for cardiovascular disease in men with and without evidence of ischaemic heart disease, we have examined the relationship separately in men with no evidence of myocardial ischaemia, in men with evidence short of a myocardial infarction and in men with definite myocardial infarction (Fig. 11.1). In men free of any evidence of ischaemic heart disease, after adjusting for the major cardiovascular risk factors and breathlessness, increasing severity of calf pain was significantly associated with an increase in risk of both major heart attacks and cardiovascular mortality. Moderate and severe calf pain were associated with a 50% and 80% increase in risk of heart attacks respectively. Those who reported calf pain, moderate or severe, were at about a two-fold increase in risk of cardiovascular mortality. In men with evidence of myocardial ischaemia short of a myocardial infarction, although the overall risk of heart attack and cardiovascular death was higher, calf pain had no prognostic value. In men with definite myocardial infarction, only severe calf pain was associated with an increase in heart attacks and cardiovascular death. The increase was not significant, possibly because of small numbers.

Calf Pain and Stroke. Fig. 11.2 shows the relative risk of stroke (fatal plus non-fatal) in men with no evidence of ischaemic heart disease, after adjustment for the

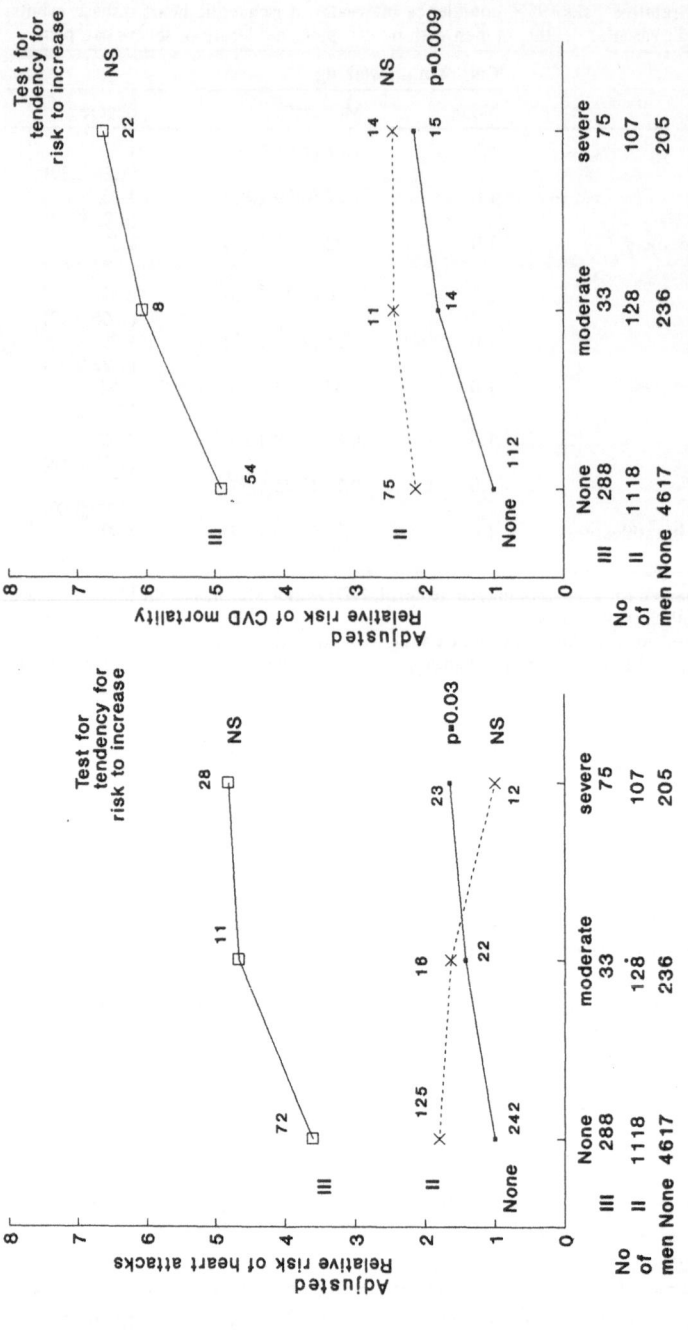

Fig. 11.1. Adjusted relative risk of major ischaemic heart disease events or cardiovascular (CVD) deaths in men with no calf pain, moderate or severe calf pain according to presence and severity of ischaemic heart disease.
Relative risk adjusted for age, smoking, cholesterol, HDL-cholesterol, systolic blood pressure, physical activity, social class, diabetes and breathlessness.
II, men with pre-existing ischaemic heart disease short of myocardial infarction.
III, men with definite evidence of myocardial infarction.

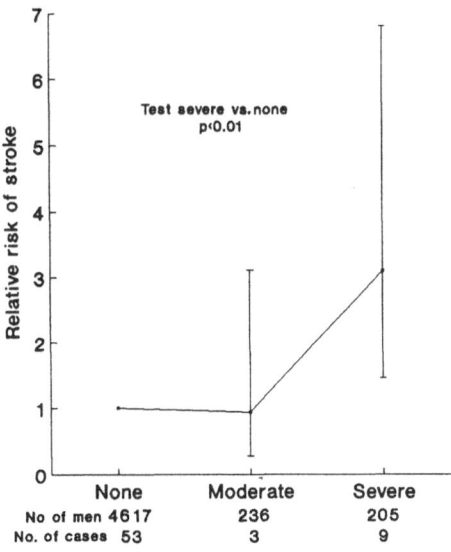

Test severe vs.none
p⁰0.01

No of men   4617   236   205
No. of cases   53   3   9

Fig. 11.2. Adjusted relative risk of stroke in men with severe, moderate or no calf pain and no evidence of ischaemic heart disease. Relative risk adjusted for age, smoking, cholesterol, HDL-cholesterol, systolic blood pressure, physical activity, social class, diabetes and breathlessness.

major risk factors and breathlessness. Men with severe calf pain on walking had a three-fold increase in the risk of stroke (p<0.01), an association far stronger than that seen for heart attack (fatal plus non-fatal) or for cardiovascular death (Fig. 11.1). In men with evidence of ischaemic heart disease at screening there was no association between calf pain and the risk of stroke.

Discussion: Calf Pain on Walking

Prevalence

The 11.5% prevalence of calf pain on walking (intermittent claudication) in these middle-aged British men may seem high when compared with the other major prospective study of middle-aged men in Great Britain, the Whitehall Study of civil servants, in which 1.8% satisfied the researchers' criteria for intermittent claudication (Davey Smith et al. 1990). Their criteria, based on the London School of Hygiene and Tropical Medicine (LSHTM) questionnaire, required that not only should the men develop calf pain while walking, but in addition "that they had not developed such pain when standing still or sitting, that they had stopped walking or slowed down when pain developed, and that the pain had then usually disappeared within 10 minutes". In addition, they had to report that "the pain had never disappeared while walking" to achieve a "probable" (most definite) status, otherwise they were classified as "possible". In the present study, the men were simply asked questions about calf muscle pain on walking and were classified as none, moderate or severe on this basis. In a recent editorial comment on the Whitehall and other studies, Criqui (1990) has stated that defining claudication as *any exercise calf pain* not present at rest (i.e. possible or probable claudication) doubled the sensitivity for large vessel peripheral arterial disease (Criqui et al. 1985a) and doubled the relative risk for subsequent mortality (Criqui et al. 1985b). Using our simple criteria, men aged 40–59 years in the Whitehall Study had a 12.4% prevalence of leg pain on walking, but the age-standardised coronary heart disease mortality in the group not satisfying their strict criteria for "probable" or "possible" claudication did not differ significantly from that observed in men with no leg pain (G. Davey Smith, 1991, personal communication). Criqui (1990) draws an analogy between intermittent claudication and angina, in which it has been demonstrated that the risk of coronary heart disease mortality for those reporting *any chest pain on exertion* was higher than that for those whose symptoms met the WHO (Rose) criteria for angina. In the BRHS men, we have shown the same rate of heart attack in men with chest pain on exertion who do not satisfy all the strict criteria as in those who do (Cook and Shaper 1989b).

Risk Factors

As peripheral vascular disease and ischaemic heart disease are both clinical manifestations of atherosclerosis, it is not surprising that their relationship to risk factors is similar. However, it is not identical, and some specific comparisons will be made with the Whitehall Study as this is the only other large prospective study in Britain which deals with the subject in detail. In the present study, systolic blood pressure, body mass index, social class, HDL-cholesterol, cigarette smoking status, alcohol intake, diabetes mellitus, physical activity status, breathlessness and FEV1 are all significantly associated with calf pain on walking in the univariate analysis. No relationship was observed with serum total cholesterol or with serum triglyceride concentrations and there was no relationship with diastolic blood pressure although the mean diastolic pressure was higher in men with severe calf pain on walking than in the other men. In the Whitehall Study, no significant relationships were observed with blood pressure or with glucose intolerance. Our data are similar to the Framingham findings, except for their observation that relative weight was inversely associated with the occurrence of intermittent claudication (Kannel et al. 1970). The data from our study also suggest that risk factor status in men with moderate calf pain on walking is more nearly akin to that seen in men with severe calf pain on walking than to men with no such pain.

The association between claudication and blood cholesterol has been very inconsistent between studies, but several have noted higher levels of serum triglycerides (Greenhalgh et al. 1971; Hughson et al. 1978b). The lower HDL-cholesterol concentration in our men with calf pain, with similar levels in both moderate and severe cases, may be associated with the high prevalence of cigarette smoking which is known to be associated with a decrease in HDL-cholesterol concentration. Physical activity was strongly and inversely associated with calf pain but this may be the result of calf pain inhibiting physical activity rather than a lack of physical activity predisposing to calf pain. Most studies (Criqui et al. 1985b; Davey Smith et al. 1990) have shown the increased risk of cardiovascular events/mortality in subjects with claudication to be independent of the cardiovascular risk factors and this is also observed in our study. However, "independence" is used in the statistical sense, and only indicates that once claudication is present, its relationship to cardiovascular outcome is no longer entirely dependent on the level of current risk factors.

Ischaemic Heart Disease. The prevalence of all forms of evidence of ischaemic heart disease is increased in men with moderate and severe calf pain on walking. This is most striking for recall of a doctor-diagnosis of ischaemic heart disease, for combined angina and possible myocardial infarction on the WHO (Rose) questionnaire and for definite myocardial ischaemia and definite myocardial infarction on electrocardiogram. Overall, about 45% of the men with calf pain on walking have firm evidence of ischaemic heart disease. In the Whitehall Study, there was no relationship between intermittent claudication and electrocardiographic evidence of myocardial ischaemia, which is very suprising and not consistent with most other studies.

Outcome

In the present study, calf pain on walking was strongly associated with increased risk of heart attacks, stroke and cardiovascular mortality. Both event rates and mortality rates increased with increasing severity of the calf pain. Severe calf pain on walking was associated with a two-fold risk of heart attacks and a nearly three-fold risk of stroke and cardiovascular mortality after adjustment for age. Moderate calf pain on walking was also associated with a significant but smaller increase in risk of heart attack and cardiovascular mortality but not of stroke.

As in other studies (Kannel et al. 1970; Hughson et al. 1978b; Reunanen et al. 1982), calf pain on walking (intermittent claudication) is strongly associated with evidence of ischaemic heart disease. In the Finnish study (Reunanen et al. 1982), the association between claudication and cardiovascular mortality was accounted for by indicators of ischaemic heart disease. In the present study, after adjustment for evidence of ischaemic heart disease, calf pain was no longer significantly associated with the risk of heart attacks but was still associated with an increased risk of stroke and cardiovascular mortality.

When the association between calf pain and walking and outcome was examined separately in men with and without evidence of ischaemic heart disease it became evident that calf pain on walking was a significant predictor of heart attack, stroke and cardiovascular death *only* in men with no evidence of underlying ischaemic heart disease. This is not to say that the relative risk of heart attack or cardiovascular death did not increase as the presence and degree of calf pain increased in men with previous definite myocardial infarction (Fig. 11.1) but that the finding did not achieve statistical significance.

In our study, severe calf pain on walking carried a similar risk of heart attack and cardiovascular death to that observed in men without calf pain who had evidence of ischaemic heart disease short of a definite myocardial infarction. The Whitehall Study similarly found claudication to have the same predictive power of cardiovascular death as angina present on the WHO (Rose) questionnaire. In an American study of 567 subjects followed for 4 years, large peripheral artery disease measured by non-invasive tests was more strongly associated with mortality when subjects with a history of cardiovascular disease were excluded (Criqui et al 1985b). Calf pain in the presence of myocardial ischaemia was not significantly associated with risk of heart attack or cardiovascular death.

Stroke. Few studies have closely examined the relationship between claudication and stroke. The Whitehall Study showed that a strong association was present between claudication and stroke in men with no evidence of ischaemic heart disease at screening, with "possible" and "probable" claudication showing a three-fold and five-fold increase respectively. Despite the markedly different criteria used for classification, the magnitude of relationships is similar to the three-fold increase in stroke risk in men with severe calf pain in this study. Altogether, it appears that claudication is an important risk indicator for stroke in men with no evidence of ischaemic heart disease.

Breathlessness. In our earlier studies we have demonstrated the importance of breathlessness as an early indicator of ischaemic heart disease risk in the absence of angina or ECG evidence (Cook and Shaper 1988, Cook and Shaper 1989a). In the present study breathlessness has been shown to be a significant predictor of ischaemic heart disease, particularly in men with no evidence of ischaemic heart disease at screening. Breathlessness not only reflects respiratory function and the effects of cigarette smoking, it also mirrors myocardial function and to some degree, physical fitness and aerobic capacity. It is therefore a subtle precursor of a number of disorders and may indeed be a major marker for many specific conditions – cardiac, respiratory and metabolic. Adjustment for breathlessness in addition to adjustment for the major risk factors for ischaemic heart disease (Table 11.4) had considerably diminished the risk of heart attack in men and severe calf pain on walking, with less effect on cardiovascular mortality and no effect on risk of stroke. This strongly suggests a major association between breathlessness and ischaemic heart disease and this could be its most specific relationship.

Summary

Calf pain on walking was present in 11.5% of 6807 men aged 40–59 years drawn from general practices in 21 British towns (The British Regional Heart Study); pain was moderate in 5.8% and severe in 5.7%. Increasing severity of calf pain was significantly associated with a number of recognised risk factors for cardiovascular disease and with the prevalence of ischaemic heart disease. Over 9.5 years of follow-up there were 548 heart attacks and 121 strokes (fatal and non-fatal), and 325 deaths from all cardiovascular causes (268 ischaemic heart disease, 29 strokes and 28 other cardiovascular causes). The rates of heart attack, stroke and cardiovascular mortality increased significantly with increasing severity of calf pain. After adjustment for the major risk factors including breathlessness and pre-exisiting heart disease, the association with heart attacks was reduced to non-significance. The association with cardiovascular mortality, comprised mainly of ischaemic heart disease deaths, remained significant even after these adjustments, indicating a stronger relationship between calf pain and fatal heart attacks than between pain and non-fatal attacks. Severe calf pain was associated with a nearly two-fold increase in risk of stroke even after adjustment for risk factors, pre-existing ischaemic heart disease and left ventricular hypertrophy.

The relationship was then examined separately in men with and without evidence of ischaemic heart disease. In men without evidence of ischaemic heart disease, severe calf pain was associated with a significant increase in heart attacks, cardiovascular mortality and with a nearly three-fold increase in risk of stroke, even after adjusting for the risk factors and breathlessness. In men with evidence of ischaemic heart disease no significant association was seen although there was a suggestive trend in men with previous heart attack.

Calf pain on exercise, particularly if severe, is an early indicator of risk of heart attacks, stroke and cardiovascular mortality, independent of evidence of ischaemic heart disease. The routine use of a simple questionnaire on calf pain on walking would add considerably to our ability to predict the risk of a cardiovascular event, in clinical situations and in epidemiological surveys.

Acknowledgement. The British Regional Heart Study is a British Heart Foundation Research Group and also receives support from the Department of Health and The Chest, Heart and Stroke Association.

References

Cook DG, Shaper AG (1988) Breathlessness, lung function and the risk of heart attack. Eur Heart J 9:1215–22

Cook DG, Shaper AG (1989a) Breathlessness, angina pectoris, and coronary artery disease. Am J Cardiol 63:921–4.

Cook DG, Shaper AG (1989b) Using the WHO (Angina) questionnaire in cardiovascular epidemiology. Int J Epidemiol 18:607–13

Criqui MH (1990) Peripheral arterial disease and subsequent cardiovascular mortality: a strong and consistent association. Circulation 82:2246–47

Criqui MH, Fronek A, Klauber MR, Barret-Connor E, Gabriel S (1985a) The sensitivity, specificity and predictive value of traditional clinical evaluation of peripheral arterial disease: results from a non-invasive testing in a defined population. Circulation 71:516–21

Criqui MH, Coughlin SS, Fronek A (1985b) Non-invasively diagnosed peripheral arterial disease as a predictor of mortality: results from a prospective study. Circulation 72:768–73

Davey Smith G, Shipley M, Rose G (1990) Intermittent claudication, heart disease risk factors and mortality: The Whitehall Study. Circulation 82:1925–31

Greenhalgh RM, Rosengarten DS, Meivart I, Lewis B, Colnan IS, Martin P (1971) Serum lipid and lipoproteins in peripheral vascular disease. Lancet ii:947

Hughson WG, Mann JI, Tibbs DJ, Woods HF, Walton I (1978a) Intermittent claudication: factors determining outcome. Br Med J 293: 377–9

Hughson TG, Mann JI, Garrod A (1978b) Intermittent claudication: prevalence and risk factors. Br Med J 293:379–81

Jelnes R, Gaardsting O, Hougaard Jensen K, Baekgaard N, Tonnesen KH, Schroeder T (1986) Fate in intermittent claudication: outcome and risk factors. Br Med J 293:1137–9

Kannel WB, Skinner JJ, Schwartz MJ, Shurtleff D (1970) Intermittent claudication: incidence in the Framingham Study. Circulation 51:875–83

Reunanen A, Takkunen H, Aromaa A (1982) Prevalence of intermittent claudication and its effect on mortality. Acta Med Scand 211:249–56

Rose G (1962) The diagnosis of ischaemic heart pain and intermittent claudication in field surveys. Bull WHO 27:645–58

Rose G, McCartney P, Reid DD (1977) Self-administration of a questionnaire on chest pain and intermittent claudication. Br J Prev Soc Med 31:42–48

Shaper AG, Pocock SJ, Walker M, Cohen NM, Wale CJ, Thomson AG (1981) British Regional Heart Study: cardiovascular risk factors in middle-aged men in 24 towns. Br Med J 283:179–186

Shaper AG, Cook DG, Walker M, MacFarlane PW (1984) Prevalence of ischaemic heart disease in middle-aged British men. Br Heart J 51:595–605

Shaper AG, Wannamethee G (1991) Physical activity and ischaemic heart disease in middle-aged British men. Br Heart J (submitted for publication).

Thelle DS, Shaper AG, Whitehead TP, Bullock DG, Ashby D, Patel I (1983) Blood lipids in middle-aged British men. Br Heart J 49:205–13

Walker M, Shaper AG (1984) Follow-up of subjects in prospective studies based in general practice. J R Coll Gen Pract 34:365–70

12 Smoking

J.T. POWELL

Introduction

Although the first taste of tobacco in Britain has been attributed to Sir Walter Raleigh, cigarette smoking was not a common habit until the late nineteenth century. This coincided with the introduction of cigarette making machines into Britain during the Crimean War. Perhaps unwittingly, the mood elevating properties of nicotine were widely used firstly during the Crimean War and then during both world wars, with extensive smoking amongst military personnel. More recently the Iraqi soldiers entrenched in the Kuwait desert still had their cigarettes even though no food was left – cigarettes also help abate hunger. The smoking habit was largely confined to men until the end of the Second World War and the smoking histories of women probably only date from the 1940s onwards. By the early 1960s almost half the adult population of Britain smoked, the habit still being more popular amongst men than women. The number of smokers had reduced to about one-third of the adult population by the beginning of this decade, women now constituting half the smoking population.

The health hazards associated with smoking were not recognised immediately and it was not until the 1950s that they became a cause for concern. Smoking is a risk factor for lung cancer, other cancers of the gastrointestinal tract, chronic bronchitis and obstructive airways disease and all cardiovascular diseases. The harmful effects of smoking have been associated with three principal products of tobacco combustion, carbon monoxide, nicotine and tar, although hundreds of other potentially harmful components have been identified in smaller amounts among the combustion products. The tobacco companies have responded to the increasing evidence of the dangers of smoking by the introduction of filter cigarettes in the 1950s and continuing further modifications to reduce the tar yield and nicotine content of cigarettes. The tar content of cigarettes has been reduced from about 35 mg to about 15 mg per cigarette over the past 50 years.

Any discussion of how smoking-related diseases are caused must consider such cigarette history. The Framingham study provided one of the earliest indications that smoking was a risk factor for aortic aneurysms (Hammond and Garfinkel 1969). The prevalence of aortic aneurysms appears to be increasing (Melton et al. 1984; Fowkes et al. 1989) with, in the 1990s, males outnumbering females by 4:1, whereas in the early 1970s males outnumbered females by 8:1. Smoking as a risk factor for aortic aneurysms would appear to depend on a long pack-year history, with women now developing more aneurysms, having been exposed to a lifetime (40 or more years) of smoking. Reduction of the tar content of cigarettes would not appear to have reduced the risk of aortic aneurysm. Neither has reduction of the tar content decreased the risk of coronary heart disease associated with smoking (Higgenbotham et al. 1982).

Smoking: The Risk Factor for Peripheral Vascular Disease

Smoking has been considered as a risk factor for atherosclerosis and all forms of cardiovascular disease: haemorrhagic and thrombotic stroke, coronary artery disease, aortic aneurysm and peripheral vascular disease (Hammond and Garfinkel 1969; Kannel and Shurtleff 1973; Nomura et al. 1974; Kannel et al. 1976; Doll and Peto 1976; Abbott et al. 1986). Of all these manifestations of atherosclerosis smoking has been particularly associated with peripheral vascular disease, occlusive lesions in the aorto-distal circulation (Fowkes 1989). The Framingham study remains one of the few prospective surveys of the risk of intermittent claudication developing in smokers (Kannel and Shurtleff 1973). In such surveys the presence of intermittent claudication can be recognised by a questionnaire which is highly specific (Rose 1962). In the Framingham study after 16 years of follow-up, intermittent claudication developed in 84/2318 (3.6%) men and 44/2865 (1.5%) women (Kannel and Shurtleff 1973). Heavy smokers (>20 cigarettes/day) had three times the risk of developing intermittent claudication compared with non-smokers. Intermittent claudication is seldom considered to be a fatal disease, which gives greater credence and reliability to retrospective studies than is true for a potentially fatal disease such as myocardial infarction. Several retrospective studies reporting that more than 95% of patients with peripheral vascular disease have an extensive smoking history (Juergens et al. 1960; Lord 1965; Hughson et al. 1978a) support the findings of the Framingham study.

Intermittent claudication is the mildest symptom of peripheral vascular disease and those stalwarts with mild claudication may not even seek medical attention. In contrast the more severe manifestations of peripheral vascular disease, rest pain and gangrene, always command medical attention: more than 95% of patients presenting to the Charing Cross Vascular Surgical Service with rest pain or gangrene have a long smoking history. When a patient presents with symptoms of intermittent claudication the presence of peripheral vascular disease may be confirmed by objective test, measurement of the ankle-brachial systolic pressure ratio; where this ratio is less than 0.9 there is good evidence for a distal occlusive lesion.

The ankle pressure together with pulse waveform have been used to assess the prevalence of peripheral arterial occlusions amongst the working population of Basle: prevalence increased with age and in the age range 55–65 years there was evidence for distal arterial occlusions in 6% of men but unfortunately no significant association with smoking history was made (Widmer et al. 1964). In this study, reporting in 1964, arterial occlusions were rare amongst women: this may be a partial reflection of the small number of Swiss women with a long smoking history at the time of the survey. A later Danish study also used ankle blood pressure to determine the prevalence of peripheral vascular disease amongst 666 subjects born in 1914 (Schroll and Munck 1981). These Danish subjects were examined at the age of 50 years in 1964 and again 10 years later at the age of 60 years. Intermittent claudication, diagnosed from the Rose questionnaire (Rose 1962), was observed in 5.8% of the men and 1.3% of the women at the age of 60 years but a total of 14% of the subjects had an ankle-brachial pressure index of less than 0.9 in one or both legs (Schroll and Munck 1981). In this study an estimate of the total weight of tobacco consumed was significantly associated with the presence of peripheral vascular disease. Again the small percentage of women affected by intermittent claudication may reflect the much longer smoking history of men than women, a consideration often overlooked.

The ankle-brachial pressure index after 1 minute of defined treadmill exercise may provide an even more sensitive test of presymptomatic peripheral vascular disease than the ankle-brachial pressure index at rest (Laing and Greenhalgh 1983). In a pilot study the use of such an exercise test has provided an indication that such presymptomatic disease may be present in 10% of middle-aged men, all of whom had a long smoking history (Baxter et al. 1988). Other evidence suggests that the early atherosclerotic lesion remains asymptomatic in about 65% of patients for 10 years prior to the onset of claudication (Baxter et al. 1988).

Peripheral vascular disease, the development of intermittent claudication, rest pain or gangrene, is typically a disease of the seventh and eighth decades, the mean age at presentation at Charing Cross Hospital during the period 1985–1990 was 66.2 years for men and 68.0 years for women, with men outnumbering women by 2:1 (Fig. 12.1). As women start to accumulate smoking histories comparable with those of men they become an increasing proportion of the

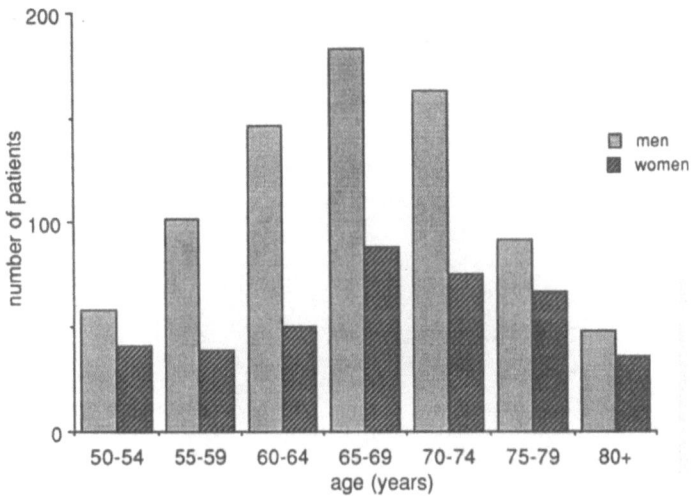

Fig. 12.1. Age–sex distribution of patients presenting with intermittent claudication.

patients presenting with symptomatic peripheral vascular disease. Since most of these patients (92%) have a greater than 20 pack-year smoking history, they may be fortunate to have survived the risk of lung cancer, fatal coronary heart disease or stroke. Smoking appears to be a particularly potent risk factor for peripheral vascular disease. Diabetes is another risk factor for peripheral vascular disease, but here the pattern of arterial disease is often distinctive with disease in the very distal and smallest arteries. For non-diabetics with intermittent claudication, disease is more commonly femoro-popliteal or aorto-iliac, femoro-popliteal disease accounting for 80–90% and more proximal aorto-iliac disease for 10%–20% of all claudication (Whittemore et al. 1985). Angiograms showing the different types of disease are shown in Fig. 12.2. Other risk factors which have been weakly associated with peripheral vascular disease in retrospective studies are hypertriglyceridemia and increased plasma viscosity (Greenhalgh et al. 1971; Dormandy et al. 1973).

In contrast other strong risk factors different from smoking are associated with coronary heart disease and stroke. For coronary heart disease these include increased plasma concentrations of cholesterol and fibrinogen, hypertension,

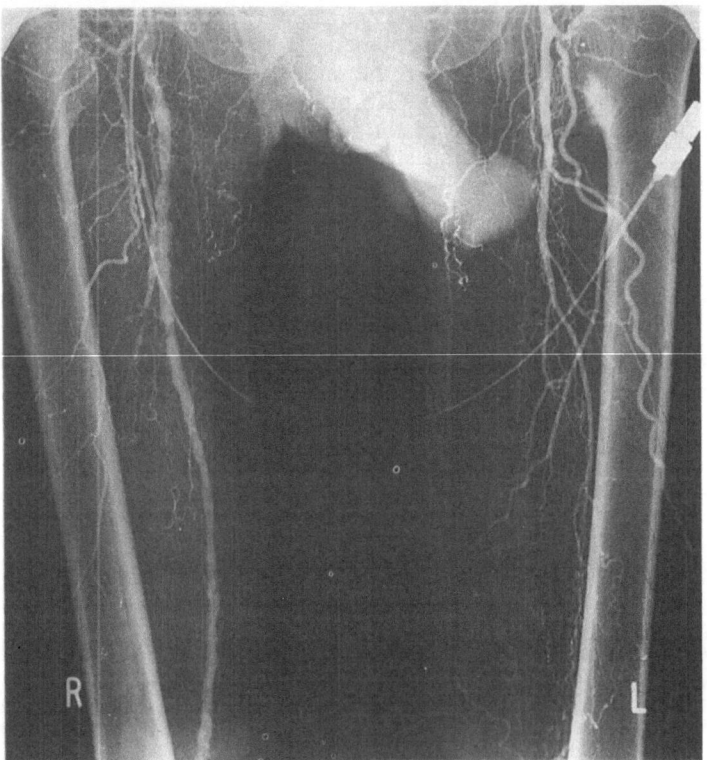

Fig. 12.2. a Angiogram illustrating aorto-iliac disease causing claudication.

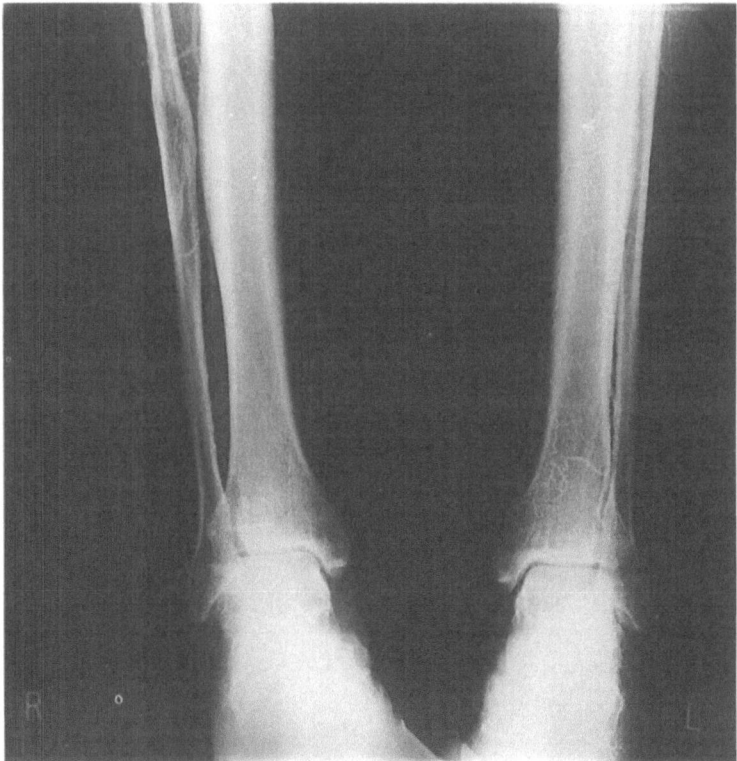

Fig. 12.2. b Angiogram illustrating femoropopliteal disease causing claudication.

family history of heart disease and obesity (Kannel et al. 1971; Gordon et al. 1981; Larsson et al. 1984; Meade et al. 1986; Durrington et al. 1988). Further, the risk of coronary heart disease associated with smoking appears to decline with age: in a prospective study of 34 000 British doctors the risk of coronary heart disease was four times higher in smokers than non-smokers between the ages of 45 and 54 years, only 1.5 times higher in smokers between the ages of 55 and 64 years, but smokers over 65 years appeared to have no significant increase in risk of coronary heart disease (Doll and Hill 1964). A study of British postal workers showed that the presence of an ischaemic ECG was just as common in non-smokers as smokers (Reid et al. 1966). Other studies also have indicated that giving up smoking in those over 65 years had little effect on the development of coronary heart disease (Gordon et al. 1974). The most important risk factor for stroke is hypertension, although obesity and increased levels of plasma fibrinogen are other considerations (Kannel et al. 1970; Heyman et al. 1971; Wilhelmsen et al. 1984; Welin et al. 1987). Smoking emerges as a relatively weak risk factor for stroke (Nomura et al. 1974; Abbott et al. 1986). The physical disabilities associated with completed stroke are likely to mask the development of intermittent claudication. Chronic bronchitis and chronic obstructive airways disease are other diseases strongly associated with smoking whose presence also could mask the development of intermittent claudication.

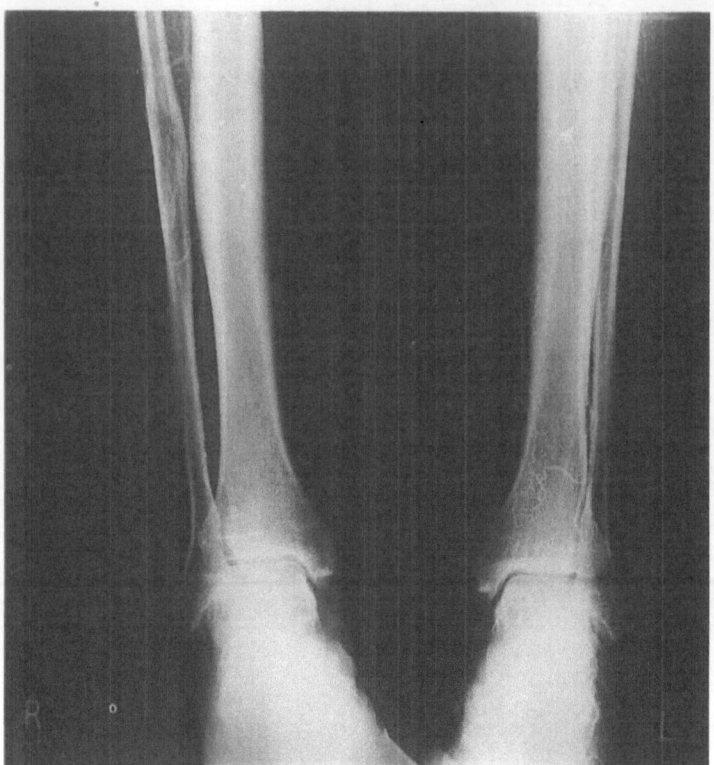

Fig. 12.2. c Angiogram illustrating distal disease in a diabetic.

Autopsy evidence also supports the particular association of cigarette smoking with peripheral vascular disease. In a detailed autopsy study of 1320 men, 25–64 years of age, the extent of abdominal aortic atherosclerosis was much more strongly associated with smoking than with atherosclerosis of the coronary arteries (Strong and Richards 1976). In each decade the extent of abdominal aortic atherosclerosis was greatest in the heaviest smokers (more than 25 cigarettes per day) in whom raised atherosclerotic lesions covered twice as much of the aortic intimal surface as in non-smokers. In contrast heavy smoking was associated with only a modest 5%–10% increase in the total area of raised atherosclerotic lesions in the three coronary arteries (Strong and Richards 1976).

Therefore, by exclusion, peripheral vascular disease is the disease of the healthier smokers because those with other risk factors such as hypercholesterolaemia, hypertension or alpha-1-antitrypsin deficiency are likely to have succumbed to coronary heart disease, stroke or chronic obstructive airways disease at an earlier age. This hypothesis also is concordant with the information indicating that hypercholesterolaemia and hypertension are only weak risk factors for peripheral vascular disease. This exclusion principle defines smoking as the most important, and perhaps only significant, risk factor for the development of peripheral vascular disease in the elderly population. Smoking may interact with other minor risk factors including the upright posture of humans. Arm claudication indicating occlusive disease in the arteries of the arm is very rare and is associated with catheterisation injury or disease in the supply vessels at the aortic arch. It also follows from this exclusion principle that the mechanism(s) whereby smoking promotes atherosclerosis should be most clearly evident from the study of its effects on the peripheral arteries and peripheral vascular disease. The association of intermittent claudication with a lifetime of smoking would lead to the neglect of passive smoking as a risk factor but perhaps those patients presenting with intermittent claudication without history of diabetes or smoking may have been the victims of passive smoking.

How Does Smoking Cause Peripheral Vascular Disease?

Both the tobacco companies and the Exchequer would like to be convinced that the safe cigarette can be produced. Smokers are ingenious and smoke in such a manner as to optimise or to titrate their nicotine level to the required amount. The reduction in the nicotine content of cigarettes, from 2 to about 1.25 mg per cigarette has failed, however, to effect a reduction in the nicotine extracted from each cigarette by the dedicated smoker: plasma nicotine values show a very limited fluctuation when smokers change to cigarette brands with a changed nicotine content (Gori and Lynch 1985). Other components might be varied more successfully to reduce or remove the active aetiological agent(s) associated with lung cancer, coronary heart disease, etc. There is no compelling evidence to associate any particular product of tobacco combustion and inhalation with peripheral vascular disease. The exclusion principle, that peripheral vascular disease is a disease that manifests in the healthier smokers can be challenged to investigate whether particular tobaccos, modes of smoking or other environmental factors promote the risk of developing peripheral vascular disease. Hence we have established a case–control study which recruits all current smokers newly referred to the Vascular Surgical Service at Charing Cross Hospital with proven peripheral vascular disease (ankle-brachial systolic pressure index less than 0.9) as cases and recruits age–sex matched current smokers without symptomatic peripheral vascular disease, as determined by the Rose questionnaire (Rose 1962), as controls. Prior diagnosis of lung cancer, chronic obstructive airways disease or coronary heart disease are exclusion criteria for the recruitment of controls. Interim analysis to check the adequate matching of cases and controls has demonstrated no significant difference in the number of cigarettes currently smoked, tar brand currently used, manner of smoking (depth of inhalation), amount of each cigarette smoked or serum cotinine (measure of nicotine intake) between cases and controls (Table 12.1).

There have been assumptions that patients with peripheral vascular disease are smokers of the higher tar cigarettes but such assumptions are unfounded (Table 12.1). Further, there are no indications that the tar yield of cigarettes influences the development of peripheral vascular disease. The mutagenic potential of tobacco products has been considered to reside in the tar fraction with its

Table 12.1. Characteristics of elderly smokers with and without peripheral vascular disease

	Cases of peripheral vascular disease	Controls
Number	136	244
Mean age (years)	64.5	62.4
% male	65	56
Cigarettes/day	19.3 (13.2)	17.6 (11.7)
Lowest tar (%)	43	40
Low-middle tar (%)	56	59
<½ cigarette smoked (%)	16	11
Deep inhalation (%)	22	25
Moderate inhalation (%)	42	41
Carboxyhaemoglobin (g/l)	4.0 (2.1)	3.7 (2.2)
Cotinine (nmol/l)	460 (326)	407 (336)

The numerical values are given as mean (standard deviation). None of the smoking-related parameters listed show a significant difference between cases and controls.

numerous aromatic and polycyclic hydrocarbons. Mutagens interact with and damage the genetic material of cells. At a molecular level smokers have been shown to have an increased amount of aromatic hydrocarbons adducted to lung cell and heart cell DNA (Phillips et al. 1988). Such adducts of aromatic hydrocarbons to DNA could contribute to the changes in cell growth and behaviour that characterise the cells of lung cancer and perhaps also underlie cellular changes in coronary heart disease. The ascending aorta of smokers has few such adducts (Randerath et al. 1989) and our own work has failed to demonstrate significant quantities of aromatic hydrocarbon DNA adducts in the abdominal aorta of smokers. The mechanisms whereby smoking causes lung cancer and peripheral vascular disease are probably very different.

Nicotine has many physiological effects on the cardiovascular system, including increasing the pulse rate and cardiac contractility, both of which increase myocardial oxygen demand, and through catecholamine release the potentiation of peripheral vasoconstriction. There are also indications that nicotine directly affects the health of endothelial cells. Opening of endothelial cell junctions and subendothelial oedema can be observed in the umbilical arteries of children born to smoking mothers (Asmussen and Kjeldsen 1975). Both smoking and nicotine infusion have been demonstrated to increase arterial permeability in dogs (Allen et al. 1988; Allen et al. 1989). The upright posture of man results in increased hydrostatic pressure in the leg arteries. Damaged endothelial cell junctions in those arteries exposed to highest hydrostatic pressure permit the greatest influx of serum components into the subendothelial space. The presence of hypertension would exacerbate such a pressure dependent process to increase the permeability of the endothelium and provides a mechanism whereby hypertension is a risk factor for peripheral vascular disease (Hughson et al. 1978a; Schroll and Munck, 1981). The endothelial cells of the distal arteries therefore may be particularly vulnerable to the disruptive effects of nicotine on the arterial endothelium. Such endothelial damage is probably the initiating mechanism for atherosclerotic change (Ross 1986). If nicotine is the aetiologic agent associated with the development of peripheral vascular disease there is little prospect of a safer cigarette to reduce the prevalence of intermittent claudication, critical ischaemia and gangrene, since the smoker successfully maximises and adjusts the nicotine yield to maintain their nicotine level.

The yield of carbon monoxide from cigarettes is also relatively constant, although filter cigarettes may deliver more carbon monoxide than untipped cigarettes. For most cigarettes the concentration of carbon monoxide in smoke is about 4%, with each fully smoked cigarette delivering about 15 mg of carbon monoxide. The extent of inhalation and rapidity of each puff determine the amount of carbon monoxide uptake into the blood, where it can be measured as carboxyhaemoglobin (COHb). In a cross-sectional study the risk of subjects having either coronary heart disease or intermittent claudication was increased about twenty-fold in those with a high COHb level (greater than 5%) compared with those whose COHb level was less than 3% (Wald et al. 1973). Increased carbon monoxide levels reduce the oxygen saturation of the blood and can promote the development of cardiac arrhythmias, but these are secondary phenomena. There are other indications from animal studies that carbon monoxide also increases the permeability of the arterial endothelium (Allen et al. 1988), although the effect may be less important than the damage caused by nicotine.

Current evidence might indicate that nicotine, the addictive component of tobacco, is the most likely culprit in initiating endothelial damage in peripheral vascular disease. There are thousands of products of the combustion of tobacco, many of which are inhaled and absorbed into the bloodstream where they could damage the blood–arterial interface. The tar fraction, carbon monoxide and nicotine only are those that have been considered in greatest detail.

Smoking and the Progression of Peripheral Vascular Disease

Smoking could also be a risk factor for the progression of peripheral vascular disease. Progression to disabling claudication or critical ischaemia may necessitate vascular reconstruction or amputation. Vascular reconstruction can often provide dramatic relief of symptoms and prevent a threatened amputation. The outcome of vascular reconstruction also could be influenced by smoking.

Certainly smoking is the most important risk factor for the development of peripheral vascular disease but it has also been associated with a more rapid deterioration of patients with peripheral vascular disease. In a study of 60 patients with intermittent claudication the 33 patients who continued smoking heavily had an increased incidence of adverse events over 4 years compared with those who reduced their consumption of cigarettes or stopped smoking (Hughson et al. 1978b). The adverse events included onset of intermittent claudication in the other limb, onset of rest pain, vascular reconstruction, amputation, myocardial infarction, stroke and death. Once the vessel wall has been damaged, disturbances due to smoking in platelet function, clotting mechanisms and prostanoid metabolism can exacerbate and extend the pathology at the blood–arterial wall interface. Currently we have no firm evidence to show that stopping smoking improves the prognosis for the limb affected by intermittent claudication or that stopping smoking prevents the onset of claudication in the other leg. We have collected detailed information on the progression of peripheral vascular disease in 900 patients attending the Charing Cross Arterial Clinic between April 1985 and December 1990, giving over 2000 years of patient follow-up, evenly split between those who continued smoking and those who gave up smoking within 3 months of confirmation of the presence of peripheral vascular disease. Analysis is not sufficiently complete to present this important information here but one aspect of the study deserves emphasis: giving up smoking needs to be confirmed by an objective test. "Stop smoking" are probably words often heard by the patient. Sometimes these words fall upon deaf ears, sometimes the patient gives up smoking, sometimes they cannot, sometimes they get no help, sometimes the addiction to nicotine is so powerful that they do not want to give up smoking. Unfortunately many patients deceive the clinician about their smoking habit and patient self-reporting can provide unreliable information about smoking status (Sillett et al. 1978; Wiseman et al. 1989). Objective assessment of smoking habit therefore becomes necessary.

Four objective estimates of smoking status are available: Expired carbon monoxide, plasma carboxyhaemoglobin, serum cotinine (a long-lived metabolite of nicotine) and serum thiocyanate (derived from inhaled cyanide). Of these, carbon monoxide and carboxyhaemoglobin provide evidence of smoking within the last few hours whereas serum cotinine ($t\frac{1}{2}$ about 17 hours) and serum thiocyanate ($t\frac{1}{2}$ about 6.5 days) may give a more reliable index of current smoking habits. The use of such markers has clearly demonstrated that continued smoking reduces the chances of successful vascular reconstruction. Carboxyhaemoglobin was the marker used in a small study which included aortic grafts and both vein and prosthetic distal grafts (Greenhalgh et al. 1981). More recently both carboxyhaemoglobin and thiocyanate have been used to investigate smoking habit in a larger number of patients entered into a multicentre trial for femoro-popliteal reconstruction. Use of these markers indicated that 25% of patients were less than truthful when they reported having given up smoking (Wiseman et al. 1989). Small surprise then that many studies have provided ambiguous results about the benefits of giving up smoking after vascular reconstruction. The use of smoking markers to discriminate continuing smokers has demonstrated the disadvantages of smoking after vascular reconstruction. In smokers only 63% of femoro-popliteal vein grafts remained patent 1 year after bypass, whereas 84% of grafts were patent in non-smokers, p<0.02. (Fig. 12.3). The results for prosthetic grafts were very similar.

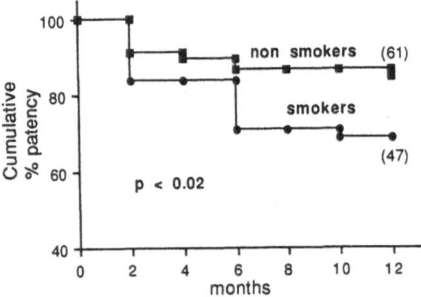

Fig. 12.3. The effect of smoking on the patency of femoro-popliteal vein grafts. Patients with thiocyanate levels greater than 70 μmol/l, 6 months after bypass, are assumed to be continuing smokers, those with thiocyanate levels less than 70 μmol/l are considered to be non-smokers. The number of patients is given in parentheses.

Conclusions: Stop Smoking

With the general ageing of the population, peripheral vascular disease assumes more importance than hitherto. All evidence suggests that the message to be conveyed is unchanging: STOP SMOKING. Clearly it is of interest to the tobacco companies to discern how smoking causes arterial disease. The exclusion principle that I have proposed here, "peripheral vascular disease is the disease of the healthier smokers", this population having no other major risk factors for pulmonary or other cardiovascular disease, indicates that peripheral vascular disease provides the simplest model in which to evaluate how smoking initiates arterial damage. The initiation and propagation of arterial disease may have different causative agents and mechanisms. The study of patients with peripheral vascular disease together with Buerger's disease patients may also provide valuable insights into how smoking damages the vasculature. Buerger's disease often presents more floridly than peripheral vascular disease, with the upper extremities also being affected. Here stopping smoking provides the cure. Stopping smoking may also prevent or retard the progression of peripheral vascular disease. Until we have the benefit of more information, helping patients with peripheral vascular disease to stop smoking is the first and most important therapeutic intervention we can offer.

Acknowledgement. Original work in this chapter was supported by the British Heart Foundation and the Tobacco Products Research Trust.

References

Abbott RD, Yin Y, Reed DM, Yano K (1986) Risk of stroke in male cigarette smokers. N Engl J Med 315:717–20

Allen DR, Browse NL, Rutt DL et al. (1988) The effect of cigarette smoke, nicotine and carbon monoxide on the permeability of the arterial wall. J Vasc Surg 7:139–52

Allen DR, Browse NL, Rutt DL (1989) Effects of cigarette smoke, carbon monoxide and nicotine on the uptake of fibrinogen by the canine arterial wall. Atherosclerosis 77:83–8

Asmussen I, Kjeldsen K (1975) Intimal ultrastructure of human umbilical arteries. Circ Res 36:579–89

Baxter K, Wiseman SA, Powell JT, Greenhalgh RM (1988) A pilot study of a screening test for peripheral arterial disease in middle-aged men: fibrinogen as a possible risk factor. Cardiovasc Res 22:300–2

Doll R, Hill AB (1964) Mortality in relation to smoking: ten years observations of British doctors. Br Med J i:1395–1410, 1460–7

Doll R, Peto R (1976) Mortality in relation to smoking: 20 years observations on male British doctors. Br Med J ii:1525–36

Dormandy JA, Hoare E, Colley J et al. (1973) Clinical, haemodynamic, rheological and biochemical findings in 126 patients with intermittent claudication. Br Med J iv:576–81

Durrington PN, Hunt L, Ishola M, Arrol S (1988) Apolipoproteins (a), AI and B and parental history in men with early onset ischaemic heart disease. Lancet i:1070–3

Fowkes FGR (1989) Aetiology of peripheral atherosclerosis. Br Med J 298:405–6

Fowkes FGR, MacIntyre CCA, Ruckley CV (1989) Increasing incidence of aortic aneurysm in England and Wales. Br Med J 298:33–5

Gordon T, Kannel WB, McGee D, Dawber TR (1974) Death and coronary attacks in men after giving up cigarette smoking: a report from the Framingham Study. Lancet ii:1345–8

Gordon T, Kannel WB, Castelli WB, Dawber TR (1981) Lipoproteins, cardiovascular disease and death. Arch Intern Med 141:1128–31

Gori GB, Lynch CJ (1985) Analytical cigarette yields as predictors of smoke bioavailability. Regul Toxicol Pharmacol 5:314–26

Greenhalgh RM, Lewis B, Rosengarten DS et al. (1971) Serum lipids and lipoproteins in peripheral vascular disease. Lancet ii:947–50

Greenhalgh RM, Laing SP, Cole PV, Taylor CTW (1981) Smoking and arterial reconstruction. Br J Surg 68:605–7

Hammond EC, Garfinkel L (1969) Coronary heart disease, stroke and aortic aneurysm: factors in etiology. Arch Environ Health 19:167–82

Heyman A, Karp HR, Heyden S et al. (1971) Cerebrovascular disease in the biracial population of Evans County, Georgia. Arch Intern Med 1971 128:949–55

Higgenbotham T, Shipley MS, Rose G (1982) Cigarettes, lung cancer and coronary heart disease: the effects of inhalation and tar yield. J Epidemiol Comm Health 36:113–17

Hughson WG, Mann JI, Garrod A (1978a) Intermittent claudication: prevalence and risk factors. Br Med J i:1379–81

Hughson WG, Mann JI, Tibbs DV et al. (1978b) Intermittent claudication: factors determining outcome. Br Med J i:1377–9

Juergens JL, Barker NW, Hines EA (1960) Arteriosclerosis obliterans: a review of 520 cases with special reference to pathogenic and prognostic factors. Circulation 21:188–95

Kannel WB, Wolf PA, Verter J, McNamara PA (1970) Epidemiologic assessment of the role of blood pressure in stroke: The Framingham Study. JAMA 214:301–10

Kannel WB, Castelli WP, Gordon T, McNamara PM (1971) Serum cholesterol, lipoproteins and the risk of coronary heart disease: The Framingham Study. Ann Intern Med 74:1–12

Kannel WB, Shurtleff D (1973) The Framingham study, cigarettes and the development of intermittent claudication. Geriatrics 28:61–8

Kannel WB, McGee D, Gordon T (1976) A general cardiovascular risk profile: the Framingham study. Am J Cardiol 38:46–51

Laing SP, Greenhalgh RM (1983) The detection and progression of asymptomatic peripheral arterial disease. Br J Surg 70:628–30

Larsson B, Svardsudd K, Welin L et al. (1984) Abdominal adipose tissue distribution, obesity and risk of cardiovascular disease: 13-year follow-up of participants of men born in the year 1913. Br Med J 288:1401–4

Lord JW (1965) Cigarette smoking and peripheral atherosclerotic occlusive disease. JAMA 191: 249–51

Meade TW, Mellow S, Brozovic M et al. (1986) Haemostatic function and ischaemic heart disease: principal results of the Northwick Park Heart Study. Lancet ii:533–7

Melton LJ, Bickerstaff LK, Hollier LM et al. (1984) Changing incidence of abdominal aortic aneurysms: a population-based study. Am J Epidemiol 120:379–86

Nomura A, Comstock GW, Kuller L, Tonascia JA (1974) Cigarette smoking and strokes. Stroke 5:483–6

Phillips DH, Hewer A, Martin CN et al. (1988) Correlation of DNA adduct levels in human lung with cigarette smoking. Nature 336:790–2

Randerath E, Miller RH, Mittal D et al. (1989) Covalent DNA damage in tissues of cigarette smokers as determined by a 32[P]-postlabelling assay. J Natl Cancer Inst 81:341–7

Reid DD, Holland WW, Humerfelt S, Rose G (1966) A cardiovascular survey of British postal workers. Lancet i:614–18

Rose GA (1962) The diagnosis of ischaemic heart pain and intermittent claudication in field surveys. Bull WHO 27:645–58

Ross R (1986) The pathogenesis of atherosclerosis – an update. N Engl J Med 1986 314:488–500

Schroll M, Munck O (1981) Estimation of peripheral atherosclerotic disease by ankle blood pressure measurements in a population study of 60-year-old men and women. J Chron Dis 34:261–9

Sillett RW, Wilson MB, Malcolm RE, Ball KP (1978) Deception amongst smokers. Br Med J ii: 1185–6

Strong JP, Richards ML (1976) Cigarette smoking and atherosclerosis in autopsied men. Atherosclerosis 23:451–76

Wald N, Howard S, Smith PG, Kjeldsen K (1973) Association between atherosclerotic diseases and carboxyhaemoglobin levels in tobacco smokers. Br Med J i:761–765

Welin L, Svardsudd K, Wilhemsen L et al. (1987) Analysis of risk factors for stroke in a cohort of men born in 1913. New Engl J Med 317:521–6

Whittemore AD, Couch NP, Mannick JA (1985) Treatment of arterial occlusive disease of the lower extremities. Ann Rev Med 36:505–14

Widmer LK, Greensher A, Kannel WB (1964) Occlusion of peripheral arteries. Circulation 34:261–9

Wilhelmsen L, Svardsudd K, Korstan-Bengsten K et al. (1984) Fibrinogen as a risk factor for stroke and myocardial infarction. New Engl J Med 311:501–5

Wiseman S, Kenchington G, Dain R et al. (1989) Influence of smoking and plasma factors on the patency of femoropopliteal vein grafts. Br Med J 299:643–6

Wittek, Richard; Heuberger, Ralf et al. (19...) ... ...
... ...

... ... (198.) ... ...

... ...

... ...

... ...

13 Lipids: Outstanding Questions

M.F. OLIVER

There have been impressive, fundamental and far-reaching advances in our understanding of the relations of lipids to the causes and management of vascular diseases over the last 30 to 40 years and a short review of unresolved or outstanding questions should be prefaced with a brief summary of this progress.

Areas of Agreement

An unassailable body of evidence exists to indicate that raised blood cholesterol causes atheroma in the aorta and coronary arteries in many experimental species and, specifically, in man. In general, the higher the concentration the greater the extent of atheromatous involvement. Raised low density lipoproteins (LDL) and reduced high density lipoproteins (HDL) are separately and jointly responsible.

A second fact, very strongly supported by many epidemiological and experimental studies, is that a high dietary intake of saturated fat is a leading cause of high blood cholesterol and of coronary heart disease.

A third area of agreement is that inheritance is a potent determinant of blood cholesterol/lipoprotein concentrations and lipoprotein receptor activity; indeed, genetic influences probably determine more than half of the plasma lipoprotein concentrations.

Fourth – but finally for the moment – is the impressive congruity of the benefit of reducing hypercholesterolaemia. All five major primary prevention trials (Dayton et al. 1969; Report from the Committee of Principal Investigators 1978; Turpeinen et al. 1979; Lipid Research Clinics Coronary Prevention Trial 1984; Frick et al. 1987) have shown that it is possible to reduce the incidence of coronary heart disease in men with initially high plasma concentrations of cholesterol and LDL, and there should no longer be any doubt about the need for aggressive treatment in such men. This is an important economic and therapeutic advance.

The strength of all these relationships is much weaker for carotid and cerebral artery disease, and also for peripheral lower limb disease, than for coronary disease. But the biochemical composition of partly or completely occlusive lesions in these arterial systems is similar to that of aortic and coronary artery lesions. The well-documented fact that lesions in the coronary arteries dominate clinically those in comparable sized arteries is more related to the crucial role of the myocardium in sustaining life than to any major metabolic differences in the nature of lesions elsewhere. The myocardium is continually exercising – often doubling its rate and workload – in contrast to the other arterial systems which are metabolically at comparative rest each night. Thus, the demands for chemical energy by the myocardium are more critically dependent on the maintenance and adjustment of blood flow, through neurohormonal influences, than in other arterial systems.

Outstanding Questions

There are so many unresolved issues that it is certain that the basic and clinical scientists of the forthcoming decade will have plenty to study (Table 13.1). Some of these have been reviewed recently (Grundy 1990). The zeal of many health educationalists for lowering everyone's cholesterol, backed by the US National Cholesterol Education Program (Lenfant 1986), carries the danger of closing the book on the cholesterol question and, with the commercially based enthusiasm of pharmaceutical companies, problems which the scientific community should address with urgency may be submerged or overlooked. These need to be identified clearly and are discussed below.

Table 13.1. Some unresolved problems concerning lipids and atheroma

How does raised plasma cholesterol cause atheroma?
How does cholesterol leave the arterial wall?
Can reduction of hypercholesterolaemia cause significant regression of atheroma?
How is measurement of atheroma regression best undertaken?
How far down should cholesterol be reduced?
Why does reduction of hypercholesterolaemia not also decrease total mortality?
Does reduction of hypercholesterolaemia increase non-cardiac mortality?
Why is the relation of cholesterol to vascular disease different in women from men?
Why does the relation of cholesterol to vascular disease decrease with age?
What are the pathogenic roles of triglycerides and fatty acids?
What are the relationships between lipid abnormalities and thrombosis?

Cholesterol and Atheroma

While the relationship between high plasma LDL and atheroma is undoubted, the proportion and characteristics of LDL taken through the arterial endothelium is unknown and variable. Is the uptake by macrophages of LDL important, or is most or all of the LDL taken up in an oxidised form (Steinberg et al. 1989) by the modified acetyl-LDL receptor present on these cells? How do reactive oxygen radicals arise locally and from what cell types? Is oxidised LDL present in plasma? To what extent are these processes amenable to control by antioxidant vitamins or drugs? How might lipid uptake by macrophages and their conversion to foam cells have a chemotactic effect on circulating monocytes?

How do "scavenger" lipoproteins, such as HDL, remove cholesterol from the arterial wall? While transfer enzyme systems are well described, little seems to be known about the kinetics or influences which lead to removal of cholesterol from deposits in the arterial wall. What is the role in this regard of phospholipids such as phosphatidyl choline (Stein et al. 1975) acting as acceptors for apo A-I? Can crystalline cholesterol be mobilised?

Regression of Atheroma

A particularly important issue is whether it is possible to cause regression of cholesterol-rich atheromatous deposits in coronary or other arteries by lowering LDL cholesterol or raising HDL cholesterol. One small controlled clinical trial (Duffield et al. 1983) in advanced femoral atherosclerosis, using cholestyramine to lower elevated lipids, reported non-progression of lesions over a 13-month period between femoral angiograms. Another (Olsson et al. 1990), comprising 20 treated and 25 control hyperlipidaemic men, showed regression of femoral lesions in nearly half of those given fenofibrate and nicotinic acid (see later). Two trials, one using cholestyramine (Brensike et al. 1984) and another using a high polyunsaturated/saturated diet (Arntzenius et al. 1985), also showed non-progression in coronary arteries between coronary angiograms. More recently, there have been three trials of the effects of lipid lowering on coronary atherosclerosis, as assessed by angiography. These are encouraging by indicating that a small degree of actual regression may be achievable over a 2-year period (Blankenhorn et al. 1987; Brown et al. 1990): the longest of these trials was 8 years and, using partial ileal bypass to reduce hypercholesterolaemia, it showed the greatest degrees of regression (Buchwald et al. 1990). More and larger well-controlled trials are needed.

Another unresolved problem is how to interpret changes which may occur in the arterial wall during regression trials (De Feyter et al. 1991). Is it more important, for example, in relation to the function of the coronary arteries in supplying blood to the myocardium to achieve a 50% regression of a 30% occlusive lesion (one which is not normally associated with clinical symptoms) or a 20% regression of a 70% occlusive lesion (usually regarded as clinically important)? It is more likely that the former would be able to be demonstrated, while the latter may be the really important test of the regression hypothesis. And will successful regression, as distinct from non-progression, necessarily be associated with restoration of the arterial wall to a more or less normal state? Successful repair of arterial lesions presumably would be associated with an increase in platelet–fibrin thrombus formation and this might carry an increased risk of small emboli passing downstream into the tissues supplied.

Many regard the current regression trials as surrogates for clinical trials, because of the expense, large numbers and duration of clinical trials. But regression trials and clinical trials address different questions. It is entirely possible for regression or non-progression of atheroma to be demonstrated yet, because of continuing or even worsening thrombus formation, no improvement occurs in the incidence of clinical events. Both are needed, therefore.

A related issue is to ensure that the pharmaceutical world is not too hasty in interpreting what would appear to be a positive result from regression trials as an indication that the clinical manifestations of heart disease will regress to the same extent.

Cholesterol as a Risk Factor

There are alternative views. One states that elevated plasma cholesterol concentrations, and particularly LDL cholesterol, are the *sine qua non* for the development of atheroma, particularly the clinical manifestation of coronary heart disease, and effectively that everyone in developed nations has concentrations too high. This argument is even advanced in explanation of the fact that the commonest cause of death in individuals with the lowest concentration of plasma cholesterol is coronary heart disease since it is argued that even these individuals have levels unacceptably high. Those who hold this view believe that everyone should have their plasma cholesterol lowered to levels of 4 mmol/l or less.

The alternative view is that the body requires for cell membrane homeostasis a "balanced" plasma cholesterol concentration. There is a J-shaped relationship between cholesterol, total mortality and an increase in the incidence of cancer and other diseases at the lowest end of the normal distribution. It has been proposed that most of this relates to incident cancer and is a result of this process (Epstein 1991); yet there are now several studies showing that low cholesterol predicts the development of cancer eight, ten and more years later. Also, there is a direct relationship between low cholesterol and haemorrhagic (not thrombotic) stroke (Iso et al. 1989; Yano et al. 1989).

If a large proportion of the cholesterol in the blood of a given individual is genetically determined, cellular homeostasis may also be individually modulated. It is not inconceivable that the maintenance of normal biological membrane function, in terms of immune resistance, for example, may demand in some individuals a higher concentration of cholesterol in the plasma than the exponents of the first view would find acceptable.

Reduction of Hypercholesterolaemia

The five major primary prevention trials (Dayton et al. 1969; Report from the Committee of Principal Investigators 1978; Turpeinen et al. 1979; Lipid Research Clinics Coronary Prevention Trial 1984; Frick et al. 1987) and some of the secondary prevention trails indicate that in men reduction of *high* plasma cholesterol levels is associated with reduction of non-fatal myocardial infarction; but the degree of reduction of cardiac death is less impressive, possibly because of the numbers recruited in these trials have been insufficient to have the power to show an effect. There are two secondary prevention studies (the Coronary Drug Project (Canner et al. 1986) and the Swedish Secondary Prevention Trial (Carlson and Rosenhamer 1988) which have reported reduction of coronary heart disease mortality and both used nicotinic acid. This has long been known to cardiologists as a vasodilator and, since both studies had a majority of patients with myocardial infarction, it is not impossible that the improved mortality was related to decreased left ventricular work rather than reduction of cholesterol. We need to keep an open mind about this point.

None of these trials have shown any reduction in total mortality (Oliver 1988). While none had the power to examine this question adequately, we should be careful not to dismiss the problem and the logical step is to wait for the completion of new clinical trials with such power. Clinical trials now being conducted using HMG-CoA reductase inhibitors, which will lower cholesterol to a far greater extent than the first generation trials, may show a reduction in total mortality.

A related and potentially serious issue is the possibility that lowering hypercholesterolaemia increases non-cardiac mortality (Oliver, 1991). This has occurred in several primary and some secondary prevention trials. One meta-analysis of six primary prevention trials (Muldoon et al. 1990) has reported an overall odds ratio of 1.76 ($p<0.004$) for non-disease mortality and in another of six secondary prevention trials (Rossouw et al. 1990) the odds ratio for non-cardiac non-cancer deaths was 2.10 ($p<0.01$). What does this mean? There are three possibilities: a drug effect, chance, or a biological effect. It now seems very unlikely that the increase in non-cardiac mortality is due to a drug effect; it has occurred in trials using four drugs – clofibrate, cholestyramine, gemfibrozil (admittedly, another fibric acid) and nicotinic acid – and a high polyunsaturated/saturated fat diet. The play of chance can seldom be eliminated and one must recognise that none of the trials individually was designed to test the question of whether reduction of hypercholesterolaemia might increase non-cardiac mortality. It has recently been pointed out that the confidence limits around the adverse finding of an increase in non-cardiac mortality are so narrow as to make chance unlikely. One reason, given that chance is the likely explanation, is that the increase in mortality is spread over a wide variety of diseases and that no plausible explanation for this can be advanced. Unfortunately, this is also a powerful argument in favour of the most serious of the explanations: namely, the consequnces of long-term impairment of the normal biologic functions of cell membranes as a result of prolonged reduction of membrane cholesterol.

Cholesterol and Women

In contrast to men, there is a weak relationship in women between LDL and HDL cholesterol, on the one hand, and coronary heart disease mortality on the other: the correlation coefficient between total cholesterol and coronary heart disease mortality in 19 countries with accurate death certification (WHO standards) is +0.67 for men and a non-significant +0.24 for women (Simons 1986). There is a very much weaker relationship between plasma cholesterol as a predictor of coronary heart disease in women in comparison with men and, effectively, it disappears by the age of 50. Why is this? One explanation may be that the levels of LDL/HDL needed to produce extensive coronary atheroma do not occur in women until after the menopause and that it may be necessary for such concentrations to be present for 15–20 years to produce coronary heart disease. The demonstration of any relationship between lipoproteins and coronary heart disease in women would then be difficult because of the confounding and diluting factors of other diseases in women in their seventies.

None of the clinical trials have been conducted in women and it is a false extrapolation to assume that the benefits regarding non-fatal myocardial infarction shown by treatment of middle-aged hypercholesterolaemic men will apply to women. This is an important issue in so far as more than half of the adult population are women. Screening programmes and health education programmes directed vigorously at women may be uneconomic and unjustified.

It does seem fairly clear, however, that oestrogen replacement therapy has reduced the incidence of coronary heart disease strikingly (Ross et al.1990). It has yet to be shown whether hormone replacement therapy (opposed oestrogens, as distinct from oestrogens alone) have such a beneficial effect.

Age and Cholesterol

The relation between cholesterol and vascular disease decreases in both sexes with advancing age. This may be because most of the cholesterol-related deaths have already occurred but also because other diseases become proportionately more common. It has yet to be demonstrated that changing the lifestyle of elderly people or lowering their cholesterol reduces their risk of developing coronary heart disease or other vascular disease. Perhaps it would be less intrusive for the quality of their lives if they were left alone.

Other Lipids as Risk Factors

The extent to which raised serum triglycerides and VLDL are or are not imporant in the pathogenesis of vascular disease is far from resolved (Grundy 1990). Yet, the lipoprotein moiety specifically related inversely to VLDL – namely HDL – is clearly also inversely related to coronary heart disease. There is good evidence that hypertriglyceridaemia is positively correlated with coronary heart disease on a univariate basis, but other influences such as HDL concentrations are more powerful correlates in multivariate analysis.

Hypertriglyceridaemia appears to be relatively more pathogenic for femoral than coronary artery disease. How might hypertriglyceridaemia be important? Is it a result of an excess of large very low density lipoproteins – if so why? Or is it a function of the fatty acid esterification of VLDL and triglycerides? Or is insulin resistance a common denominator? If raised plasma triglycerides are important, might the adverse effect not be due to a prothrombotic or antifibrinolytic action?

Lp (a) appears to be one of the most powerful predictors of coronary heart disease (Editorial 1991) It is independent of the common lipoprotein moieties, structurally, genetically, clinically and epidemiologically. Raised Lp (a), or one of its isoforms, may well be an explanation of the not uncommon finding of a young patient presenting with coronary heart disease without any of the orthodox risk factors. Yet we know little about the reasons for its pathogenicity, such as its binding to plasminogen receptors or how to modify plasma concentrations of apo (a) or Lp (a).

Apolipoprotein E polymorphism, particularly the E4 allele (Mahley 1985), also needs more investigation in order to provide perspective regarding cholesterol absorption from the gut and its relationship to coronary heart disease.

Fatty Acids

Individual fatty acids have individual effects and much more research will be needed to put these in perspective (Oliver, 1989; Grundy 1990; Oliver 1991).

While it is established in general that saturated fatty acids favour the development of coronary atheroma and coronary heart disease, not all have the same actions. Myristic is prothrombotic but palmitic acid, for example, may have a beneficial effect through an antithrombotic action.

Unsaturated fatty acids also have different actions. Oleic acid not only lowers LDL, but raises HDL and lowers blood pressure. Linoleic acid – the principal omega 6 fatty acid in food – also lowers LDL but has no effect on HDL. The extent to which oleic acid has a "better" effect on lipoproteins than linoleic acid has yet to be resolved. It now appears clear that some of the polyenes, notably eicosapentaenoic acid (EPA), actually increases LDL cholesterol. Yet EPA, the principal fish fatty acid normally available, probably has antithrombotic action, and may favour atheroma regression.

Diets rich in saturated fatty acids are by definition deficient in essential fatty acids and more studies will be required to identify the importance of linoleic acid (the principal essential fatty acid) deficiency in communities and individuals with coronary disease. Linoleic acid and other long chain polyenes are potentially unstable and are liable to undergo peroxidation. This may become important as emerging evidence is indicating an inverse relationship between coronary heart disease and naturally-occurring antioxidants, such as vitamin E and vitamin C (Riemersma 1991). A relative deficiency of essential fatty acids and antioxidants would favour lipid peroxidation and free-radical formation and may determine the rate of oxidation of LDL.

Lipids and Thrombosis

One of the most important unresolved areas is the interrelationship between different lipid moieties and thrombosis/fibrinolysis. While we know that fibrinogen (Meade 1988), factor V.II and possibly plasminogen activator inhibitor are related to raised triglyceride levels, insufficient detail is yet available concerning the nature of this relationship and, specifically, the relationship of individual fatty acids to plasma factors controlling thrombosis (Nordoy and Goodnight 1990). This is true also for the interrelationship of lipids and platelets. Better control of atheroma and vascular diseases may come from understanding of this interrelationship. It is a much neglected area possibly because of the dominance of studies of cholesterol-related lipoproteins, until recently, and the lack of appropriate methodology.

Education

Atheromatous vascular disease is not going to disappear as a result of lowering cholesterol and treating abnormal lipoproteins, and those who claim that it will are either naive or ignorant of the pathogenesis of the disease. Therefore, much needs to be done to educate the public correctly and it is the responsibility of those close to the field of lipids and vascular disease to provide the right balance. The areas of agreement should be clearly and consistently presented. Excessive claims, false expectations and nihilism must all be resisted.

It is also necessary to educate our own profession. There are four risk factors, not three. They are smoking, raised cholesterol, raised blood pressure and physicians. Cardiologists and physicians in many countries still need to be convinced of the need for aggressive action to lower high blood cholesterol. One reason for their casual interest is that their duty is to diagnose and treat advanced disease, and many are overwhelmed by the load of clinical problems. Another is scepticism. Others include a legitimate suspicion of public campaigns promoted by self-appointed health educationalists and of the increasing intrusion into clinical judgement by pharmaceutical companies.

The balance of knowledge favours aggressive action in young and middle-aged men with unequivocal hypercholesterolaemia. Physicians should be more alert to high risk families and encourage opportunistic screening for raised cholesterol levels. They should also be aware of the unsettled questions presented in this brief review.

References

Arntzenius AC, Kromhout D, Barth JD et al. (1985) Diet, lipoproteins and the progression of coronary atherosclerosis. The Leiden Intervention Trial. N Engl J Med 312:805–11

Blankenhorn DH, Nessim SA, Johnson RL, Sanmarco ME, Azen SP, Cashin-Hemphill L (1987) Beneficial effects of combined colestipolniacin therapy on coronary atherosclerosis and coronary venous bypass grafts. JAMA 3233:40

Brensike JF, Levy RI, Kelsey SF et al. (1984) Effects of therapy with cholestyramine on progression of coronary atherosclerosis: results of the NHLBI Type II Coronary Intervention Study. Circulation 69:313–24

Brown G, Albers JJ, Fisher LD et al. (1990) Regression of coronary artery disease as a result of intensive lipid-lowering therapy in men with high levels of apolipoprotein B. N Engl J Med 323:1289–98

Buchwald H, Varco RL, Matts PJ et al. (1990) Effect of partial ileal bypass surgery on mortality and morbidity from coronary heart disease in patients with hypercholesterolaemia: report of the Programme on Surgical Control of Hyperlipidaemias (POSCH). N Engl J Med 323:946–55

Canner PL, Berge KG, Wenger NK et al. (1986) Fifteen-year mortality in Coronary Drug Project patients: long-term benefit with niacin. J Am Coll Cardiol 8:1245–55

Carlson LA, Rosenhamer G (1988) Reduction of mortality in the Stockholm Ischaemic Heart Disease Secondary Prevention Study by combined treatment with clofibrate and nicotinic acid. Acta Med Scand 223:405–18

Dayton S, Pearce ML, Hashimoto S, Dixon WJ, Tomiyasu U (1969) A controlled trial of a diet high in unsaturated fat in preventing complications of atherosclerosis. Circulation 39/40 (Suppl II):II–1–63

De Feyter PJ, Serruys PW, Davies MJ, Richardson P, Lubsen J, Oliver MF (1991) Quantitative coronary angiography to measure prevention and regression of coronary atherosclerosis. Circulation (in press)

Duffield RG, Lewis B, Miller NE, Jamieson CW, Brunt JN, Colchester AC (1983) Treatment of hyperlipidaemia retards progression of symptomatic femoral atherosclerosis. A randomised controlled trial. Lancet ii:639–42

Editorial (1991) Lipoprotein (a). The Lancet 337:397–8

Epstein FH (1991) The effect of coronary heart disease prevention on the prevention of non-cardiovascular diseases. Proc. 7th International Meeting on Atherosclerosis and Cardiovascular Diseases, Bologna 1991. Kluwer, Dordrecht (in press)

Frick MH, Elo O, Haapa K et al. (1987) Helsinki Heart Study: primary prevention trial with gemfibrozil in middle-aged men with dyslipidaemia. N Engl J Med 317:1237–45

Grundy SM (1990) Cholesterol and coronary heart disease. JAMA 264:3053–9

Iso H, Jacobs DR, Wentworth D, Neaton JD, Cohen JD (1989) Serum cholesterol levels and six-year mortality from stroke in 350 977 men screened for MRFIT. N Engl J Med 320:904–10

Lenfant C (1986) A new challenge for America: the National Cholesterol Education Programme. Circulation 73:855–6

Lipid Research Clinics Coronary Prevention Trial (1984) I. Reduction in incidence of coronary heart disease. II. The relationship of reduction in incidence of coronary heart disease to cholesterol lowering. JAMA 251:351–74

Mahley RW (1985) Atherogenic lipoproteins and coronary artery disease. Circulation 72:943–8

Meade TW (1988) Lipids, coagulability and ischaemic heart disease. In: Suckling KE, Groot PHE (eds) Hyperlipidaemia and atherosclerosis. Academic Press, New York, pp. 103–16

Muldoon MF, Manuck SB, Matthews KA (1990) Lowering cholesterol concentrations and mortality: a quantitative review of primary prevention trials. Br Med J 301:309–14

Nordoy A, Goodnight SH (1990) Dietary lipids and thrombosis: relationships to atherosclerosis. Arteriosclerosis 10:149–63

Oliver MF (1988) Reducing cholesterol does not reduce mortality. J Am Coll Cardiol 12:814–17

Oliver MF (1989) Cigarette smoking, polyunsaturated fats, linoleic acid and coronary heart disease. Lancet 335:1241–3

Oliver MF (1991) New nutritional aspects of coronary heart disease: polyunsaturated fatty acids and anti-oxidants. In: Anderson RH, Yacoub M (eds) Royal Brompton reviews in diseases of the heart and lung. Oxford. Butterworth–Heinemann, Oxford

Oliver MF (1991) Might treatment of hypercholesterolaemia increase non-cardiac mortality? Lancet 337:1529–31

Olsson AG, Ruhn G, Erikson U (1990) The effect of serum lipid regulation and the development of femoral atherosclerosis in hyperlipidaemia: a non-randomized controlled study. J Intern Med 227:381–91

Report from the Committee of Principal Investigators (1978) A cooperative trial in the primary prevention of ischaemic heart disease using clofibrate. Br Heart J 40:1069–118

Riemersma RA, Wood DA, MacIntyre CCA et al. (1991) Risk of angina and plasma concentrations of vitamins A, C and E and carotene. The Lancet 337:1–5

Ross RK, Pike MC, Mack TM, Henderson BE (1990) Oestrogen replacement therapy and cardiovascular disease. In: Drife JO, Studd JWW (eds) HRT and Osteoporosis. Springer-Verlag, London, pp. 209–22

Rossouw JE, Lewis B, Rifkind BM (1990) The value of lowering cholesterol after myocardial infarction. N Engl J Med 16:1112–20

Simons LA (1986) Interrelations of lipids and lipoproteins with coronary artery mortality in 19 countries. Am J Cardiol 5D:5G–10G

Stein Y, Glangeaud MC, Fainaru M, Stein O (1975) The removal of cholesterol from aortic smooth muscle cells in culture and Landschutz ascites cells by fractions of human high-density apolipoprotein. Biochim Biophys Acta 380:106–18

Steinberg D, Parthasarathy S, Carew TE, Khoo JC, Witztum JL (1989) Beyond cholesterol: Modifications of low density lipoprotein that increase its atherogenicity. N Engl J Med 320:915–24

Turpeinen O, Karvonen MJ, Pekkarinen M et al. (1979) Dietary prevention of coronary heart disease: The Finnish Mental Hospital Study. Int J Epidemiol 8:99–118

Yano K, Reed DM, MacLean CJ (1989) Serum cholesterol and hemorrhagic stroke in the Honolulu Heart Program. Stroke 20:1460–65

14 Lipids: Epidemiology

G.C. LENG and F.G.R. FOWKES

Smoking and blood lipids are probably the factor which have been most intensively studied in the aetiology and pathogenesis of peripheral vascular disease. Investigations on blood lipids have paralleled those in ischaemic heart disease. Initial work in the 1960s and 1970s concentrated on serum cholesterol and triglycerides, and subsequently, to a lesser extent, on the lipoprotein fractions, particularly HDL cholesterol. More recently, attention has focused on apolipoproteins and fatty acids. The relationship between lipid levels in the blood and coronary artery disease has been studied extensively. In this chapter we examine the evidence based only on studies of peripheral vascular disease, and conclude by considering the question "Is there any major difference in the lipid profiles between peripheral vascular disease and ischaemic heart disease?"

Possible Role of Lipids in Atherogenesis

Atheromatous lesions are plaques of intimal thickening produced primarily by the deposition of lipids and the formation of fibrous tissues. Lipid deposits are first seen in childhood and adolescence, when they are described as "fatty streaks". Microscopically these consist of accumulations of lipid droplets, predominantly cholesterol ester, in intimal smooth muscle cells and in aggregates of macrophages lying beneath the endothelium. As these lesions progress they develop into raised patches of intimal thickening due to further deposition of lipid deep in the intima, and to the formation of fibrous tissue more superficially. This "simple" plaque may become "complicated" by the deposition of calcium salts, and by the breakdown of the fibrous layer to produce "ulceration of the plaque", often precipitating thrombus formation and vessel occlusion.

Fat represents the most concentrated source of energy in the average Western diet. It accounts for 30%–40% of total calorie intake, and this is consumed mainly as triglyceride. The digestion and absorption of fat from the small intestine is very efficient, resulting in the formation of water-soluble chylomicrons which pass into the systemic circulation via the thoracic duct. Chylomicrons consist predominantly of triglyceride (Table 14.1) with an outer crust of cholesterol and lecithin. The core triglyceride is removed by the action of lipoprotein lipase, an enzyme found on cell

Table 14.1. The composition of human plasma lipoproteins

Constituent	Percentage weight of constituents in each lipoprotein			
	Chylomicrons	VLDL	LDL	HDL
Cholesterol	7	19	48	27
Triglyceride	84	54	6	4
Phospholipid	7	18	23	24
Protein	2	9	23	45

Data derived from Smith (1987).

membranes throughout the body, leaving chylomicron "remnants" which are usually rapidly removed by the liver. However, if their removal is delayed, as in subjects with coronary heart disease, then these remnants appear to be particularly atherogenic, possibly by inducing cholesterol ester deposition (Packard and Shepherd 1990). The other main carrier of triglyceride in plasma is very low density lipoprotein (VLDL) which is produced primarily by the liver (Fig. 14.1). The triglyceride core is again degraded by lipoprotein lipase, leaving an intermediate density lipoprotein (IDL); this is hydrolysed in the liver to produce low density lipoprotein (LDL) which consists predominantly of cholesterol (Castelli 1986).

LDL is the main carrier of cholesterol to hepatic and peripheral tissues. It interacts with a specific receptor site on cell membranes to produce endocytosis of the lipoprotein, and then hydrolysis of its component parts to release cholesterol for the cells' needs. When sufficient sterol is present in a cell, synthesis of the LDL receptor is suppressed (Lipid Research Clinics Program 1984). However, in situations where LDL levels are raised, for example by excessive dietary intake or genetically, then LDL is redirected to receptor-independent routes, possibly involving the monocyte–macrophage series which play a key role in early atherogenesis (Goldstein and Brown 1978).

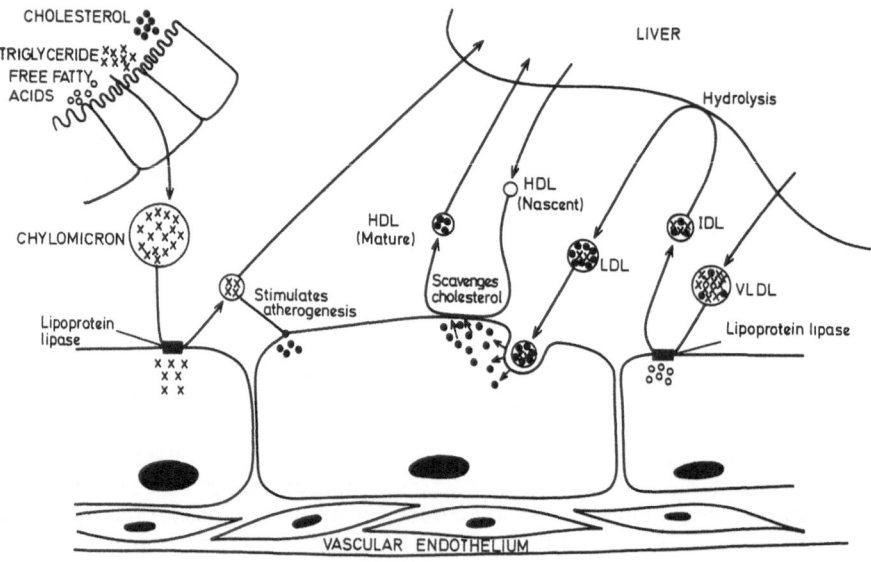

Fig. 14.1. Circulation of lipoproteins and possible roles in atherogenesis.

HDL is secreted in "nascent" form from the liver and ileum. It then picks up cholesterol from peripheral tissues and arterial walls and returns it to the liver. It therefore acts in opposition to LDL, as mediator of "reverse cholesterol transport" (Gotto 1989).

The complex interactions involved in lipoprotein metabolism require the presence of several proteins, known as apolipoproteins (apo). These have several different functions, including lipid binding, binding to cell-surface receptors, activation of enzymes involved in lipoprotein metabolism, and cholesterol ester transfer (Smith 1987).

Each low density lipoprotein contains one molecule of apolipoprotein B, and a small proportion also includes another equally large protein, apolipoprotein (a), which is chemically linked to apo B. This subfraction of LDL is known as lipoprotein (a), and has been identified in much higher concentrations in those with a family history of early onset coronary heart disease. The structure of apo (a) is very similar to that of plasminogen, and although its precise function is unknown, it may contribute to thrombus formation by binding fibrinogen (Editorial 1988a; Hegele 1989; Editorial 1991).

Lipids and Peripheral Vascular Disease: Epidemiological Studies

Triglycerides

Table 14.2 shows that elevated levels of serum triglycerides have been found in patients with peripheral vascular disease in numerous hospital-based case–control studies. Only one study (Bradby et al. 1978) found no significant increase in triglyceride levels. The findings have been so consistent over the years that triglyceride has been put forward as being a specially important risk factor for atherosclerosis affecting peripheral arteries. However, in the studies shown in Table 14.2, no account has been taken either of the relationship with other lipids, or the independent association of triglycerides with disease. Also, in some studies the control groups were not properly matched for age and sex.

In cross-sectional studies of peripheral vascular disease in the community some, but not all, studies show a univariate association between triglycerides and peripheral vascular disease (measured by intermittent claudication, the ankle-brachial pressure index or a combination of clinical features) (Table 14.3). However, when multivariate analyses were also conducted to adjust for the influence of other risk factors including lipids, no independent relationships could be found between triglycerides and the occurrence of peripheral vascular disease (Schroll and Munck 1981; Pomrehn et al. 1986; Fowkes et al. 1992). Indeed, in the Edinburgh Artery Study we found a highly significant univariate relationship of triglycerides with disease (p<0.001) which was not sustained on multivariate analysis.

In the Basle study (da Silva and Widmer, 1980) and in the Glostrup Study (Schroll and Munck 1981) the longitudinal relationship between serum triglycerides and the future occurrence of peripheral vascular disease has been examined.

Table 14.2. Clinical case–control studies of serum triglycerides and peripheral vascular disease

Reference	Triglycerides elevated in cases compared with controls	Controls age- and sex-matched
Angquist et al. 1982	*	Yes
Bliss et al. 1972	**	No
Bradby et al. 1978	NSD	No
Cardia et al. 1990	**	Yes
Greenhalgh et al. 1971	**	Yes
Greenhalgh et al. 1981	***	Yes
Jacobsen et al. 1984	***	Yes
Meerloo and Billimoria 1979	*	Partly
Pilger et al. 1983	**	Partly
Schrade et al. 1960	**	Partly
Sirtori et al. 1975	**	Yes
Skrede and Kvarstein 1975	*** (not females)	Yes
Trayner et al. 1980	*	Yes
Vyden et al. 1975	*	No

* = $p < 0.05$; ** = $p < 0.01$; *** = $p < 0.001$; NSD, No significant difference.

Table 14.3. Community cross-sectional studies of serum triglycerides and peripheral vascular disease

Reference	Population	Indicators of peripheral vascular disease	Positive association of triglycerides with disease	Independent association of triglycerides with disease
Hughson et al. 1978	M & F 50–69 years	Claudication	*	
de Backer et al. 1979 (Belgian Study)	M 40–55 years	ABPI	None	
Schroll and Munck 1981 (Glostrup Study)	M & F 50 years	ABPI Claudication+ABPI	* *	None
Reunanen et al. 1982 (Finnish Soc. Ins. Study)	M & F 30–59 years	Claudication	None	
Pomrehn et al. 1986 (Lipid Research Clinics Study)	M & F 20+ years	Claudication Exercise test ABPI	NK * None	None
Gofin et al. 1987 (Jerusalem Lipid Research Clinics Study)	M & F 25+ years	ABPI Claudication+ABPI	None None	None
Hale et al. 1988 (Dunedin Program)	M & F 65+ years	Claudication	***	
Fowkes et al. 1992 (Edinburgh Artery Study)	M & F 55–74 years	Claudication ABPI	*** ***	None None

NK = significance not known; * = $p < 0.05$; ** = $p < 0.01$; *** = $p < 0.001$.

Among 2759 men in the Basle study, 174 developed asymptomatic evidence of peripheral vascular disease during a five-year period, and of these 102 had elevated baseline levels of triglyceride ($p<0.001$). In the Glostrup Study a fasting serum triglyceride at 50 years of age was correlated with the ankle-brachial pressure index 10 years later (Spearman rank r value $= -0.1$, $p<0.05$). However, this relationship disappeared on multivariate analysis adjusting for sex, smoking, blood pressure, and serum cholesterol.

Thus, current evidence would suggest that the relationship of triglycerides with peripheral arterial disease is comparable to that with ischaemic heart disease – a strong univariate association disappearing on multivariate analysis (Hulley et al. 1980). In the Edinburgh Artery Study, we confirmed that this was mostly due to the correlation of triglycerides with non-HDL cholesterol ($r = 0.48$) and HDL cholesterol ($r = -0.45$) (Fowkes et al. 1992). However, these results do not necessarily mean that triglycerides are not important in aetiology. Indeed, in examining the association of triglycerides in multivariate logistic regressions at various cut-off points of the ABPI, we found that triglycerides were independently associated at lower levels (≤0.7), suggesting that triglycerides could be a risk factor for severe disease independent of a relationship with other lipids.

It has also been suggested that atherosclerosis may be related more to a slower rate of removal of triglycerides (particularly in the very low density lipoprotein fraction) than to elevated levels of serum triglyceride. In male patients with coronary artery disease, the fractional removal rate of intravenously injected triglyceride emulsion (Intralipid) has been shown to be lower than in age-matched controls (Tollin et al. 1984). Also, similar differences were found between cases of peripheral arterial disease and controls ($p<0.02$) (Angquist et al. 1982). This could be related to physical inactivity because high exercise levels have been shown to be associated with more rapid clearance of intravenously injected triglyceride emulsion (Ericsson et al. 1982).

Cholesterol

The majority of case–control studies conducted on hospital patients have found higher levels of serum cholesterol in cases than in controls (Schrade et al. 1960; Greenhalgh et al. 1971; Bliss et al. 1972; Skrede and Kvarstein 1975; Lipinska et al. 1979; Trayner et al. 1980; Greenhalgh et al. 1981; Angquist et al. 1982; Pilger et al. 1983; Rühling et al. 1989; Cardia et al. 1990). However, in some studies, significantly elevated levels are not found in cases (Dormandy et al. 1973; Sirtori et al. 1974; Vyden et al. 1975; Bradby et al. 1978); or only in women (Meerloo and Billimoria 1979). Cholesterol levels have also been shown to be related to the severity of disease (Jacobsen et al. 1984; Davignon et al. 1977). Discrepancies in the results between studies may be due to many causes, but an important factor may be poor age- and sex-matching of control groups. Fig. 14.2 shows the mean serum cholesterol levels in cases and controls from eight studies in which matching was satisfactory. Although there was substantial overlap in 95% confidence intervals in two studies (Sirtori et al. 1975; Vyden et al. 1975), the mean cholesterol levels were higher in cases than in controls in all the studies.

Cross-sectional studies in the community have also produced variable results. Significant elevations of serum cholesterol were found in cases in Finland (Reunanen et al. 1982; Schroll and Munck 1981), the London Whitehall Study

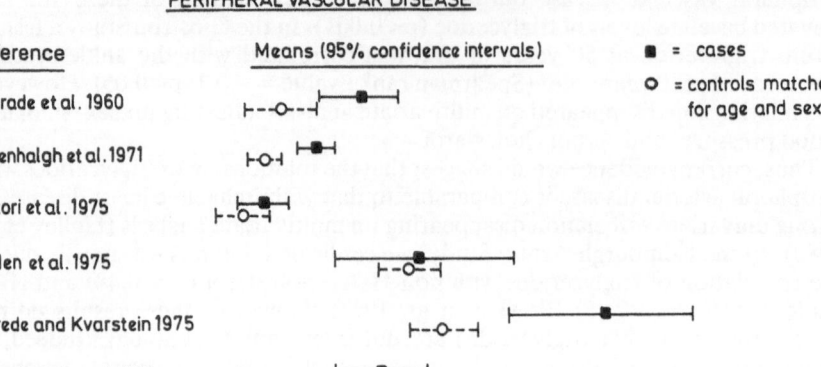

Fig. 14.2. Serum cholesterol in case–control studies of peripheral vascular disease.

(Davey Smith et al. 1990), Israel (Gofin et al. 1987) and the Edinburgh Artery Study (Fowkes et al. 1992) but not in Belgium (De Backer et al. 1979), Oxford (Hughson et al. 1978) or the Lipid Research Clinics Study (Pomrehn et al. 1986). Positive associations have however been found in the two longitudinal studies relating serum cholesterol at entry to later occurrence of peripheral arterial disease. Serum cholesterol was found to be a significant, but weak risk factor for intermittent claudication after 26 years of follow-up in Framingham (Kannel and McGee 1984). Similarly, in the Danish study, serum cholesterol at age 50 years was found to correlate with ankle-brachial pressure indices 10 years later (Spearman Rank $r = -0.18$, $p < 0.001$) and to be independently related (Schroll and Munck 1981). In the Edinburgh and Dunedin cross-sectional studies (Fowkes et al. 1992; Hale et al. 1988) the relationship between cholesterol and peripheral vascular disease also persisted after adjustment for other risk factors, but this was not the case in the Israeli study (Gofin et al. 1987). However, on balance there would appear to be sufficiently strong evidence to indicate that serum cholesterol is a "risk factor" for peripheral vascular disease.

HDL Cholesterol

Reduced levels of HDL cholesterol in hospital cases of peripheral vascular disease, compared with controls, have almost without exception (Pilger et al.

1983) been found in recent years (Kirstein and Olsson 1979; Lipinska et al. 1979; Trayner et al. 1980; Angquist et al. 1982; Franceschini et al. 1982; Rühling et al. 1989) although significant differences have been noted between men (Bihari-Varga et al. 1981) and women (Bradby et al. 1978). HDL cholesterol levels would appear to be related inversely to the severity of peripheral vascular disease (Beach 1979; Jacobsen et al. 1984) and the reduction in levels may be related to the amount of cigarette smoking and physical inactivity (Bihari-Varga et al. 1981).

HDL cholesterol levels have not been studied widely in community studies. In one report from the Lipid Research Clinics Prevalence Study (Pomrehn et al. 1986), men with a history of claudication or leg pain on exercise testing had lower mean levels of HDL cholesterol than those without. Whereas in another report on a different population from the Lipid Research Clinics Study of peripheral vascular disease, which included some asymptomatic subjects, HDL cholesterol levels did not differ from those without peripheral vascular disease. Likewise, the Israeli study found no significant association between HDL cholesterol and the ankle-brachial pressure index (Gofin et al. 1987). In the Edinburgh Artery Study, we found a strong inverse association between HDL cholesterol levels and intermittent claudication ($p<0.001$) and a less significant association with the ankle-brachial pressure index ($p<0.1$) (Fowkes et al. 1992). However, the relationships did persist on multivariate analyses adjusted for other lipids, obesity, smoking, diabetes, and alcohol consumption. In the cohort phase of the Edinburgh Artery Study, we intend to observe the relationship between HDL cholesterol and peripheral vascular disease 5 years later.

Apolipoproteins

Pilger et al. (1983) have suggested that apolipoproteins rather than lipid or lipoprotein levels may be a better discriminator of cases of peripheral vascular disease and normal subjects. In their case–control study on 88 male subjects, they found that the cases had significantly higher levels of apo B than the controls ($p<0.001$) and although apo A-I and A-II levels were only slightly and non-significantly lower in the cases, apo A-I/apo B ratios were much lower ($p<0.001$) such that the apo A-I/apo B ratio was a better discriminator of peripheral vascular disease than LDL or HDL cholesterol. In other case–control studies however, apo B has either not been elevated in any lipoprotein subfraction (Carlson et al. 1986), has been elevated in VLDL and not in LDL subfractions (Franceschini et al. 1982) or has been significantly lower in cases of peripheral vascular disease (Bradby et al. 1978). Apo A-I and apo A-II have been shown to be reduced in cases of peripheral vascular disease (Bradby et al. 1978; Angquist et al. 1982) although in another study no differences were found in apo A-I and A-II high density lipoprotein cholesterol (Franceschini et al. 1982) between cases and controls.

Thus, the differences between apolipoprotein levels in peripheral vascular disease patients and controls are not wholly consistent, and this is probably due to relatively few studies, small numbers of subjects in some (Angquist et al. 1982; Franceschini et al. 1982; Carlson et al. 1986), and the methods of selecting cases and controls according to lipid status. There is little doubt that apo B levels are related to the occurrence of coronary heart disease (Editorial 1988a), and certainly there is no consistent evidence to suggest that the same does not hold true for peripheral vascular disease. Several studies have suggested that the circulating level of apo B is a better predictor of premature coronary heart disease than is plasma cholesterol concentration (Avogaro et al. 1979; Campeau et al. 1984; Durrington et al. 1986). Indeed the plasma concentration of apo B would appear to be under strong genetic control (Rajput-Williams et al. 1988). In one study on 205 patients with peripheral vascular disease (Monsalve et al. 1988), significant differences occurred at the apo B locus in patients compared to a healthy population but, interestingly, no differences occurred in allele frequencies between patients with peripheral vascular disease alone and those with known coexisting coronary or carotid disease.

Lipoprotein (a) is an independent risk factor for myocardial infarction or coronary death (Rosengren et al. 1990), but there is little, if any, research on this lipoprotein or its apolipoprotein, apo (a), in peripheral vascular disease.

Essential Fatty Acids

Essential fatty acids are derived from two parent molecules, linoleic acid (18:2, n–6), and alpha-linolenic acid (18:3, n–3), where n–6 and n–3 represent the position of double bonds from the methyl end of the molecule. As man cannot synthesise either of the two, they are essential components of the diet, most commonly found in vegetable and marine oils. They are required for the normal growth and function of all tissues.

The activity of linoleic acid depends upon its conversion to other highly unsaturated long-chain fatty acids. An important rate-limiting step in this pathway is the desaturation of linoleic acid to gamma-linolenic acid, followed by conversion to the biologically active dihomo-gamma-linolenic and arachidonic acids. However, some people are now known to be unable to perform this rate-limiting desaturation step, and are therefore effectively lacking in this essential fatty acid (Abraham et al. 1990). Dietary sources of gamma-linoleic acid are rare, but relatively large quantities (9%) can be found in evening primrose oil.

Deficiencies of linoleic acid in plasma cholesterol esters have been found in patients with coronary, aortic and peripheral arterial disease (Kingsbury et al. 1969). Linoleic acid levels were also found to be reduced in the adipose tissue of subjects with coronary artery disease, probably reflecting a reduced long-term dietary intake (Wood et al. 1984). More recently, Ishikawa et al. (1989) found a decrease in LDL-cholesterol and apolipoprotein B levels following a 16-week administration of evening primrose oil in a controlled trial of hypercholesterolaemic patients. This evidence suggests that deficiencies in the n–6 series of essential fatty acids may contribute to the risk of developing vascular disease, but whether these fatty acids can promote regression of established atheroma is still unknown.

Cold water fish contain fatty acids of the n–3 series derived from marine plankton, in particular eicosapentaenoic acid (20:5, n–3) and docosahexaenoic acid (22:6, n–3). These fatty acids compete with arachidonic acid (20:4, n–6) which is the main precursor of prostaglandin, thromboxane and prostacyclin synthesis in those eating a standard Western diet. They completely replace arachidonic acid when the consumption of fish oil is increased (Thorngren and Gustafson 1981). The net functional result of these changes is difficult to predict, although both anti-inflammatory and antithrombotic effects might be expected (Editorial 1988b).

The possible importance of marine oils in preventing atherosclerosis was first noticed in population studies in Greenland, where eskimos were found to have a much lower than expected incidence of myocardial infarction (Kromann and Green 1980). Recent controlled trials have shown that fish oil reduces plasma triglyceride levels by approximately 25% (Thorngren and Gustafson 1981), although the effect on cholesterol is inconclusive (Hornstra 1989; Smith et al. 1989). Fish oil also favourably alters certain haemostatic variables, in particular fibrinogen, blood viscosity and red cell deformability (Hornstra 1989), all of which are thought to be related to the development of atherosclerosis. A major clinical trial by Burr et al. in 1989 produced a 29% reduction in 2-year all-cause mortality in patients with ischaemic heart disease who had been advised to eat fatty fish. Fish oil may also decrease re-stenosis rates following percutaneous coronary angioplasty (Hornstra 1989). However, administration of n–3 fatty acids was shown to have no effect on peripheral vascular disease as assessed by Doppler ultrasound techniques (Woodcock et al. 1984), although this was a short study in patients with advanced arterial disease. Thus the role of fish oil in modifying the natural history of atherosclerosis is still not clear, particularly as there have been no long-term intervention studies.

Lipid Peroxides

Lipid peroxides are formed by the peroxidation of polyunsaturated fatty acids, and for many years it has been known that the severity of aortic atherosclerosis is correlated with the concentration of lipid peroxides in the aortic wall (Harland et al. 1971). Also, animals maintained on atherogenic diets develop an increase in serum and aortic lipid peroxide concentrations (Goto 1982). Thus, it has been suggested that lipid peroxides may be important in the development of athero-sclerosis. In one recent study, plasma lipid peroxide levels were found to be significantly higher in both cases of peripheral arterial disease and ischaemic heart disease than in control patients with evidence of atherosclerosis (Stringer et al. 1989). In a case–control study, part of the Edinburgh Artery Study, we have also found that lipid peroxides were elevated in cases (Fowkes et al. unpublished observations), but further analysis is required to determine the extent to which this is independent of lipid levels and other confounding factors. Vitamin E is a major antioxidant which protects polyunsaturated fatty acids from peroxidation, high levels of which have been shown to be associated with a decreased risk of coronary heart disease (Riemersma et al. 1991) and could therefore conceivably also have a protective effect in the development of peripheral atherosclerosis.

Effect of Modifying Lipid Profiles on Peripheral Vascular Disease

Numerous clinical and epidemiological studies have demonstrated that elevated serum cholesterol and probably also triglyceride are important risk factors for peripheral vascular disease. Some abnormal lipoprotein patterns such as occur in familial hypercholesterolaemia, are known to be genetic in origin, whilst others result from disorders of renal, liver and metabolic function. However, the majority of elevated lipid levels in the population probably result from a complex interaction between dietary and social factors such as smoking and exercise (Bolton-Smith et al. 1991).

The medical management of hyperlipidaemia initially involves dietary modification to reduce fat intake, and increase the ratio of polyunsaturated fats. If this fails, drug therapy is instituted, and occasionally in severe cases of familial hypercholesterolaemia partial ileal bypass surgery is performed to reduce the absorption of fat. An increase in moderate-intensity exercise can also produce favourable lipoprotein changes, in both healthy subjects and patients with vascular disease (Haskell 1986). Cigarette smoking has also been shown to be associated with substantially lower levels of high density lipoprotein (Criqui et al. 1980).

Few studies have looked at the effects of lowering lipid levels on peripheral vascular disease, and in general these have examined asymptomatic subjects with hyperlipidaemia rather than claudicants. However, the one randomised controlled trial of lipid-lowering therapy (diet and drugs) in claudicants did show significantly less disease progression in the treatment group after a mean period of 19 months (Duffield et al. 1983). Other studies have also shown either regression or reduced progression of peripheral vascular disease following the administration of lipid-lowering drugs (Table 14.4).

These results suggest that effective lipid-lowering therapy favourably influences the natural history of peripheral atherosclerosis in patients with hyperlipoproteinaemia. However, very large, population-based studies would be required to determine whether dietary changes are successful in preventing the development of peripheral atheroma and its complications.

Peripheral Vascular Disease and Ischaemic Heart Disease

The relationship between blood lipid concentrations and peripheral vascular disease would appear from the evidence presented to be very similar to that for ischaemic heart disease. Serum triglycerides have a strong univariate relationship with peripheral vascular disease which tends not to persist on multivariate analysis. The picture with serum cholesterol is somewhat mixed, but on balance the evidence points to an elevated serum cholesterol associated with peripheral vascular disease. HDL cholesterol has been less extensively studied but shows a typical inverse association, although this has not been examined in longitudinal studies. The few studies examining apolipoproteins, essential fatty acids, and lipid peroxides do not show any consistent evidence, suggesting discrepancies with the findings in ischaemic heart disease.

Lipid profiles in peripheral vascular disease and ischaemic heart disease have been compared directly in only a few studies. In the Glostrup Study, Schroll and

Table 14.4. Effects of lipid-lowering regimens on peripheral vascular disease

Reference	Study population	Trial design	Mean period of follow-up	Intervention	Effect on lipids	Effect on peripheral vascular disease
Barndt et al. 1977	25 patients with hyperlipoproteinaemia and intermittent claudication (22–65 yrs)	Non-controlled trial	13 months	Diet and drugs	TG ↓ C ↓	↑ Regression on angiography only in 36% with lower lipid levels
Duffield et al. 1983	24 patients with hyperlipidaemia and stable claudication (<65 yrs)	Randomised controlled trial	19 months	Diet and drugs	TG ↓ C ↓ LDL ↓ HDL ↑	→ Progression ↑ Regression on angiography
Koivisto and Leinonen 1988	37 patients with familial hypercholesterolaemia (av.44 yrs)	Non-randomised controlled trial	10 years	Ileal bypass (control:diet and drugs)	C ↓ LDL ↓ HDL ↑	No difference in claudication or ABPI
Erikson et al. 1988	62 men with hyperlipidaemia: 2 with symptoms of intermittent claudication (35–65 yrs)	Non-randomised controlled trial	18 months	Fenofibrate and nicotinic acid	TG ↓ C ↓	↑ Regression on angiography
Olsson et al. 1990	45 asymptomatic men with hyperlipidaemia (35–65 yrs)	Non-randomised controlled trial	12 months	Diet, nicotinic acid and fenofibrate	TG ↓ C ↓ LDL ↓ HDL ↑	→ Progression ↑ Regression on angiography

TG, triglyceride; C, cholesterol; LDL, low density lipoprotein cholesterol; HDL, high density lipoprotein cholesterol.

Munck (1981) stated that blood lipids (cholesterol, triglycerides) were "more pronounced risk factors for peripheral artery disease than for coronary heart disease after the age of fifty". But in the Framingham Study after 16 years of follow-up (Gordon and Kannel 1972), serum cholesterol had the same degrees of association with intermittent claudication, coronary heart disease, and atherosclerotic brain infarction. One problem in studies such as this, however, is that comparisons may be being made between differing underlying severities of atherosclerosis in the total vascular system rather than comparisons of atherosclerosis at different sites. In the Edinburgh Artery Study, we overcame this problem to some extent by carrying out multiple logistic regressions of lipids and other risk factors on three separate measures of both symptomatic and asymptomatic peripheral vascular disease and ischaemic heart disease (Fowkes et al. 1992). We found no significantly consistent differences in the lipid profiles for peripheral vascular disease and heart disease. Thus, these direct comparisons do not provide any evidence suggesting that peripheral vascular disease is any different from ischaemic heart disease in this respect.

Therefore, it is likely that dietary changes, cholesterol screening in high risk groups, and other interventions to alter lipids and prevent ischaemic heart disease may well have a spin-off effect in contributing to the prevention of peripheral vascular disease.

References

Abraham RD, Riemersma RA, Elton RA, Macintyre C, Oliver MF (1990) Effects of safflower oil and evening primrose oil in men with a low dihomo-gamma-linolenic level. Atherosclerosis 81:199–208

Angquist KA, Johnson O, Ericsson M, Tollin C (1982) Serum lipoproteins and apolipoproteins AI and AII in male patients with peripheral arterial insufficiency. Acta Chir Scand 148:675–8

Avogaro P, Bittolo BG, Cazzolato G, Quinii GB (1979) Are apolipoproteins better discriminators than lipids for atherosclerosis? Lancet i:901–3

Barndt R, Blankenhorn DH, Crawford DW, Brooks SH (1977) Regression and progression of early femoral atherosclerosis in treated hyperlipoproteinaemic patients. Ann Intern Med 86:139–46

Beach KW (1979) The correlation of arteriosclerosis obliterans with lipoproteins in insulin dependent and non-insulin dependent diabetes. Diabetes 28:836–40

Bihari-Varga M, Szekely J, Gruber E (1981) Plasma high density lipoproteins in coronary, cerebral and peripheral vascular diseases. The influence of various risk factors. Atherosclerosis 40:337–45

Bliss BP, Kirk CJC, Newall RG (1972) Abnormalities in glucose tolerance, lipid and lipoprotein levels in patients with atherosclerotic peripheral vascular disease. Angiology 23:69–75

Bolton-Smith C, Woodward M, Smith WCS, Tunstall-Pedoe H (1991) Dietary and non-dietary predictors of serum total and HLD-cholesterol in men and women: results from the Scottish Heart Health Study. Int J Epidemiol 20:95–104

Bradby GV, Valente AJ, Walton KW (1978) Serum high density lipoproteins in peripheral vascular disease. Lancet ii:1271–4

Burr ML, Fehily AM, Gilbert JF et al. (1989) Effects of changes in fat, fish and fibre intakes on death and myocardial infarction: diet and reinfarction trial (DART) Lancet ii: 757–61

Campeau L, Enjalbert M, Lesperance J et al. (1984) The relation of risk factors to the development of atherosclerosis in saphenous vein bypass grafts and the progression of disease in the native circulation. N Engl J Med 311:1329–32

Cardia G, Grisorio D, Impedovo G, Lillo A, Regina G (1990) Plasma lipids as a risk factor in peripheral vascular disease. Angiology 41:19–22

Carlson LA, Pauciullo P, Ekland B, Johansson J (1986) Modified composition of very low density lipoproteins in peripheral vascular disease: evidence for low cholesteryl ester but normal apolipoprotein B content of small VLDL. In Ventura A, Crepaldi G, Senin U (eds) Extracoronary atherosclerosis. Karger, Basel, pp. 31–4 (Monographs in atherosclerosis 14)

Castelli WP (1986) The triglyceride issue: a view from Framingham. Am Health J 112:432–7

Criqui MH, Wallace RB, Heiss G et al. (1980) Cigarette smoking and plasma high-density lipoprotein cholesterol. The Lipid Research Clinics Program prevalence study. Circulation 62: (Suppl 4, pt 2) IV70–IV76

da Silva A, Widmer LK (1980) Occlusive peripheral artery disease. Early diagnosis, incidence, course, significance. Hans Huber, Bern

Davey Smith G, Shipley MJ, Rose G (1990) Intermittent claudication, heart disease risk factors, and mortality. The Whitehall Study. Circulation 82:1925–31

Davignon J, Lussier-Calan S, Ortin-George M et al. (1977) Plasma lipids and lipoprotein patterns in angiographically graded atherosclerosis of the legs and in coronary heart disease. Can Med Assoc J 116:1245–50

De Backer IG, Kornitzer M, Sobolski J, Denolin H (1979) Intermittent claudication – epidemiology and natural history. Acta Cardiol (Brux) 34:115–24

Dormandy JA, Hoare E, Colley J, Arrowsmith DE, Dormandy TL (1973) Clinical, haemodynamic, rheological and biochemical findings in 126 patients with intermittent claudication. Br Med J iv:576–81

Duffield RGM, Lewis B, Miller NE, Jamieson CW (1983) Treatment of hyperlipidaemia retards progression of symptomatic femoral atherosclerosis. A randomized controlled trial. Lancet ii:639–42

Durrington PN, Hunt L, Ishola M, Kane J (1986) Serum apolipoproteins A1 and B and lipoproteins in middle-aged men with and without previous myocardial infarction. Br Heart J 56:206–12

Editorial (1988a) Apolipoprotein-B and atherogenesis. Lancet i:1141–2

Editorial (1988b) Fish oil. Lancet i:1081–3

Editorial (1991) Lipoprotein (a). Lancet 337:397–8

Ericsson M, Johnson O, Tollin C et al. (1982) Serum lipoprotein, apolipoprotein and intravenous fat tolerance in young athletes. Scand J Rehab Med 14:209–12

Erikson U, Helmius G, Hemmingsson A, Ruhn G, Olsson AG (1988) Repeat femoral angiography in hyperlipidaemic patients. A study of progression and regression of atherosclerosis. Acta Radiol 29:303–9

Fowkes FGR, Housley E, Riemersma RA et al. (1992) Smoking, lipids, glucose intolerance, and blood pressure as risk factors for peripheral atherosclerosis compared to ischaemic heart disease in the Edinburgh Artery Study. Am J Epidemiol (in press)

Franceschini G, Bondioli A, Mantero M et al. (1982) Increased apoprotein B in very low density lipoproteins of patients with peripheral vascular disease. Arteriosclerosis 2:74–84

Gofin R, Kark JD, Friedlande Y et al. (1987) Peripheral vascular disease in a middle-aged population sample. The Jerusalem Lipid Research Clinic Prevalence Study. Isr J Med Sci 23:157–67

Goldstein JL, Brown MS (1978) Familial hypercholesterolaemia: pathogenesis of a receptor disease. Johns Hopkins Med J 143:8–16

Gordon T, Kannel WB (1972) Predisposition to atherosclerosis in the head, heart and legs. JAMA 221:661–6

Goto Y (1982) Lipid peroxides as a cause of vascular disease. In: Yagi K (ed) Lipid peroxides in biology and medicine. Academic Press, New York, pp. 295–303

Gotto AM (1989) The HDL hypothesis: Gathering the evidence. Proceedings of the XI Congress of European Society of Cardiology, Nice

Greenhalgh RM, Lewis B, Rosengarten DS et al. (1971) Serum lipids and lipoproteins in peripheral vascular disease. Lancet ii:947–50

Greenhalgh RM, Laing SP, Cole PV, Taylor GW (1981) Smoking and arterial reconstruction. Br J Surg 68:605–7

Hale WE, Marks RG, May FE et al. (1988) Epidemiology of intermittent claudication: evaluation of risk factors. Age Aging 17:57–60

Harland WA, Gilbert JD, Steel G, Brooks CJW (1971) Lipids of human atheroma, Part 5. The occurrence of a new group of polar sterol esters in various stages of human atherosclerosis. Atherosclerosis 13:239–46

Haskell WL (1986) The influence of exercise training on plasma lipids and lipoproteins in health and disease. Acta Med Scand [Suppl] 711:25–37

Hegele RA (1989) Lipoprotein (a): An emerging risk factor for atherosclerosis. Can J Cardiol 5:263–5

Hornstra G (1989) Influence of dietary fish oil on arterial thrombosis and atherosclerosis in animal models and in man. J Intern Med 225 (Suppl):53–9

Hughson WG, Mann JI, Garrod A (1978) Intermittent claudication: prevalence and risk factors. Br Med J i:1379–81

Hulley SB, Rosenman RH, Bawol RD, Brand RJ (1980) Epidemiology as a guide to clinical decisions: the association between triglyceride and coronary disease. N Engl J Med 302:1383–9

Ishikawa T, Fujiyama Y, Igarashi O et al. (1989) Effects of gammalinolenic acid on plasma lipoproteins and apolipoproteins. Atherosclerosis 75:95–104

Jacobsen UK, Dige-Pedersen H, Gyntelberg F, Svendsen UG (1984) 'Risk factors' and manifestations of arteriosclerosis in patients with intermittent claudication compared to normal persons. Dan Med Bull 31:145–8

Kannel WB, McGee DL (1984) Update on some epidemiological features of intermittent claudication: The Framingham Study. J Am Geriatr Soc 33:13–18

Kingsbury KJ, Morgan DM, Stovold R, Breit CG (1969) The relationships between plasma cholesteryl polyunsaturated fatty acids, age and atherosclerosis. Postgrad Med J 45:591–601

Kirstein P, Olsson AG (1979) HDL cholesterol is low in young and increases with age in male claudicators. Atherosclerosis 33:145–8

Koivisto PVI, Leinonen H (1988) Peripheral arterial disease in heterozygous familial hypercholesterolaemia: no difference between patients with and without partial ileal bypass. Atherosclerosis 70:21–27

Kromann N, Green A (1980) Epidemiological studies in the Upernavik district, Greenland. Acta Med Scand 208:401–6

Lipid Research Clinics Program (1984) The Lipid Research Clinics Coronary Primary Prevention Trial Results. I. Reduction in the incidence of coronary heart disease. JAMA 251:351–64

Lipid Research Clinics Program Epidemiology Committee (1979) Plasma lipid distribution in selected North American populations: the Lipid Research Clinics Program prevalence study. Circulation 60:427–39

Lipinska I, Lipinski B, Gurewich V (1979) Lipoproteins, fibrinolytic activity and fibrinogen in patients with occlusive vascular disease and in healthy subjects with a family history of heart attacks. Artery 6:254–64

Meerloo JM, Billimoria JD (1979) High density lipoprotein cholesterol levels in peripheral vascular disease and in women on oral contraception. Atherosclerosis 33:267–9

Monsalve MV, Young R, Wiseman SA et al. (1988) DNA polymorphism of the gene for apolipoprotein B in patients with peripheral arterial disease. Atherosclerosis 70:123–9

Olsson AG, Ruhn G, Erikson U (1990) The effect of serum lipid regulation on the development of femoral atherosclerosis in hyperlipidaemia: a non-randomized controlled study. J Intern Med 227:381–90

Packard CJ, Shepherd J (1990) Cholesterol, lipoproteins and atherosclerosis. Vasc Med Review 1:91–8

Pilger E, Pristautz H, Pfeiffer KH, Kostner GM (1983) Retrospective evaluation of risk factors for peripheral atherosclerosis by stepwise discriminant analysis. Arteriosclerosis 3:57–63

Pomrehn P, Duncan B, Weissfeld L et al. (1986) The association of dyslipoproteinaemia with symptoms and signs of peripheral arterial disease. The Lipid Research Clinic Prevalence Study. Circulation 73 (Suppl 1):100–7

Rajput-Williams J, Knott TJ, Wallis SC et al. (1988) Variation of apolipoprotein-B gene is associated with obesity, high blood cholesterol levels, and increased risk of coronary heart disease. Lancet ii:1442–6

Reunanen A, Takkunen H, Aromaa A (1982) Prevalence of intermittent claudication and its effect on mortality. Acta Med Scand 211:249–56

Riemersma RA, Wood DA, Macintyre CCA, Elton RA, Gey KF, Oliver MF (1991) Risk of angina pectoris and plasma concentrations of vitamins A,C,E and carotene. Lancet 337:1–5

Rosengren A, Wilhelmsen L, Eriksson E, Risberg B, Wedel H (1990) Lipoprotein (a) and coronary heart disease: a prospective case–control study in a general population sample of middle-aged men. Br Med J 301:1248–51

Rühling K, Zabel-Langhennig R, Till U, Thielmann K (1989) Enhanced net transfer of HDL cholesteryl esters to Apo B containing lipoproteins in patients with peripheral vascular disease. Clin Chim Acta 184:289–96

Schrade W, Böehle E, Biegler R (1960) Humoral changes in arteriosclerosis. Investigations on lipids, fatty acids, ketone bodies, pyruvic acid, lactic acid, and glucose in the blood. Lancet ii:1409–16

Schroll M, Munck O (1981) Estimation of peripheral arteriosclerotic disease by ankle blood pressure measurements in a population study of 60-year-old men and women. J Chron Dis 34:261–269

Sirtori CR, Biasi G, Vercellio G, Agradi E, Malan E (1975) Diet lipids and lipoproteins in patients with peripheral vascular disease. Am J Med Sci 268:325–32

Skrede S, Kvarstein B (1975) Hyperlipidaemia in peripheral atherosclerotic arterial disease. Acta Chir Scand 141:333–40

Smith EB (1987) Relationship between lipids and atherosclerosis. In: Bloom AL, Thomas DF (eds) Haemostasis and thrombosis. Churchill Livingstone, Edinburgh, pp. 679–96

Smith P, Arnesen H, Opstad T, Dahl KH, Eritsland J (1989) Influence of highly concentrated n–3 fatty acids on serum lipids and haemostatic variables in survivors of myocardial infarction receiving either oral anticoagulants or matching placebo. Thromb Res 53:467–74

Stringer MD, Görög PG, Freeman A, Kakkar VV (1989) Lipid peroxides and atherosclerosis. Br Med J 298:281–4

Thorngren M, Gustafson A (1981) Effects of 11-week increase in dietary eicosopentaenoic acid on bleeding time, lipids and platelet aggregation. Lancet ii:1190–93

Tollin C, Ericsson M, Johnson O et al. (1984) Decreased removal of triglycerides from the blood – a mechanism for the hypertriglyceridaemia in male patients with coronary artery disease. Int J Cardiol 5:185–92

Trayner IM, Mannarino E, Clyne CAC, Thompson GR (1980) Serum lipids and high density lipoproteins in peripheral vascular disease. Br J Surg 67:497–9

Van Gaal L, Rillaerts E, Creten W, De Leeuw I (1988) Relationship of body fat distribution pattern to atherogenic risk factors in NIDDM. Preliminary results. Diabetes Care II:103–6

Vyden JK, Thorner J, Nagasawa K et al. (1975) Metabolic and cardiovascular abnormalities in patients with peripheral arterial disease. Am Heart J 90:703–8

Wood DA, Butler S, Riemersma RA et al. (1984) Adipose tissue and platelet fatty acids and coronary heart disease in Scottish men. Lancet ii:117–121

Woodcock BE, Smith E, Lambert WH et al. (1984) Beneficial effect of fish oil on blood viscosity in peripheral vascular disease. Br Med J 288:592–4

15 Blood Pressure

C.J. BULPITT

This chapter reviews the association between vascular disease of the arteries of the lower limbs (peripheral vascular disease) and blood pressure. Associations between peripheral vascular disease and blood pressure are examined in three situations: the presence of intermittent claudication, absent foot pulses and a low systolic pressure in the leg. Information from cross-sectional studies is reported to assess the association between established disease and blood pressure. Longitudinal prospective studies are examined to determine whether or not a high blood pressure precedes the development of peripheral vascular disease. We will also consider the strength of any association, whether or not it is causal and whether it is related most closely to systolic or diastolic pressure.

Measurement of Peripheral Vascular Disease

We will first consider what is meant by peripheral vascular disease. For practical and ethical reasons, peripheral vascular disease cannot be studied by angiography in the general population. The following markers have therefore been frequently employed: the presence of intermittent claudication, the absence of foot pulses and a reduced systolic pressure at the ankle. Intermittent claudication has either been determined using the Rose questionnaire (1968) or clinically as in the Framingham study where claudication was considered present when "there was intermittent cramping calf pain induced by walking a while, related to the rapidity of the pace and the grade, and relieved in minutes by rest" (Kannel and McGee 1985). Nonpalpable pedal pulses in the Framingham study were the dorsalis pedis and posterior tibialis on the right and left sides "considered nonpalpable when pulses were absent at either location" (Abbott et al. 1990). The poor repeatability of these observations is well recognised (Meade et al. 1968).

The systolic pressure in the ankle is usually measured using the Doppler method. The method has been used satisfactorily but varies between the posterior tibial and dorsalis pedis arteries (Gofin et al. 1987) and also with the systolic pressure in the brachial artery. In order to allow for the latter variability the leg: brachial ratio is usually calculated. A ratio less than 0.90 is observed in about 5% of the middle-aged population (Schroll and Munck 1981) and therefore this has arbitrarily been taken to indicate the limit of normality. Unfortunately this ratio is bound to be negatively related to brachial blood pressure and observations that the higher the brachial artery pressure the lower the ratio (Gofin et al. 1987; Scholl and Munck 1981) cannot prove that hypertension produces peripheral vascular disease.

It must also be admitted that these three measures of peripheral vascular disease do not overlap to any large extent (Gofin et al. 1987; Scholl and Munck 1981). There is therefore considerable uncertainty over the diagnosis of peripheral vascular disease. An additonal question is whether or not peripheral vascular disease can increase peripheral vascular resistance by such an extent that brachial artery pressure increases. This problem would be expected to increase diastolic pressure more than systolic but patients with both legs amputated have higher systolic pressures when over the age of 50 (Labouret et al. 1983) and systolic pressure may rise as a consequence rather than a cause of atherosclerosis obliterans (Levenson et al. 1982). Owing to this association more credence will have to be given to longitudinal rather than cross-sectional studies of the relationship between blood pressure and peripheral vascular disease.

Relationship between Intermittent Claudication and Blood Pressure

In a cross-sectional study reported in 1966, Reid and colleagues reported a brachial systolic pressure 10 mmHg higher and diastolic 5 mmHg greater in 15 subjects who were questionnaire positive for intermittent claudication. In 1978 Hughson and colleagues studied 54 community cases of intermittent claudication and 108 controls. Systolic pressure was 17 mmHg higher and diastolic 5 mmHg greater in the cases. Higher brachial pressures of 13 mmHg (systolic) and 7 mmHg (diastolic) have been confirmed in the Glostrup study (Schroll and Munck 1981) but not in the Jerusalem Lipid Research Clinic Prevalence study (Gofin et al. 1987). Hale and colleagues (1988) also reported that systolic pressure was higher in a cross-sectional study of patients reporting intermittent claudication although they did not say how much higher. After adjusting for age and sex they stated that diastolic blood pressure was not related to intermittent claudication. Their subjects were all over the age of 65 years at which age diastolic blood pressure is considered to be falling in the average subject and the pressures may have been higher previously.

In the Whitehall study 18 388 men answered a questionnaire on intermittent claudication at entry to the study and at that time 0.8% and 1% had probable and possible claudication respectively. Intermittent claudication was not related to either systolic or diastolic pressure at presentation. Nevertheless during 17 years of follow-up a three-fold excess death rate from both coronary heart disease and stroke mortality was observed (Smith et al. 1990).

Reunanen et al. (1982) reported a relative risk of 2.5 for mortality in patients with intermittent claudication, reduced to 2.2 by adjusting for blood pressure, serum cholesterol and smoking. This was a large study of 5738 men and 5224 women and as in the Whitehall study systolic blood pressure was not related to the presence of intermittent claudication at first examination.

In the longitudinal prospective studies more evidence has accrued to relate blood pressure to the later development of intermittent claudication. In the 18-year Framingham follow-up (Shurtleff 1974) a systolic pressure of ≥180 mmHg compared with ≤119 mmHg was associated with a 2.3-fold increase in incidence of intermittent claudication in men and a 4.5-fold increase in women. Fig. 15.1 shows that the 26-year follow-up (Kannel and McGee 1985) reveals 2.7-fold and 5.2-fold increases for this level of systolic pressure in men and women respectively. In contrast, if we compare the corresponding quintiles for diastolic pressure (≥95 mmHg compared with ≤75 mmHg), there was no increase of incidence in men in the 18-year follow-up and a 1.7-fold increase after 26 years (Fig. 15.1). In women the corresponding increases with diastolic ≥95 mmHg were 2.7-fold and 2-fold for 18- and 26-year follow-up respectively. Definite hypertension (blood pressure >159/94 mmHg) was associated with a 2.5 (men) and 3.8 (female) excess incidence of intermittent claudication in the 20-year follow-up (Kannel and Stokes 1985). This contrasts with a 5.8 and 6.1 increase in stroke for men and women respectively. Blood pressure predicted coronary heart disease and cerebrovascular accident much better than intermittent claudication.

Of two other longitudinal studies of the incidence of intermittent claudication, the Glostrup study grouped claudication, absent pulses and a leg/arm systolic

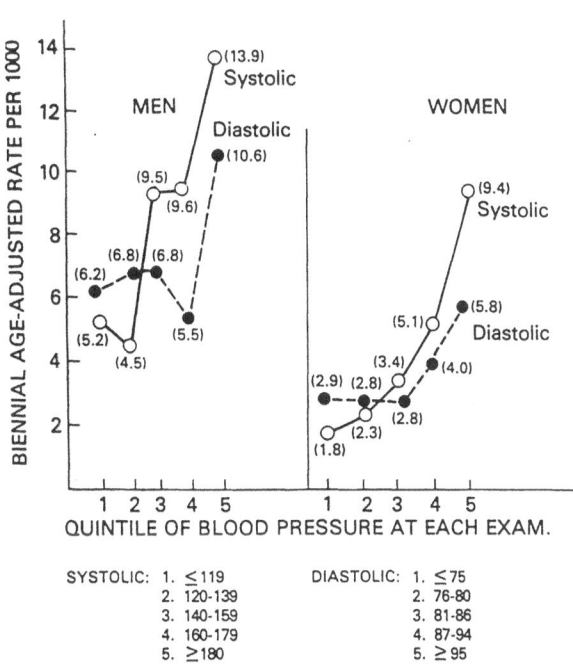

Fig. 15.1. Risk of intermittent claudication by systolic and diastolic blood pressure. Subjects aged 35–84 years. Trends significant at p<0.001 (data from 26-year follow-up of the Framingham Study). Reproduced with permission from Kannel WB, McGee DC (1985) J Am Geriatr Soc, 33:16.

index <0.90 together, to define reported cases. There was an initial higher systolic and diastolic pressure in these subjects but the use of the index makes this difficult to interpret. In the Basle longitudinal study, peripheral vascular disease at baseline was not significantly related to systolic pressure at that moment (da Silva et al. 1979). The incidence of peripheral vascular disease was not reported.

Lack of Foot Pulses and Brachial Pressure

A poor outcome in terms of cardiovascular disease has been reported for non-palable pedal pulses in the Framingham study (Abbott et al. 1990). A 1.5–2.0-fold increase occurred in men for coronary heart disease, strokes and cardiac failure and a 1.2–2.0-fold increase in women. There was a 2.5-fold increase in men and women with both diabetes and absent pulses. The data have not yet been presented for the combination of hypertension and non-palpable pedal pulses. In the cross-sectional Jerusalem Lipid Research Clinics study (Gofin et al. 1987), systolic pressure was negatively associated with absent pulses.

Low Systolic Pressure in the Foot and Brachial Pressure

Jacobsen and colleagues (1984) studied 53 patients with intermittent claudication and a leg systolic pressure at least 20 mmHg below arm pressure. The patients were compared with 106 community controls. Systolic pressure in the arm averaged 10 mmHg higher in the cases and diastolic pressure averaged 2 mmHg *lower*. Similarly Criqui et al. (1989) defined peripheral vascular disease as either an abnormal leg: arm ratio, an abnormal large vessel flow velocity or previous surgery for peripheral arterial disease. In men with moderate or severe peripheral vascular disease in this cross-sectional analysis, systolic pressure was increased by 9 mmHg and diastolic by 2 mmHg. In women with moderate peripheral vascular disease systolic and diastolic pressure were both 2 mmHg lower than in controls but with severe peripheral vascular disease systolic pressure was 16 mmHg higher and diastolic 2 mmHg higher.

Causal Association?

This overview of the association between peripheral vascular disease and blood pressure in the arm suggests an association between systolic pressure and the development of arterial lesions in the lower limbs. The evidence for such association is weak for diastolic pressure and we must ask whether or not the relationship between a high systolic pressure and peripheral vascular disease is likely to be causative or not.

Proof of prevention with active antihypertensive treatment cannot be demonstrated. Peripheral vascular disease, although not very rare, is an infrequent occurrence in randomised placebo controlled trials of antihypertensive treatment in subjects without peripheral vascular disease. In the MRC trial of 18 000 subjects with mild to moderate hypertension (Miall and Greenberg 1987), there were 638 cardiovascular events but less than 2% were attributed to causes other than myocardial infarction or stroke. We are not informed if any of the events were for peripheral vascular disease. In the Prevention of Atherosclerotic Complications with Ketanserin Trial (1989), 3899 patients with intermittent claudication and a leg: arm systolic pressure ≤ 0.85 were randomised either to ketanserin or placebo. Forty-six per cent of patients in this trial had hypertension (systolic ≥ 160 mmHg or diastolic ≥ 95 mmHg) and as ketanserin has antihypertensive properties this trial tested whether the relationship between peripheral vascular disease and blood pressure could be modified. Seventeen above ankle amputations were necessary in the ketanserin group and 32 in the placebo group. This supports the concept that peripheral vascular disease and hypertension are causally linked but ketanserin also inhibits serotonin induced platelet aggregation and improves haemorrheological properties.

Systolic pressure also increases as aortic compliance decreases. The decrease in aortic compliance results from stiffening of this artery through loss of elasticity. Such an occurrence is likely to be non-causally associated with arterial disease of the arteries of the lower limbs.

In conclusion, arm systolic pressure is associated with peripheral vascular disease but this may not be a causal relationship. The association with diastolic pressure remains to be proven but a high diastolic pressure could contribute to a stiffened aorta and peripheral vascular disease, and the subsequent fall in diastolic pressure as aortic compliance decreases may negate an original positive relationship between diastolic pressure and peripheral vascular disease. A causal relationship between a high blood pressure and arterial disease of the lower limbs remains unproven but appears to be likely.

References

Abbot RD, Brand FN, Kannel WB (1990) Epidemiology of some peripheral arterial findings in diabetic men and women: Experiences from the Framingham study. Am J Med 88:376–81

Criqui MH, Browner D, Fronek A et al. (1989) Peripheral arterial disease in large vessels is epidemiologically distinct from small vessel disease. An analysis of risk factors. Am J Epidemiol 129:1110–19

da Silva A, Widmer LK, Ziegler HW, Nissen C, Schweizer W (1979) The Basle Longitudinal Study: report on the relation of initial glucose level to baseline ECG abnormalities, peripheral artery disease, and subsequent mortality. J Chron Dis 32:797–803

Gofin R, Kark JD, Friedlander Y et al. (1987) Peripheral vascular disease in a middle-aged population sample. The Jerusalem Lipid Research Clinic Prevalence Study. Isr J Med Sci 23:157–67

Hale WE, Marks RG, May FE, Moore MT, Stewart RB (1988) Epidemiology in intermittent claudication: evaluation of risk factors. Age Aging 17:57–60

Hughson WG, Mann JI, Garrod A (1978) Intermittent claudication: Prevalence and risk factors. Br Med J i:1379–81
Jacobsen UK, Dige-Pedersen H, Gyntelberg F, Svendsen UG (1984) 'Risk factors' and manifestations of arteriosclerosis in patients with intermittent claudication compared to normal persons. Dan Med Bull 31:145–8
Kannel WB, McGee DL (1985) Update on some epidemiologic features of intermittent claudication: The Framingham Study. J Am Geriatr Soc 33:13–8
Kannel WB, Stokes J III (1985) Hypertension as a cardiovascular risk factor. In: Bulpitt CJ (ed) Handbook of hypertension, vol. 6, Epidemiology of hypertension. Elsevier, Amsterdam
Labouret G, Achimastos A, Benetos A, Safar M, Housset E (1983) L'hypertension arterielle systolique des amputes traumatiques. Presse Med 12:1349–50
Levenson JA, Simon AC, Safar Me, Fiessinger JN, Housset EM (1982) Systolic hypertension in arteriosclerosis obliterans of the lower limbs. Clin Exp Hypertens 4:1059–72
Meade TW, Gardner MJ, Cannon P, Richardson PC (1968) Observer variability in recording the peripheral pulses. Br Heart J 30:661–5
Miall WE, Greenberg G (1987) On behalf of The Medical Research Council's Working Party on Mild to Moderate Hypertension. Mild hypertension: Is there pressure to treat? Cambridge University Press, Cambridge
Prevention of Atherosclerotic Complications with Ketanserin Trial Group (1989) Prevention of atherosclerotic complications: controlled trial of ketanserin. Br Med J 298:424–30
Reid DD, Holland WW, Humerfelt S, Rose G (1966) A cardiovascular survey of British postal workers. Lancet i:614–8
Reunanen A, Takkunen H, Aromaa A (1982) Prevalence of intermittent claudication and its effect on mortality. Acta Med Scand 211:249–56
Rose GA, Blackburn H (1968) Cardiovascular survey methods. WHO, Geneva.
Schroll M (1981) Blood pressure as a cardiovascular risk factor in a 10-year prospective study of men and women born in 1914 and examined in 1964 and 1974 in Glostrup. Dan Med Bull 28:154–64
Schroll M, Munck O (1981) Estimation of peripheral arteriosclerotic disease by ankle blood pressure measurements in a population study of 60-year-old men and women. J Chron Dis 34:261–9
Shurtleff D (1974) An epidemiological investigation of cardiovascular disease. In: Kannel WB, Tavia Gordon T (eds) The Framingham Study. DHEW publication
Smith GD, Shipley MJ, Rose G (1990) Intermittent claudication, heart disease risk factors, and mortality. The Whitehall study. Circulation 82:1925–31

16 Diabetes Mellitus

R.J. JARRETT

The major difference, apart from frequency, between peripheral vascular disease in diabetic and non-diabetic individuals is the common accompaniment in the former of microvascular disease and neuropathy both of which affect the risk of amputation. It would probably be desirable to treat insulin dependent (IDDM) and non-insulin dependent diabetes mellitus (NIDDM) separately, but few studies discriminate between them, so this review will generally deal with both types of diabetes with specific references to IDDM or NIDDM where possible and appropriate.

Diabetes and Clinical Manifestations of Peripheral Vascular Disease

Intermittent Claudication

The only general population-based study is that in Framingham, based on relatively small numbers of diabetic subjects (Kannel and McGee 1979). For those with diabetes age-adjusted incidence rates over 20 years were 12.6 per 1000 person-years (men) and 8.4 per 1000 person-years (women). For those without diabetes, rates were 3.3 and 1.3 per 1000 person-years, respectively. The Israeli Ischaemic Heart Disease Study (Herman et al. 1977) in male civil servants over a 5-year follow-up reported a doubling of incidence rates in both previously diagnosed diabetic men and in those diagnosed by glucose tolerance test in the survey.

A study in Finland (Siitonen et al. 1986) compared the prevalence of intermittent claudication in newly-diagnosed subjects with NIDDM and controls randomly selected from the same population base. Although prevalence was higher in the diabetics the difference did not reach statisical significance, although again numbers were small, particularly when divided by gender. However, in the five years following the original observations, the age-adjusted incidence of intermittent claudication was significantly higher both in diabetic men (20.3%) and women (21.8%) compared with non-diabetic men (8.0%) and women (4.2%) (Uusitupa et al. 1990).

In another general population-based Finnish study (Reunanen et al. 1982) of persons aged 30–59 years, the age-adjusted prevalence of intermittent claudication was 3.4 times higher in diabetic men and 5.7 times higher in diabetic women.

Pulse Deficits

The Rochester, Minnesota, population study included 1073 diabetics aged 30 years or more and free of peripheral vascular disease at the time of diagnosis (Melton et al. 1980). The incidence of pulse deficits was 21.3/1000 person-years for men and 17.6/1000 person-years for women. There was no comparison with a control population.

The Kristianstad, Sweden, survey (Nilsson et al. 1967) compared the prevalence of pulse deficits in controls, short-term diabetics (average duration 1.5 years) and long-term diabetics (average duration 20 years). Above the age of 40 years, pulse deficits were more frequent in both groups of diabetics, but a consistent effect of duration was only apparent in women.

Lower Extremity Amputations

Amputations in diabetics may be due to neuropathy and infection as well as or in addition to peripheral vascular disease. Population-based data are few and are subject to reservations as they mostly depend upon routinely collected data, principally hospital discharge data, which represent events and not patients and may underrepresent diabetes, which is often not recorded on the form. One study which has avoided these problems is that in the Pima Indian population in Arizona (Nelson et al. 1988). This population has the highest recorded incidence of diabetes which is almost exclusively NIDDM. Between 1972 and 1984, from a study population of 4399 subjects, lower extremity amputations were performed in 84 patients, 80 of whom had NIDDM. Incidence rates were compared with those estimated by Most and Sinnock (1983) for six US counties (Table 16.1) using hospital discharge data. The age–sex adjusted rate in diabetic Pima Indians was 24.1/1000 person-years compared with 6.5/1000 person-years in the known diabetics reported by Most and Sinnock. To what extent the difference is real or apparent, due in part to underreporting in the six country series, is speculative. Differences in amputation rates may also, in part, reflect differences in health care (Waugh 1988).

Another difference between the two studies which might affect the comparison was the type of amputation. In the Pima Indians 84% were amputations of toes compared with 30.5% in the six countries study.

In the Pima Indians rates were significantly higher in men and were independently associated both with age and with duration of diabetes. Rates were also higher in men in the six countries study, the overall male/female ratio being 1.4.

Waugh (1988) used hospital discharge data plus several other data sources and was able to estimate person-based rates of major amputations (excluding toes) for the Tayside Area (Scotland) for 1980–1982. Depending on the data source, the proportion of diabetic amputees varied from 25%–35% and the rates (per

Table 16.1. Incidence rates of lower extremity amputations (per 1000 person-years) for Pima Indian diabetics and US diabetics in the six county study. Rates refer to hospitalisations

Age (years)	Sex	Pima	US
5–44	M	14.9	2.7
	F	6.3	1.0
45–64	M	29.4	6.3
	F	14.6	3.8
65+	M	57.5	13.2
	F	22.2	9.1

Sources: Most and Sinnock (1983); Nelson et al. (1988).

10 000 per year) were 46–70 times higher in diabetics, the denominator for the latter itself being an estimate.

Diabetes and Pathological Manifestations of Peripheral Vascular Disease

Atheroma

In the International Atherosclerosis Project (Robertson and Strong 1968) diabetes was inconsistently associated with the extent of fatty streaks in the abdominal aorta, but was significantly associated with the extent of raised lesions. In the WHO collaborative autopsy study (Ždanov and Vihert 1976) the extent of atherosclerotic lesions in the aorta tended to be greater in diabetics, particularly in those dying below age 50 years. In an autopsy study of persons dying above the age of 60 years, Sternby (1968) found no clear excess of aortic atherosclerotic lesions in diabetics, but did observe that lesions in the internal iliac and femoral arteries were more severe in diabetics. More frequent involvement of arteries below the knee in diabetics has been shown in pathological studies of amputation specimens and by arteriographic and ultrasound investigation in intact limbs (Strandness et al. 1964; Conrad 1967; Haimovici 1967).

Medial Calcification

Calcification of the arterial media (Monckeberg's sclerosis) is a focal calcification, particularly of medium-sized arteries, which can be identified using radiography. Previous clinical studies have shown an association with diabetes (Ferrier 1964) and this has been confirmed in a prospective study in Pima Indians (Everhart et al. 1988) (Table 16.2). There is a strong association between the prevalence of neuropathy and that of medial calcification (Edmonds et al. 1982; Everhart et al. 1988) and it is possible that neuropathy predisposes to or causes the calcification

Table 16.2. Age-adjusted prevalence (per cent) of medial calcification in Pima Indians

Site	Men		Women	
	Diabetic	Non-diabetic	Diabetic	Non-diabetic
Feet	37.9	20.8	24.6	9.3
Calf	28.9	17.1	19.4	7.7
Thigh	25.7	15.6	15.7	4.4
Hand	22.9	8.8	14.1	5.3
Any	39.5	24.7	27.7	10.7

Source: Everhart et al. (1988)

(Goebel and Füssl 1983). In the Pima Indians, medial calcification was a strong predictor of subsequent amputation (Nelson et al. 1988).

Risk Factors for Peripheral Vascular Disease in Diabetes

There are few prospective studies in which putative risk factors have been examined in relation to the various manifestations of peripheral vascular disease. In the University Group Diabetes Program (UGDP) study (Kreines et al. 1985) a clinical trial of patients with NIDDM, the incidence rates of amputations, intermittent claudication and non-palpable dorsalis pedis pulses, respectively, were recorded up to 14 years of follow-up. Calcification was also recorded, but no distinction made between intimal and medial varieties and will not be further considered here. Neither baseline blood pressure nor total serum cholesterol were related to any of the several manifestations of peripheral vascular disease. Baseline serum triglyceride was significantly related to intermittent claudication in men. Cigarette smoking was not recorded at the baseline examination. No difference in incidence rates was observed between the three randomised treatment groups despite substantially better blood glucose control in that assigned to the variable insulin regime.

In the Pima Indian prospective study (Nelson et al. 1988) the incidence of amputation was analysed in relation to baseline and subsequent factors. Of these, significant associations occurred for medial arterial calcification, retinopathy, nephropathy, absence of patellar tendon reflexes, impaired great toe vibration – perception threshold and degree of fasting and 2-hour post-load glycaemia. Serum cholesterol, blood pressure, age, and absence of Achilles tendon reflexes were not significant predictors. The Finnish study referred to earlier also looked at risk factors (Uusitupa et al. 1990). In the combined sexes the incidence of intermittent claudication was significantly related to baseline values of fasting plasma insulin, serum total cholesterol, VLDL cholesterol, total, LDL and VLDL triglycerides (all positive) and HDL cholesterol (negative). Cigarette smoking was also significantly more common in those who developed claudication.

In their prevalence study the only significant association in the newly diagnosed diabetics was of low HDL cholesterol and absent foot pulses in diabetic men (Siitonen et al. 1986).

Welborn et al. (1984) reported a study using stepwise logistic regression analysis for the association of risk variables in a population estimated to be 70% of available diabetics in areas of Western Australia. For peripheral vascular disease in IDDM no variables showed significant associations. In NIDDM significant associations were found for age, plasma creatinine, glycated haemoglobin, age at onset and the difference between diastolic blood pressure erect and supine. In this study peripheral vascular disease comprised intermittent claudication and/or pulse deficits.

Paisey et al. (1984) studied 503 clinic-based Mexicans with NIDDM. Peripheral vascular disease (intermittent claudication with or without pulse deficits) was significantly related to duration of diabetes and to cigarette smoking, but not to blood pressure or plasma lipids.

Janka et al. (1980) studied a German clinic-based population which included both types of diabetes. They used Doppler ultrasonic techniques to identify the presence and position of arterial disease and divided this into proximal (femoro-popliteal) and distal (below-knee) varieties. After age adjustment, proximal disease was significantly associated with blood pressure, duration of diabetes, serum cholesterol and serum triglycerides. However, the peripheral variety was associated only with blood pressure and duration of diabetes, although additionally with blood glucose levels.

The WHO Multinational Study (West et al. 1983) included clinic – and population – based cohorts of diabetics ages 35–54 years. The prevalence of peripheral vascular disease (amputation and/or intermittent claudication) was significantly associated with duration of diabetes and age, but not with blood pressure and plasma lipids.

Conclusions

From the foregoing it can be seen that population-based and prospective data are sparse. Furthermore, most data relate to people with NIDDM specifically or to mixed populations which are likely to be predominately composed of people with NIDDM. In NIDDM the data do support the conventional belief that peripheral vascular disease due to atherosclerosis is more common than in non-diabetics. However, intermittent claudication has been reported as a predictor of subsequent diabetes in two prospective studies (Medalie 1979; Wilson et al. 1986) and it may be that, as for coronary heart disease, NIDDM tends to occur in people at increased risk for peripheral atherosclerosis (Jarrett 1989). Nevertheless diabetes may have an effect of its own because arterial sclerosis and below-knee occlusions are more frequent in NIDDM.

In the six counties study (Most and Sinnock 1983) in patients aged <45 years (and therefore likely to have IDDM) 56.5% of amputations were of toes, which may indicate neuropathy as the principal underlying cause. However, there are convincing data indicating a strong association between renal disorder and excess coronary heart disease associated with atherosclerosis in this type of diabetes (Jarrett 1989), and so it is likely that peripheral vascular disease would also be more frequent in such patients, if not in those who preserve normal renal function.

References

Conrad MC (1967) Large and small artery occlusion in diabetics and non-diabetics with severe vascular disease. Circulation 36:83–91

Edmonds ME, Morrison N, Laws JW, Watkins PJ (1982) Medial arterial calcification and diabetic neuropathy. Br Med J 284:928–30

Everhart JE, Pettitt DJ, Knowler WC, Rose FA, Bennett PH (1988) Medial arterial calcification and its association with mortality and complications of diabetes. Diabetologia 31:16–23

Ferrier TM (1964) Radiologically demonstrable arterial calcification in diabetes mellitus. Aust Ann Med 13:222–8

Goebel FD, Füssel HS (1983) Monckeberg's sclerosis after sympathetic denervation in diabetic and non-diabetic subjects. Diabetologia 24:347–50

Haimovici H (1967) Patterns of arteriosclerotic lesions of the lower extremity. Arch Surg 95:918–22

Herman JB, Medalie JH, Goldbourt U (1977) Differences in cardiovasular morbidity and mortality between previously known and newly-diagnosed adult diabetics. Diabetologia 13:229–34

Janka HU, Standl E, Mehnert H (1980) Peripheral vascular disease in diabetes mellitus and its relation to cardiovascular risk factors: screening with the Doppler ultrasonic technique. Diabetes Care 3:207–13

Jarrett RJ (1989) Epidemiology and public health aspects of non-insulin-dependent diabetes mellitus, Epidemiol Rev 11:151–71

Kannel WB, McGee DL (1979) Diabetes and cardiovascular disease. JAMA 241:2035–8

Kreines K, Johnson E, Albrink M et al. (1985) The course of peripheral vascular disease in non-insulin-dependent diabetes. Diabetes Care 8:235–43

Medalie JH (1979) Risk factors other than hyperglycemia in diabetic vascular disease. Diabetes Care 2:77–84

Melton LJ, Macken KM, Palumbo PJ, Elveback LR (1980) Incidence and prevalence of clinical peripheral vascular disease in a population-based cohort of diabetic patients. Diabetes Care 3:650–4

Most R, Sinnock P (1983) The epidemiology of lower extremity amputations in diabetic individuals. Diabetes Care 6:87–91

Nelson RG, Gohdes DM, Everhart J et al. (1988) Lower extremity amputations in NIDDM: 12-year follow-up in Pima Indians. Diabetes Care 11:8–16

Nilsson SE, Nilsson JE, Frostber N, Emilsson T (1967) The Kristianstad survey II. Acta Med Scand (Suppl) 469:1–42

Paisey RB, Arredondo G, Villalobos A, Lozano O, Guevara L, Kelly S (1984) Association of differing dietary, metabolic, and clinical risk factors with macrovascular complications of diabetes: a prevalence study of 503 Mexican type II diabetic subjects. Diabetes Care 7:421–7

Reunanen A, Takkunen H, Aromaa A (1982) Prevalence of intermittent claudication and its effect upon mortality. Acta Med Scand 211:249–56

Robertson WB, Strong JP (1968) Atherosclerosis in persons with hypertension and diabetes mellitus. Lab Invest 18:538–51

Siitonen O, Uusitupa M, Pyörälä K, Voutilainen E, Lansimies E (1986) Peripheral arterial disease and its relationship to cardiovascular risk factors and coronary heart disease in newly diagnosed non-insulin-dependent diabetics. Acta Med Scand 220:205–12

Sternby NH (1968) Atherosclerosis in a defined population. An autopsy study in Malmo, Sweden. Acta Pathol Microbiol Scand (Suppl) 194:158

Strandness DE, Priest RE, Gibbons GE (1964) A combined clinical and pathological study of diabetic and non-diabetic peripheral arterial disease. Diabetes 13:366–72

Uusitupa MIJ, Niskanen LK, Siitonen O, Voutillainen E, Pyörälä K (1990) 5-year incidence of atherosclerotic vascular disease in relation to general risk factors, insulin level, and abnormalities in lipoprotein composition in non-insulin-dependent diabetic and non-diabetic subjects. Circulation 82:27–36

Waugh NR (1988) Amputations in diabetic patients – a review of rates, relative risks and resource use. Commun Med 10:279–88

Welborn TA, Knuiman M, McCann V, Stanton K, Constable IJ (1984) Clinical macrovascular disease in caucasoid diabetic subjects: logistic regression analysis of risk variables. Diabetologia 27:568–75

West KM, Ahuja MMS, Bennett PH et al. (1983) The role of circulating glucose and triglyceride concentrations and their interactions with other risk factors as determinants of arterial disease in nine diabetic population samples from the WHO Multinational Study. Diabetes Care 3:361–9

Wilson PWF, Anderson KM, Kannel WB (1986) Epidemiology of diabetes mellitus in the elderly; the Framingham Study. Am J Med: 80 (Suppl) 5A:3–9

Ždanov VS, Vihert AM (1976) Atherosclerosis and diabetes mellitus. Bull WHO 53:547–53

SECTION IV

Social and Life Style Factors

SECTION IV

Social and Life Style Factors

17 Social Factors

C.C.A. MACINTYRE and V.D.L. CARSTAIRS

Social Factors and Health

The association between social factors and health state has been well documented over the last decade with attention focusing on social class and mortality, but with other social variables and measures of morbidity also being considered. Strong gradients are found in mortality for most causes of death (Office of Population Censuses and Surveys and Registrar General, Scotland 1986). The Longitudinal Study has identified gradients also by tenure, number of rooms, car ownership, level of qualifications (Fox and Goldblatt 1982) and unemployment (Moser et al. 1987).

Morbidity data are less susceptible to analysis since routine health records fail to capture occupational data on a comprehensive basis. Cancer registration material has nevertheless been subject to analysis with some, but not all, tumour sites exhibiting increasing rates with descent down the social scale (Leon 1988). The General Household Survey data also provides evidence of higher levels of (self-reported) long-standing and acute illness and restricted days of activity in manual as compared with non-manual groups (Office of Population Censuses and Surveys, General Household Survey, Series GHS (annual reports)).

In addition to analysis by individual social variable an area measure of deprivation derived from 1981 census data provides similar evidence as shown in the gradient for all-causes mortality for Scotland (Carstairs and Morris 1991) (Table 17.1). The deprivation score is an aggregate measure for all households in a postcode sector and is based on the combination of four variables: men unemployed, overcrowded housing (more than one person per room), households without a car and household heads in semi- or unskilled/manual occupation.

We have used this latter approach to study the relationship between social deprivation and mortality and hospital discharges due to peripheral vascular disease. Also, we have had the opportunity to study the association between morbidity in peripheral vascular disease and social factors, and to compare the associations with different aspects of social status by analysing a community study of peripheral vascular disease – the Edinburgh Artery Study.

Table 17.1. Social deprivation and mortality and hospital discharge rates due to peripheral vascular disease

	Deprivation categories[a]						
	Most affluent					Most deprived	
	1	2	3	4	5	6	7
Mortality 1980–1985 (all ages; standardised mortality ratio)							
All cause mortality	83	89	95	100	107	113	125
Peripheral vascular disease (other) (ICD 443) (*n*=1080)	61	97	102	106	97	99	132
Scottish Hospital discharge data for peripheral vascular disease (standardised discharge rate)							
All specialities (*n*=5806)	68	92	98	100	100	110	140
General medicine (*n*=2765)	67	87	93	100	104	122	141
Surgery and orthopaedics (*n*=1594)	76	97	95	106	112	83	126
All specialities males (*n*=3654)	70	90	93	98	102	120	142
females (*n*=2152)	64	96	106	104	97	91	135

[a] Key to grouping of deprivation score into categories (1: <-5, 2: $-5<-3$, 3: $-3<-1$, 4: $-1<+1$, 5: $+1<+3$, 6: $+3<+6$, 7: $>+6$).

Social Deprivation and Peripheral Vascular Disease: Mortality and Hospital Discharges

In order to examine the gradient between deprivation and mortality from peripheral vascular disease the deaths within the International Classification of Diseases (ICD) category "Other Peripheral Vascular Disease" (ICD 443) were considered over the period 1980–1985; there were 1080 deaths in this category. Standardised mortality ratios (SMRs) were calculated for deprivation divided into seven categories (Table 17.1). There was very little evidence of a trend, although the two extremes were as anticipated.

A further source of data, Scottish hospital in-patient discharge data, was then examined for evidence of a trend between discharge rates and peripheral vascular disease and deprivation. The analysis examined all discharges (*n* = 5806), other than discharges from geriatric units, over the period 1980–1982 coded to peripheral vascular disease (defined as ICD 440.2, 440.9, 443.9), as either primary or secondary diagnosis. Standardised discharge ratios (SDR) were calculated for deprivation categories using the Scottish population data classified by age and sex (Table 17.1). There was evidence of a trend with deprivation with an SDR rising from 68 in the most affluent category to 140 in the most deprived category.

This pattern was also evident when discharges from general medicine were examined. The pattern was less clear when discharges for surgery and orthopaedics were considered, although part of the explanation may be small numbers, as there were only 1594 discharges in the latter category.

Measurement of Social Factors in Edinburgh Artery Study

The Edinburgh Artery Study provided an opportunity to examine the association between morbidity from peripheral vascular disease and social status. A total of 1592 subjects aged 55–74 years from the Edinburgh population, were examined in a cross-sectional survey of peripheral vascular disease. Details have been published elsewhere (Fowkes et al. 1991 and in Chapter 8), but briefly, an age–sex stratified random sample was taken from the age–sex registers of 10 general practices chosen to cover the socioeconomic and geographic spread of the city.

One measure of peripheral vascular disease used was the ankle-brachial pressure index (ABPI), the ratio of systolic blood pressure in the ankle and the arm, although for ease of presentation the results are scaled up by 100. This non-invasive measure appears to correlate (negatively) with disease, although it has not been used extensively in studies of the general population (Fowkes 1988). The measures of social factors recorded were social class, highest educational level and deprivation.

Information for social class classification and educational levels was recorded on a questionnaire completed at a visit to the study clinic. Social class was coded using the instructions of the Office of Population Censuses and Surveys. The occupation of the husband was used for female married subjects, and retired subjects were coded according to their last main occupation. Education distinguished three levels: college or university education, other post-school qualifications, secondary or primary education only. Deprivation was determined, as previously described, from the current home address of the subject by assigning the deprivation score for the postcode sector.

The distribution of social class and deprivation in the study sample was compared with the estimates from the Edinburgh population derived from the 1981 Census. There was an underrepresentation of social classes IV and V which is also suggested by the lower deprivation index in the sample compared with the Edinburgh population (Edinburgh mean −0.110; study mean −1.63).

Statistical Analysis

The data were analysed separately for males and females due to the different interpretation which is put on social class in females. The disadvantage of this approach is that the amount of data available for examining associations is halved, but there was still sufficient data.

Multiple linear regression with ABPI as the outcome was used for the analysis. The associations between each factor and the ABPI were determined by giving estimates of the difference between the mean ABPI in each category compared with a baseline category. These estimates were adjusted for age, as this was a potential confounding factor due to the strong association between ABPI and age and possible association between age and social status.

The association with deprivation was estimated using seven categories, but since deprivation provides a quantitative measure the relationship was also summarised by the estimated change in mean ABPI associated with a unit change in deprivation. A formal test of interaction was carried out to assess whether the associations between ABPI and social factors were different in females and males.

The graphical representation of associations used estimates of population means (SAS/STAT 1988). These were derived for each social factor separately, and are estimates of the mean ABPI in each category for subjects with age adjusted to the average age of the study sample.

Ankle-Brachial Pressure Index and Social Class, Education and Deprivation

Individual Social Factors

There was no substantial difference in the distribution of social factors between males and females except that males had a slightly higher educational level (Table 17.2). The distribution of ABPI in relation to social factors in males and females is illustrated in Fig. 17.1. The mean ABPI in the sample was 103 with a standard deviation of 18. There was a tendency towards a higher ABPI in males, but this was probably due to a positive association between ABPI and height (Fowkes et al. 1991).

In males there was evidence of a trend of lower ABPI (more disease severity) with less educational achievement, higher deprivation score and lower social class although the patterns in social class V appeared out of line.

These associations were not evident in females, and tests of interaction suggested a difference between males and females (Table 17.3). Again, the social class V group appeared to have ABPIs which were higher than expected, given the level in the other social classes. This pattern with social class may have been a result of the responders in this group tending to be healthier, or alternatively since the group was very small, it could be explained by random variation.

Combinations of Social Factors

Models including two of the three social factors were examined to investigate whether different aspects of social status are described by these measures. This was done for males only due to the weak relationships noted in females. The results of various models are presented in Table 17.4.

In significance tests of association between social class and the ABPI, social class was considered as a quantitative variable (equally spaced). This imposed a linear

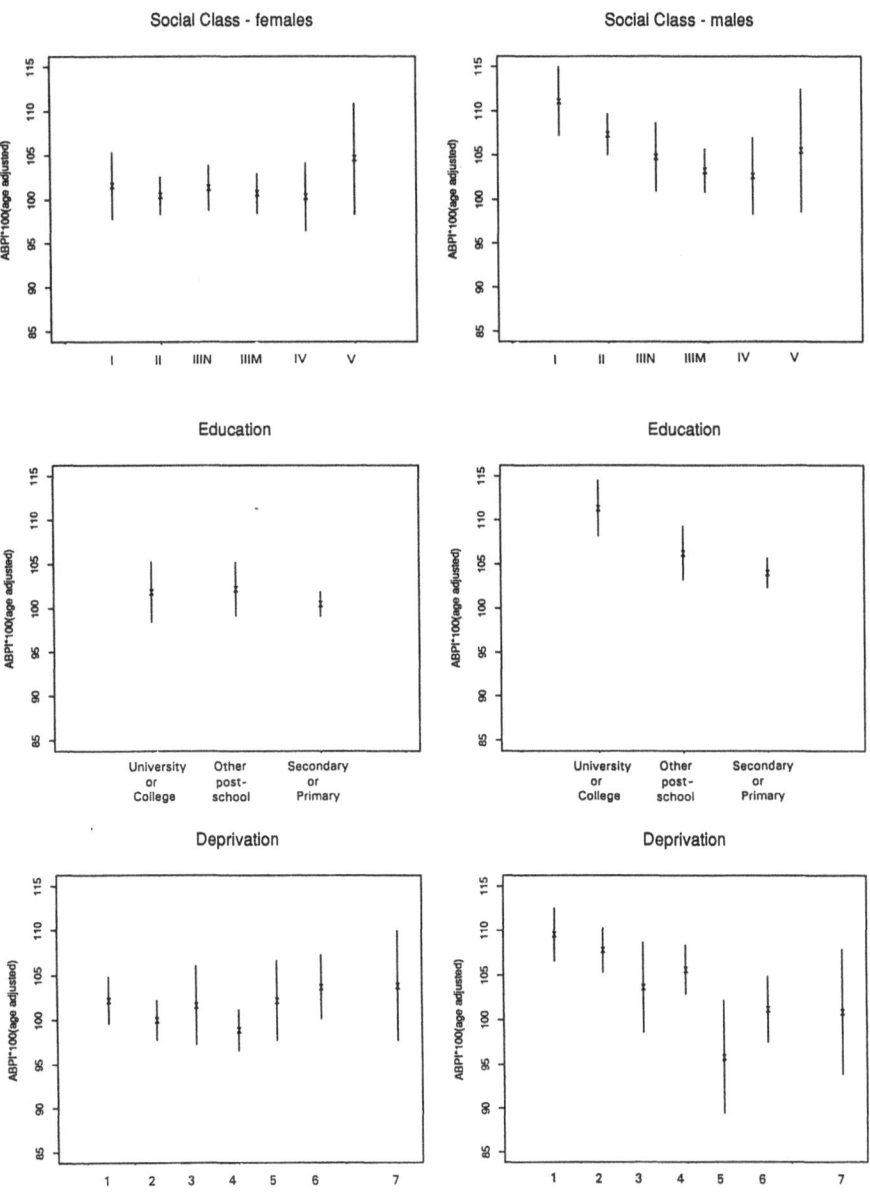

Fig. 17.1. Relationship between ankle-brachial pressure index (ABPI) and social factors in females and males.

relationship between ABPI and social class, and gave a more powerful comparison than the alternative which treats the social class categories as unordered.

As expected, adjustment of one social factor by either of the other social factors resulted in a reduction of the association with the ABPI. The association with social class was reduced considerably by either deprivation or education and

Table 17.2. Distribution of social factors in males and females in the Edinburgh Artery Study

	Females		Males	
	n	(%)	n	(%)
Social class				
I	74	(10)	94	(12)
II	240	(31)	265	(33)
III N	159	(21)	95	(12)
III M	207	(27)	245	(30)
IV	70	(9)	78	(10)
V	26	(3)	30	(4)
Total	776		807	
Highest educational level				
University/College	90	(12)	142	(18)
Other post school qualification	115	(15)	154	(19)
Primary or secondary	574	(74)	511	(63)
Total	779		807	
Deprivation category (DEPCAT)				
1 Affluent	149	(19)	154	(20)
2	215	(28)	234	(29)
3	54	(7)	55	(7)
4	201	(26)	192	(24)
5	52	(7)	35	(4)
6	83	(11)	103	(13)
7 Deprived	27	(4)	29	(4)
Total	781		802	
Deprivation score mean (standard deviation)	−1.61	(3.61)	−1.65	(3.73)

although non-significant a possible trend remained. When the association with social class was adjusted for both education and deprivation simultaneously, this possible trend almost disappeared.

The association between education and ABPI, independent of social class and deprivation, was statistically non-significant, but there was a suggestion that a college or university education was associated with higher ABPI.

There was strong evidence of an association between deprivation and ABPI which was independent of social class and education. Adjustment by social class and education reduced the change in mean ABPI associated with a unit increase in deprivation score from −0.823 to −0.678.

The model containing all three social factors suggested that the variation between ABPI and social class could be adequately explained by the two factors, highest educational level and deprivation score.

Effect of Cardiovascular Risk Factors

The extent to which the variation in the ABPI with social factors could be explained by some of the conventional cardiovascular risk factors was investigated by including deprivation and education into a regression model with smoking, diabetic status, alcohol intake, height, high density lipoprotein (HDL) cholesterol and non-HDL cholesterol.

Table 17.3. Ankle-brachial pressure index (ABPI) and social factors in males and females ($n=562$)[a]

Social class		
Estimated difference in mean ABPI (SE) between males and females of social class IIIM (male–female)		2.91 (1.68)
Estimated difference in mean ABPI (SE) compared with social class IIIM: males and females separately	*Females*	*Males*
I	1.74 (2.42)	7.93 (2.18)
II	−0.35 (1.70)	4.15 (1.59)
IIIN	0.54 (1.89)	1.42 (2.16)
IIIM	—	—
IV	−0.51 (2.51)	−0.49 (2.36)
V	3.89 (3.69)	2.41 (3.50)
		($p=0.006$)[b,c]
Highest educational level		
Estimated difference in mean ABPI (SE) between males and females with at most secondary education		3.84 (1.09)
Estimated difference in mean ABPI (SE) compared to those with at most secondary education: males and females separately	*Females*	*Males*
University/college	1.24 (2.02)	7.40 (1.69)
Other post-school qualifications	1.50 (1.82)	2.42 (1.64)
At most secondary school	—	
		($p=0.07$)[b]
Deprivation		
Estimated difference in mean ABPI (SE) between males and females with deprivation of 0		3.72 (0.98)
Estimated change in mean ABPI (SE) for a unit change in deprivation	*Females*	*Males*
	0.092 (0.177)	−0.826 (0.168)
		($p<0.001$)

[a] Based on regression models adjusted for age.
[b] Giving statistical significance of interaction between social factors and sex.
[c] Using social class as a quantitative variable.

A model was fitted with all the above risk factors and deprivation and education. The change in coefficients of deprivation and education when each of the risk factors was removed from the full model was examined in order to assess the extent to which these risk factors explained the variation in the ABPI with social status (Table 17.5). Removal of smoking from the model resulted in a substantial increase in the deprivation and education coefficients, suggesting that some of the association between deprivation and education and ABPI could be explained by smoking. There was very little effect of removing diabetes, alcohol, lipids or height from the model so there was little evidence in these data that current levels of these factors explained the difference with social status.

Explaining Associations of Social Factors and Peripheral Vascular Disease

This analysis of a very specific aspect of health state has produced evidence of association with the three social variables which is consistent with the general

Table 17.4. Ankle-brachial pressure index (ABPI) and each social factor after adjustment for the other social factors in males $(n=797)$[a]

Social class

	I	II	IIIN	IV	V	Significance level
Difference in mean ABPI (SE) from social class IIIM	7.85 (2.33)	4.08 (1.70)	1.50 (2.30)	−0.61 (2.50)	2.25 (3.74)	<0.001[b]
Adjusted for education	3.45 (3.30)	2.63 (1.94)	1.46 (2.30)	−0.53 (2.51)	2.37 (3.75)	0.24[b]
Adjusted for deprivation	5.08 (2.47)	1.97 (1.80)	0.65 (2.30)	0.02 (2.49)	4.12 (3.76)	0.17[b]
Adjusted for education and deprivation	1.29 (3.35)	0.97 (2.00)	0.61 (2.30)	0.01 (2.50)	4.09 (3.76)	0.99[b]

Highest education level	University/college	Other post-school qualification	Significance level
Difference in mean ABPI (SE) from at most secondary school education	7.36 (1.81)	2.24 (1.76)	<0.001
Adjusted for social class	5.13 (2.72)	1.04 (2.00)	0.16
Adjusted for deprivation	5.04 (1.92)	0.21 (1.85)	0.03
Adjusted for social class and deprivation	4.42 (2.71)	−0.02 (2.02)	0.20

Deprivation		Significance level
Change in mean ABPI (SE) associated with a unit change in deprivation	−0.823 (0.180)	<0.001
Adjusted for social class	−0.683 (0.208)	<0.001
Adjusted for education	−0.674 (0.197)	<0.001
Adjusted for social class and education	−0.678 (0.210)	<0.001

[a] Based on linear regression model adjusted for age.
[b] Using social class as a quantitative variable.

picture of differentials in health. The lack of gradient for mortality, while surprising, is probably because people suffering from peripheral vascular disease mostly die from other causes.

It is interesting that the gradients with social class are only demonstrated amongst the males in the study. Given that the coding of social class for females is carried out using the male spouse's occupation, this may not be surprising. However, there is also a lack of association with both highest educational level and deprivation.

An analysis of hospital discharge data separately for males and females (Table 17.1) also suggests a gradual trend in males rising from an SMR of 70 to 142, but an uneven pattern in females, giving more evidence of a different pattern in males and females. One possible explanation for this finding could be that the women have less variation in the extent of disease and so it is more difficult to demonstrate association, but there is no suggestion from the community survey

Table 17.5. Effect of cardiovascular risk factors on associations between ankle-brachial pressure index (ABPI) and social factors in males ($n=769$)

	Highest educational level: difference in mean ABPI compared to those with at most secondary education		Deprivation: change in mean ABPI for a unit increase in deprivation
	University/college	Other post-school qualification	
Model with age, deprivation, education	4.94	−0.19	−0.704
Full model with age, deprivation, education, smoking, diabetes, alcohol, height, non-HDL cholesterol, HDL cholesterol	2.27	−1.41	−0.416
Full model excluding smoking	4.03	−0.21	−0.653
,, diabetes	2.28	−1.25	−0.439
,, alcohol	2.23	−1.46	−0.431
,, height	2.77	−1.24	−0.470
,, non-HDL cholesterol	2.56	−1.67	−0.369
,, HDL cholesterol	2.45	−1.36	−0.438

that the variability in ABPI is less in women, with a standard deviation of 3.6 compared with 3.7 in males. Alternatively, it may be that the social facets of employment, formal education and material deprivation reflected in these measures are not so strongly associated with health of women.

The three social factors are all surrogates for life style factors rather than direct measures themselves. They are clearly interrelated, but education and deprivation independently explain variation in ABPI, whereas social class does not explain any more variation in addition to these other measures. A consequence of this is that social status could be studied as a two-dimensional factor rather than one summary measure contained in social class.

In order to shed some light on a biological explanation of the associations between social status and ABPI, some of the risk factors for cardiovascular disease were considered. (It was not possible to consider blood pressure due to the fact that this is part of the outcome measure ABPI and therefore would have a spuriously strong relationship with it.) Of the subset considered (smoking, diabetes, alcohol, height, HDL and non-HDL cholesterol) smoking was the only factor which appeared to contribute to the explanation. It should be borne in mind that this was a cross-sectional study and so the lack of explanation may be due to temporal alterations in life style factors, and so current levels of risk factors are not necessarily the most appropriate to consider. Data on rheological factors, diet and exercise have also been collected in the Edinburgh Artery Study, and so it will be of interest to examine how much of the variation with social status can be explained by these factors.

References

Carstairs V, Morris R (1991) Deprivation and health in Scotland. Aberdeen University Press, Aberdeen

Fowkes FGR (1988) The measurement of atherosclerotic peripheral arterial disease in epidemiological surveys. Int J Epidemiol 17:248–54

Fowkes FGR, Housley E, Cawood EHH, Macintyre CCA, Ruckley V, Prescott RJ (1991) Edinburgh Artery Study: Prevalence of asymptomatic and symptomatic peripheral arterial disease in the general population. Int J Epidemiol 20:384–92

Fox AJ, Goldblatt PO (1982) Sociodemographic mortality differentials: longitudinal study 1971–75, LS No. 1. HMSO, London

Leon D (1988) Longitudinal Study 1971–75, Social distribution of cancer, OPCS series LS No. 3. HMSO, London

Moser KA, Goldblatt PO, Fox AJ, Jones DR (1987) Unemployment and mortality: comparison of the 1971 and 1981 longitudinal study samples. Br Med J iv:86–90

Office of Population Censuses and Surveys and Registrar General, Scotland (1986) Occupational mortality, decennial supplement 1979–80, 1982–83, DS No. 4. HMSO, London

Office of Population Censuses and Surveys, General Household Survey, Series GHS (annual reports). HMSO, London

SAS/STAT (1988) User's Guide. SAS Institute Inc, Cary NC, USA, p 600

18 Diet and Alcohol

P.T. DONNAN and M. THOMSON

In this chapter the relationships between diet, alcohol and peripheral vascular disease are discussed with reference to the Edinburgh Artery Study, a cross-sectional survey in the general population. In presenting the results from this study we will concentrate on the nested case–control study within the main survey. A fuller assessment of diet and alcohol in relation to peripheral arterial disease in the general population is planned in the future.

There is an extensive literature on diet and plasma lipids, and the relationship has been examined in a consensus conference (1985). The evidence relating dietary fats to serum cholesterol and coronary heart disease has been recently reviewed (LaRosa et al. 1990). In addition, there is a limited literature on diet and coronary heart disease mortality (Morris et al. 1977; Yano et al. 1978; Gordon et al. 1981; Shekelle et al. 1981; Kushi et al. 1985). In comparison, the literature on diet and peripheral vascular disease is sparse. Recently, Katsouyanni et al. (1991) published a case–control study concerning the association between diet and peripheral vascular disease. They found that saturated fatty acids, proteins and dietary cholesterol were significantly associated with increased risk of peripheral vascular disease, while polyunsaturated fatty acids and crude fibre significantly reduced risk. They concluded that substitution of olive oil for saturated fats and/or the consumption of high levels of vegetables and fruit and other fibre containing foods may help to explain the low occurrence of atherosclerotic disease in Mediterranean countries.

Prior to the study by Katsouyanni et al. (1991) the limited publications on diet and peripheral vascular disease have been concerned with dietary interventions to alter lipid levels in patients (Hutchinson et al. 1983; Brown et al. 1984; Heller et al. 1989) and the effect of fish oil on blood viscosity in peripheral vascular disease has also been investigated (Woodcock et al. 1984).

Dietary Survey Methodology

The influence of nutrition on the genesis of chronic disease has become an important area of research, but work is hindered by the difficulty of measuring people's habitual diets. Garrow (1974), for example, has observed that "The measurement of the habitual food intake of an individual must be among the most difficult tasks a physiologist can undertake".

Range of Methods

Diet assessment methods currently available include weighed food records, food consumption with estimated weight of food, the daily (24-hour) recall method, the diet history method, and the food frequency questionnaire (FFQ). Each method may have random or systematic errors and each has advantages and disadvantages in use.

The literature on dietary survey methodology is scattered and few textbooks consider methods in detail. However, two comprehensive reviews have recently been published (Bingham 1987; Cameron and Van Staverne 1988) and these describe available diet assessment methods and their accuracy. Biological markers of food consumption and their potential validatory use are discussed by Bingham (1987) who also makes recommendations on the selection of suitable methods for clinical and epidemiological research.

Diet assessment methods used in epidemiological research into coronary heart disease have ranged from weighed records of food intake (Keys 1970) to food frequency questionaires (Tunstall-Pedoe et al. 1989). Weighed food records have usually covered a 7-day period which is not necessarily the most appropriate way to assess the habitual dietary intake of all nutrients of all individuals within a population. However, this is usually sufficient to assess an individual's current energy intake and percentage energy derived from protein, fat and carbohydrate with an acceptable degree of precision. It also places individuals correctly at the very extremes of the distribution for more variable nutrients (Thomson et al. 1988). A single 24-hour record or recall should never be used to assess dietary status or to test associations between diet and other risk factors such as serum lipids (Marr 1971; Bingham 1987).

Food Frequency Questionnaires

Wiehl and Reed (1960) suggested that a questionnaire should be used in epidemiological studies of cardiovascular disease. "If groups of individuals can be clearly differentiated by use or non-use, frequent or infrequent use of selected food . . . then such characteristics can be tested for association with disease." A number of questionnaires were subsequently developed (Hankin et al. 1975; Jain et al. 1982; Yarnell et al. 1983). Quantitative estimation of their accuracy is not possible in many cases, because food frequency methods and questionnaires are designed to assess the intake of a specific nutrient or food, rather than intake of all nutrients and, despite their routine use in epidemiology, attempts to validate them are sparse. Yarnell et al. (1983) compared the results of a food frequency questionnaire with a 7-day weighed record in a Welsh population and found correlation coefficients for dietary intake ranged from 0.27 to 0.41, and 0.75 for alcohol. Clearly, agreement between FFQ and weighed intake (WI) is very good for alcohol and moderate for other nutrients.

The same questionnaire with some modifications for the alcohol intake was used in the Scottish Heart Health Study (SHHS) of coronary heart disease risk factors (Tunstall-Pedoe et al. 1989). Bolton-Smith and Milne (1990) validated the FFQ used in the SHHS and found correlations from 0.18 for sugar to 0.46 for fat, and 0.64 for alcohol. Whenever the percentage energy from nutrients or nutrient density was calculated, the correlation increased to 0.33 for protein to 0.71 for fat, and 0.68 for alcohol, indicating the necessity of adjusting for total energy intake (Willett et al. 1985). The better agreement of the FFQ used in the SHHS compared with Wales (Yarnell et al. 1983) may have been due to the age differences, the longer time interval of weighed diet assessment (14 days compared with 7 days), and the higher percentage of non-manual workers (84%) in the Scottish sample.

These few comparisons of the food frequency questionnaire and weighed intake suggest reasonable agreement. The WI method has flaws, but is considered to be the "gold standard" for assessing dietary intake of nutrients. An advantage of the FFQ is the ease of administration in large epidemiological surveys where WI is not feasible for such large numbers.

Edinburgh Artery Study and Diet: Methods

The Edinburgh Artery Study was a cross-sectional survey of 1592 men and women aged 55–74 years, selected from the age–sex registers of 10 general practices. These general practices had catchment populations spread geographically and socioeconomically throughout the city. The subjects were invited to a university clinic to complete a questionnaire and have a comprehensive medical examination. The response rate was 65% and follow-up of a sample of non-responders did not show any significant bias. Details of the study population, recruitment, and prevalence of peripheral vascular disease have been described previously (Fowkes et al. 1991; and in Chapter 8). The questionnaire included validated questions on social class, cardio-vascular history, intermittent claudication and angina (WHO questionnaire, Rose 1962), smoking history and alcohol consumption (Duffy 1985).

Selection of Cases and Controls

Among the 1592 participants of the cross-sectional study, 153 were identified as having definite peripheral arterial disease. These cases had either (a) at least *two* of the following criteria: intermittent claudication, ABPI ≤0.9, hyperaemic reduction in ankle pressure >20%; or (b) *one* of the following criteria: ABPI ≤0.7, hyperaemia reduction in ankle pressure >35%.

Subjects without vascular disease were identified from the remaining participants, and were defined as having none of the above criteria plus no evidence of angina or previous myocardial infarction on questionnaire nor ECG evidence of ischaemia. From this group of disease-free subjects a random sample of 153 controls was selected, who were frequency matched by age and sex with the diseased subjects.

Dietary Analysis

A version of the validated food frequency questionnaire developed for the Caerphilly Heart Disease Study (Yarnell et al. 1983) and applied by Tunstall-Pedoe et al. (1989) was used with some alterations to obtain nutrient information. This was completed by all who attended the clinic in the cross-sectional study, and a nurse was available to go through the questionnaire if there were any problems of interpretation.

Calculations of energy and nutrient intakes for individuals was estimated by multiplying the nutrient content of a typical portion size of specified food item by the frequency of consumption and summing over all food items. Where participants failed to complete all questions, nutrients were still calculated providing they had given a response to foods which were major sources of a specific nutrient. Hence, some nutrients may be underestimated. Table 18.1 indicates some of the individual food items which appeared in the questionnaire and the grouping of items, which was used for some analyses.

Data were analysed for individual food items, food groups, nutrient intake, and percentage energy from nutrients (including and excluding alcohol from the total), and nutrient density.

Nutrient density is the amount of a nutrient expressed per 1000 kilocalories (4.18 MJ). In comparing cases and controls it is necessary to adjust for total energy intake (Willett et al. 1985), and age and sex.

Logistic regression was used to assess dietary factors which are significantly associated with peripheral vascular disease using the statistical package BMDP (Dixon 1988) and to assess their independence from other factors such as social class, and smoking history. Smoking history was defined in terms of "pack-years" which is a measure of the amount smoked and the number of years as a smoker. This variable along with indicators of current smokers, recent ex-smokers (stopped in last 5 years) and smokers of pipe or cigar only, were included in the models. A square root transformation of pack-years was used to give a more symmetric distribution for the analyses. As the cases and controls were frequency matched by age and sex, all analyses were stratified by these factors (Schlesselmann 1982).

Diet in Cases vs. Controls

Food Frequencies

For the purposes of this analysis the frequency of consumption of individual foods was split into three categories; less than once a week, 1–3 times per week, and 4–7 times per week. There were significant linear trends across frequency categories

Table 18.1. Main individual foods and food groups from the food frequency questionnaire

Individual food	Food groups
Wholemeal	
Brown	
White	Bread
Crispbread and crackers	
Breakfast cereals	Breakfast cereals
Beef (including minced, beefburgers)	
Beef sausages	
Lamb	
Pork/Bacon/Ham	
Pork sausages	Meat and meat products
Chicken, Turkey	
Liver	
Heart	
Meat pies and pasties	
Fish	Fish
Vegetables (including chips)	Vegetables
Pasta/rice	Pasta/rice
Biscuits	
Cakes	
Ice-cream, tarts, puddings	Sweet foods
Sweets	
Apples	
Pears	
Oranges	Fresh fruit
Bananas	
Eggs	Eggs
Sugar	Sugar
Milk	Milk
Alcohol	Alcohol
Cream/cheese	Cream/cheese
Butter	Butter
Margarine	Margarine
Cooking oil	Cooking oil
Solid cooking fat	Solid cooking fat
Soft drinks	Soft drinks

with cases consuming white bread, pork/bacon/ham, beef sausages, meat pies and pasties, and solid cooking fat more frequently than controls. There were significant linear trends across the frequency categories with controls consuming more cornflakes/muesli, bran flakes, sweets/biscuits and bananas, apples, and pears than cases.

Some caution must be applied in interpreting these results because the problem of multiple significance testing. Nevertheless, they do indicate an association between meat and meat products and solid cooking fat with cases of peripheral vascular disease and fibre with freedom from atherosclerotic disease.

The individual foods were grouped as shown in Table 18.1. The controls showed higher mean levels of bread, breakfast cereals, pasta/rice, sweet foods, fruit, alcohol, cream/cheese, and eggs. After adjusting for total energy intake and age and sex, cases consumed significantly greater amounts of meat and solid cooking fat compared with controls (both p<0.01). Those free of disease had significantly higher levels of fruit (p<0.01), alcohol (p<0.01), pasta/rice (p<0.05), breakfast cereal (p<0.05) and cream/cheese (p<0.05).

Nutrients and Energy

Table 18.2 shows that cases of peripheral vascular disease have higher intakes of fat, saturated fat, protein, starch, sugar, energy, and cholesterol and lower intakes of alcohol, cereal fibre, vegetable fibre, vitamin C and β-carotene. After adjusting for age, sex and the difference in total energy intake which would have affected the levels of the nutrients, there remained statistically significant higher levels of alcohol (p=0.007) and cereal fibre (p=0.001) in those free of atherosclerotic disease. These results were more than would be expected by chance alone in any multiple comparison.

Table 18.2. Nutrients in cases of peripheral vascular disease and controls

Nutrient	Cases		Controls		p-value[b]
	Mean	(95% C.I.)[a]	Mean	(95% C.I.)[a]	
Energy (kcal)	2051	(1955,2147)	1947	(1880,2014)	
Protein (g)	82.9	(79.3,86.6)	78.8	(76.1,81.6)	0.91
Fat (g)	83.4	(79.2,87.9)	77.9	(74.6,81.3)	0.56
Saturated fat (g)	36.6	(34.5,38.7)	33.9	(32.3,25.5)	0.46
Polyunsaturated fat (g)	11.9	(11.1,12.8)	11.9	(11.1,12.7)	0.37
Starch (g)	141.2	(133.4,149.4)	136.6	(129.1,144.4)	0.32
Sugar (g)	82.3	(76.7,88.3)	78.5	(73.9,83.3)	0.67
Alcohol (g)	3.5	(2.9,4.3)	5.0	(4.0,6.1)	0.007
Cereal fibre (g)	9.9	(9.2,10.6)	11.1	(10.4,11.9)	0.001
Vegetable fibre (g)	11.2	(10.4,11.9)	11.1	(10.5,11.7)	0.62
Linoleic acid (g)	9.5	(8.8,10.3)	9.5	(8.8,10.3)	0.37
Vitamin C (mg)	56.3	(51.8,60.8)	58.5	(54.4,62.5)	0.33
Cholesterol (mg)	331	(309,355)	314	(297,323)	0.96
Retinol (μg)	675	(607,747)	653	(600,711)	0.63
β-Carotene (μg)	2823	(2423,3290)	3069	(2705,3482)	0.34
A-Tocopherol (mg)	10.4	(9.6,11.3)	10.9	(10.0,11.8)	0.13

[a] For positively skewed variables the geometric mean is given.
[b] Cases vs. controls adjusted for age, sex, and total energy.

When alcohol was included as a component of total energy the only differences between cases and controls were a higher percentage energy from fat and saturated fat in cases of peripheral vascular disease while controls had higher percentage energy from alcohol (Table 18.3). None of these differences were statistically significant at the 5% level. When alcohol was excluded as a component of total energy there were minor differences between cases and controls in the source of total energy. In comparing the nutrient density for fibre and vitamins there was a highly significant increased geometric mean nutrient density for cereal fibre in the controls compared to the cases.

Logistic regressions (Table 18.4) indicated reduced odds of disease for increases in percentage energy from carbohydrates, polyunsaturates, alcohol and cereal fibre, while increases in fat, saturated fat or protein were associated with increased odds of disease. After adjustment for total energy intake and age and

Table 18.3. Percentage energy/nutrient density (ND)[a] in cases of peripheral vascular disease and controls

	Cases Mean (SE)[b]	Controls Mean (SE)[b]	p-value
Fat	37.8% (0.5)	36.8% (0.5)	0.13
Saturated fat	16.5% (0.3)	15.9% (0.4)	0.14
Polyunsaturated fat	5.4% (0.2)	5.6% (0.2)	0.50
Protein	16.7% (0.2)	16.5% (0.2)	0.55
Carbohydrate	43.0% (0.4)	43.4% (0.5)	0.62
Alcohol	2.4% (0.3)	3.3% (0.4)	0.06
Cereal fibre (g)	4.4 (4.1,4.7)	5.2 (4.9,5.6)	0.0008
Vegetable fibre (g)	5.7 (0.2)	5.9 (0.2)	0.57
Vitamin C (mg)	29.3 (1.4)	31.4 (1.2)	0.24
Retinol (μg)	3.3 (3.1,3.4)	3.4 (3.3,3.5)	0.16
A-tocopherol (mg)	4.7 (4.3,5.1)	5.1 (4.7,5.6)	0.12
Linoleic acid (mg)	4.2 (3.9,4.6)	4.4 (4.1,4.8)	0.44

[a] % Energy includes alcohol; ND, amount/1000 kcal for fibre and vitamins.
[b] Geometric mean and 95% confidence intervals have been given for positively skewed distributions.

Fig. 18.4. Odds ratios (OR) of peripheral vascular disease by source of energy adjusted for total energy intake and age and sex

Changes in source of energy	OR (95% confidence interval)
Percentage from carbohydrate (+10%)	0.84 (0.58,1.24)
Percentage from protein (+10%)	1.79 (0.65,4.92)
Percentage from fat (+10%)	1.41 (0.94,2.10)
Percentage from saturated fat (+10%)	1.67 (0.89,3.13)
Percentage from polyunsaturated fat (+10%)	0.75 (0.31,1.81)
Percentage from alcohol (square root, +0.1)	0.79 (0.64,0.98)*
Log$_n$ cereal fibre (+1)	0.37 (0.20,0.67)**

* $p<0.05$; ** $p<0.01$.

sex, alcohol (Odds ratio, OR) = 0.79, 95% confidence interval = 0.64, 0.98) and cereal fibre (OR = 0.37, 95% confidence interval = 0.20, 0.67) were statistically significantly associated with reduced risk of peripheral vascular disease. These differences also remained statistically significant after adjustment for social class and smoking history, indicating independent effects of alcohol and cereal fibre.

Interpreting Dietary Differences

Despite some drawbacks the food frequency questionnaire is a useful method for obtaining nutrient intake in large epidemiological surveys. The levels of nutrients found in the Edinburgh Artery Study general population (unpublished observations) were similar to those found in the Scottish Heart Health Study (Bolton-Smith and Milne 1990).

In the case–control study in Edinburgh there were significant differences in the intake of meat and meat products between cases and controls after adjustment for total energy. In terms of nutrient, cases had higher intake of total fat and saturated fat and lower intake of polyunsaturated fat compared with controls, but these differences were not statistically significant.

The differences in alcohol may be due to cases reducing their intake voluntarily or under guidance from their doctor because many will have symptoms (48%) of peripheral vascular disease or a history of angina or heart attack compared with controls. Alcohol history was not measured but only more recent intake and so we cannot infer that alcohol reduces risk of peripheral vascular disease, but it is more probable that those with peripheral vascular disease have reduced alcohol intake relative to those who are disease free.

The reduction in odds of peripheral vascular disease associated with cereal fibre is consistent with the case–control study of Katsouyanni et al. (1991) where crude fibre was found to be associated with a reduced odds of peripheral vascular disease. However, comparisons are difficult because they analysed only crude fibre, and it is not clear what proportion of this was from cereals and what proportion was from fruit and vegetables. Katsouyanni et al. (1991) found a significant association between polyunsaturated fat and reduced risk of peripheral vascular disease. While a similar result was found in our study, the higher levels of polyunsaturated fat and lower levels of saturated fat in the controls failed to reach statistical significance. Despite some drawbacks in the design of the study (Katsouyanni et al. 1991), their results are consistent with ours for peripheral vascular disease and fibre. They did not have quite the same problem of whether the diet preceded disease or disease preceded the diet because each subject was asked about their diet over a 1-year period before the onset of the disease. Of course, this increases the problem of recall bias in their study. They also used a postal questionnaire with the potential response bias that this entails.

The reduction in odds of peripheral vascular disease associated with cereal fibre remained significant after adjustment for social class, age, sex, and smoking history, and hence appears to be an independent effect. This association is similar to that found in coronary heart disease; Morris et al. (1977) found a lower incidence of coronary heart disease risk in men aged 30–67 years employed in banking and transport who had a high intake of dietary fibre from cereals. In a study of men who emigrated from Ireland to Boston and their brothers who remained in Ireland, it was found that the latter had a lower prevalence of coronary heart disease, the main dietary difference being higher intake of cereals and potatoes than their brothers in America (Trulson et al. 1964).

From the results of such epidemiological studies it is suggested that fibre intake is related to cholesterol synthesis and the development of atherosclerosis. The results presented in this chapter are consistent with an association between cereal fibre and reduction in risk of peripheral vascular disease. In comparing studies it is important that fibre is sufficiently defined, and the analyses presented in this chapter will be extended to explore the relationship between peripheral vascular disease and dietary factors in the full random sample of the general population.

References

Bingham SA (1987) The dietary assessment of individuals: methods, accuracy, new techniques and recommendations. Nutr Abstr Rev (Series A) 57, 10:705–42

Bolton-Smith C, Milne AC (1990) Food frequency v. weighed intake data in Scottish men. Proceedings of Nutrition Society, University of York, p. 35

Brown GD, Whyte L, Gee MI et al. (1984) Effects of two "lipid-lowering" diets on plasma lipid levels of patients with peripheral vascular disease. J Am Diet Assoc 84(5):546–50

Cameron ME, Van Staverne WA (1988) Manual on methodology for food consumption studies. Oxford University Press, Oxford

Consensus conference (1985) Lowering blood cholesterol to prevent heart disease. JAMA 253:2080–6

Dixon WJ (ed) (1988) BMDP statistical software manual, Vol. II. University of California Press, California

Duffy J (1985) Questionnaire measurement of drinking behaviour in sample surveys. J Official Statistics 1:229–34

Fowkes FGR, Housley E, Cawood EHH et al. (1991) Edinburgh Artery Study. Prevalence of asymptomatic and symptomatic peripheral artery disease in the general population. Int J Epidemiol 20:384–92

Garrow JS (1974) Energy balance and obesity in man. North Holland, Amsterdam

Gordon T, Kegan A, Garcia-Palmieri M et al. (1981) Diet and its relation to coronary heart disease and death in three populations. Circulation 63:500–15

Hankin JH, Rhoads GG, Glober GA (1975) A dietary method for an epidemiologic study of gastrointestinal cancer. Am J Clin Nutr 31:353–9

Heller RF, Elliott H, Bray AE, Alabaster M (1989) Reducing blood cholesterol levels in patients with peripheral vascular disease: dietitian or diet fact sheet? Med J Aust 151(10):566–8

Hutchinson K, Oberle K, Crockford P et al. (1983) Effects of dietary manipulation on vascular status of patients with peripheral vascular disease. JAMA 249 (24) 3326–30

Jain MG, Harrison L, Howe GR, Miller AB (1982) Evaluation of a self-administered dietary questionnaire for use in a cohort study. Am J Clin Nutr 36:931–5

Katsouyanni K, Skalkidis Y, Petridou E et al. (1991) Diet and peripheral arterial occlusive disease: the role of poly-, mono- and saturated fatty acids. Am J Epidemiol 133:24–31

Keys A (1970) Coronary heart disease in seven countries. Am Heart Assoc Monogr 29. Circulation (Suppl) 1:41–2

Kushi LH, Lew RA, Stane FJ (1985) Diet and 20-year mortality from coronary heart disease. The Ireland–Boston Heart Study. New Engl J Med 312(13):811–18

LaRosa JC, Hunninghake D, Bush D et al. (1990) The cholesterol facts. A summary of the evidence relating dietary fats, serum cholesterol, and coronary heart disease. Circulation 81(5):1721–33

Marr JW (1971) Individual dietary surveys: purposes and methods. Wld Rev Nutr Diet 13:105–64

Morris JN, Marr JW, Clayton DG (1977) Diet and heart: a postscript. Br Med J ii:1307–14

Rose GA (1962) The diagnosis of ischaemic heart pain and intermittent claudication in field surveys. Bull WHO 27:645–58

Schlesselmann JJ (1982) Case–control studies: design, conduct, analysis. Oxford University Press, New York

Shekelle RB, Shyrock AM, Paul O et al. (1981) Diet, serum cholesterol and death from coronary heart disease. N Engl J Med 304:65–70

Thomson M, Elton RA, Fulton M, Brown S, Wood DA, Oliver MF (1988) Individual variation in the dietary intake of a group of Scottish men. J Hum Nutr Diet 1:47–57

Trulson MF, Clancy RE, Jessop WJE, Childers RW, Stare FJ (1964) Comparisons of sibling in Boston and Ireland. J Am Diet Assoc 45:225–9

Tunstall-Pedoe H, Smith WCS, Crombie IK, Tavendale R (1989) Coronary risk factors and lifestyle variations across Scotland: Results from the Scottish Heart Health Study. Scot Med J 34:556–60

Wiehl DG, Reed R (1960) Development of new or improved dietary methods for epidemiological investigations. Am J Pub Health 50:824–8

Willett WC, Sampson L, Stampfer MJ et al. (1985) Reproducibility and validity of a semiquantitative food frequency questionnaire. Am J Epidemiol 122:151–61

Woodcock BE, Smith E, Lambert WH et al. (1984) Beneficial effect of fish oil on blood viscosity in peripheral vascular disease. Br Med J 288:592–4

Yano K, Rhoads GG, Kagan A and Tittatson J (1978) Dietary intake and the risk of coronary heart disease in Japanese men living in Hawaii. Am J Clin Nutr 31:1270–9

Yarnell JWG, Fehily AM, Milbank JE, Sweetnam PM, Walker CL (1983) A short dietary questionnaire for use in an epidemiological survey: Comparison with weighed dietary records. Hum Nutr Appl Nutr 37A:103–12

19 Personality

I.J. DEARY

Personality and Coronary Heart Disease

Health psychology is in vogue at present. A number of new journals devoted to this physical subspecialty have appeared and health psychologists are springing up in profusion. Diseases which have putative psychological risk factors, usually in the form of stress, are legion. Coronary heart disease (CHD) has a special place in this endeavour. The contribution of temperamental factors to the aetiology of heart disease is a well-researched area, and the hypothesis that personality attributes have a bearing on the state of the vasculature predates the establishment of psychosomatic medicine as a recognised area of study.

In his Lumleian Lectures on angina pectoris Osler (1910) asserted that, "It is not the delicate neurotic person who is prone to angina, but the robust, the vigorous in mind and body, the keen and ambitious man, the indicator of whose engines is always at 'full speed ahead' ". This is a remarkable anticipation of the work on heart disease and the Type A behaviour pattern which began about half a century later. There are other heralds of this work, but none is as impressive as Osler. Menninger and Menninger (1936) applied psychoanalytic methods to cases of cardiac disorder and came up with a suggested risk factor that today attracts considerable attention, namely hostility: "It would appear that heart disease and heart symptoms are (sometimes) a reflection of strongly aggressive tendencies which have been *totally* repressed". This plausible clinical guess suffered the fate of many psychoanalytic notions by becoming bound up in a bizarre chain of reasoning. Briefly, the Menningers' (1936) story was that the angina sufferer buried hostility toward his father and outwardly proferred affection. In an act of "identification" angina pain suffered by the father was felt by the son, allowing patricidal urges to be replaced by focal suicide. The authors did qualify their confidence in this explanation by stating that it was "a most tentative hypothesis".

Type A Behaviour Pattern

Friedman and Rosenman (1959) began the large research effort which has attempted to establish that coronary heart disease is more prevalent among individuals who show chronic time urgency, excessive drive, competitiveness and, to some degree hostility and frustration. This pattern of behaving has been labelled the Type A behaviour pattern (TABP). The volume of research conducted since then is huge and cannot be represented in full here. Apart from the number of papers involved, the research is complicated by the fact that different studies have used different study populations, different personality assessment techniques, different disease outcomes and different study designs.

As an introduction to the research it is worthwhile examining a few of the larger, better-designed studies in the area, and then summarising the meta-analyses in this field of research. The Western Collaborative Groups Study (Rosenman et al. 1975) followed up 3154 men between 39 and 59 years of age at intake for 8.5 years. All subjects were initially well. TABP was assessed using standardised structured interviews administered by trained interviewers. The sample contained 1589 individuals classified as Type A and 1565 men designated Type B, the more relaxed personality type. Death rate for coronary heart disease was 2.92 per 1000 person-years for the Type A group and 1.32 for the Type B group on average. In general terms Type A individuals were about twice as likely as Type B subjects to suffer from myocardial infarction (symptomatic or unrecognised), or angina pectoris. When the potentially confounding effects of parental history of coronary heart disease, cigarette smoking, blood pressure, serum triglyceride and cholesterol and alpha- and beta-lipoprotein ratios were controlled for in the TABP–CHD relationship the relative risk (odds ratio) fell only marginally, i.e. from 2.21 ($p<0.0001$) to 1.87 ($p<0.002$) in younger subjects and from 2.31 ($p<0.0002$) to 1.98 ($p<0.019$) in the older age group. Therefore, the personality–heart disease association was not caused by these other, more traditional risk factors.

Similar positive findings were also reported for the Framingham study (Haynes et al. 1980) for professional, non-clerical occupations but not for clerical or blue-collar groups. Further, it was shown that the association between behaviour pattern and coronary heart disease remained when TABP was assessed using a self-report questionnaire, the Jenkins Activity Survey (Jenkins et al. 1974). A further large prospective trial, the Multiple Risk Factor Intervention Trial (MRFIT; Shekelle et al. 1985), followed up 12 772 initially well men aged between 35 to 57 years for a mean of 7.1 years using the Jenkins Activity Survey (JAS) to estimate individual differences in TABP. No significant relationship between behaviour pattern and coronary heart disease was found. The top fifth of JAS scorers (Type A subjects) had a risk of 5.0% for coronary heart disease and the lowest quintile (Type B subjects) had a risk of 4.1%. For the 3100 men who were also given the structured interview the results were equally negative. This was despite very careful attempts to achieve reliable TABP estimates, and adequate power.

By the late 1980s studies in the field were numerous and some order was brought to an unmanageable literature by the meta-analysis conducted by Booth-Kewley and Friedman (1987). They concluded that TABP accounted for about 2% of the variance in coronary heart disease (including myocardial infarction and angina). This may appear to be a trivially small contribution, but it is of similar magnitude to other significant risk factors, and it is compatible with the Rosenman et al. (1975) estimate that Type A individuals are twice as likely to suffer from coronary heart disease as are Type B subjects. Booth-Kewley and Friedman (1987) found that scores from the Type A structured interview had a closer association with coronary heart disease indices than did scores from the self-administered JAS.

Hostility

It was further suggested by Booth-Kewley and Friedman (1987) that Type A was not a monolithic personality feature, and that not all aspects of the TABP were predictive of heart disease. Job involvement (the classic workaholic factor) was largely unrelated to coronary heart disease while speed and impatience/time urgency were more powerful predictors and, overall, competitiveness/hard-driving/aggressiveness was the most closely related to heart disease occurrence. Booth-Kewley and Friedman (1987) noted that two other personality variables emerged as potentially important predictors of heart disease and should receive more attention in research, namely hostility and anxiety. Moreover, depression was as strong a predictor of coronary heart disease as was TABP. Therefore, the true picture of the CHD-prone individual is not the time-pressured workaholic but, rather, someone who frequently has one or more negative emotions such as depression, hostility, anger, aggressive-competitiveness or frustration.

The above extensive meta-analysis was criticised by Matthews (1988) who only partly confirmed that hostility, anxiety and depression were significant predictors of coronary heart disease. TABP remained a significant predictor only when all studies were considered together, the structured interview and JAS studies considered separately failed to achieve significance. Another interesting finding in the Matthews meta-re-analysis (1988) was that the TABP–CHD relationship was highly significant in general population studies, but not in high risk studies. Matthews (1988), like Booth-Kewley and Friedman (1987) urged that hostility be investigated further as a risk factor in coronary heart disease.

Although their own study did not find positive results Helmer et al. (1991) provide a helpful and concise review of the somewhat inconsistent results in studies where the association between coronary heart disease and personality traits of hostility have been studied. In their own study they administered the Cook–Medley Hostility Inventory (derived from the Minnesota Multiphasic Personality Inventory) and the Behaviour Pattern Hostility Index (derived from audiotapes of the Type A structured interview) to 118 men and 48 women undergoing angiography. The coronary heart disease measures were measures of occlusion at angiography. Low and high scorers on the Cook–Medley Hostility Inventory had rates of significant occlusion of 75.9% and 71.2%, respectively and extreme high and low scorers on the Behaviour Pattern Index had 73.1% and 74%, respectively. None of these comparisons was significant. As Matthews (1988) suggests of other negative studies found in high risk populations, the problem here may be that there was too little variance in the coronary heart disease variable to allow for a CHD–hostility relation to be uncovered. In view of the other recent positive studies (Hecker et al. 1988; Koskenvuo et al. 1988; Hearn et al. 1989) hostility remains an important variable for study in heart disease predisposition.

Personality and Peripheral Vascular Disease

The above findings on the contribution made by temperamental factors in the occurrence of CHD are an important consideration when attempting to examine the risk factors for peripheral vascular disease. Ischaemic heart disease, arterial disease in the lower limbs and stroke are manifestations of atherosclerosis, and the three conditions frequently occur together (Fowkes 1988). Therefore, it is of interest to discover whether similar behaviour patterns constitute risk factors in stroke and in peripheral vascular disease. While there is some indication that patients suffering from stroke, especially when combined with coronary artery disease, tend toward a pressurised pattern of behaviour (Adler et al. 1971; Gianturco et al. 1974), peripheral vascular disease has received little attention.

The hypothesis that TABP predisposes to peripheral vascular disease was tested by Cottier et al. (1983). They studied three groups of 13 patients: a group with intermittent claudication; with intermittent claudication and coronary artery disease; and a control group. The control group consisted of 13 consecutive admissions to the Department of Internal Medicine, seven of whom had malignant disease. Pressurised behaviour tendencies were assessed by interview and by administration of the Bortner Personality Inventory (Bortner 1969), a 14-item self-rated test of Type A personality. Diagnosis of intermittent claudication was based on the presence of typical symptoms without pain at rest and absent distal lower limb pulses and/or typical angiographic findings. There were no significant differences between the groups on the total Type A score from the structured interview, but the scores did rank in the expected order with the control group lowest and the group with both intermittent claudication and coronary artery disease scoring highest. Within the interview there were five items related to the expression of anger and three of these tended toward significance in the expected direction, indicating that patients with intermittent claudication are poor at controlling feelings of anger. Results from the Bortner scale revealed significant differences between groups in the hypothesised direction with controls showing fewest and patients with both intermittent claudication and coronary artery disease showing most Type A characteristics. Wives' ratings of patients' Bortner scores showed almost identical mean group values and only just failed to reach significance. The correlation between Type A interview and Bortner Type A self-rating scores was 0.41 ($p < 0.01$). The authors concluded that their exploratory study offered encouragement for larger scale research into the temperamental associations of predisposition to peripheral vascular disease.

Measurement of Personality in Edinburgh Artery Study

The Edinburgh Artery Study is a large-scale attempt to examine the prevalence and risk factors of symptomatic and asymptomatic peripheral vascular disease in the general population (Fowkes et al. 1991). The age–sex registers of 10 general practices in the City of Edinburgh have provided 1592 randomly selected male and female subjects aged 55 to 74 years (a response rate of about 65% to postal invitations to attend a clinic). The prevalence of intermittent claudication was 4.5%, and 8% of the subjects had major asymptomatic disease causing a significant impairment of blood flow to the lower limbs. Subjects with major asymptomatic disease were at significantly greater risk of suffering from ischaemic heart disease than those without evidence of peripheral vascular disease. Having established that limitation of peripheral blood flow in the lower limbs is common in the general population it was of interest to discover the risk factors for peripheral vascular disease. Among the risk factors studied was personality.

The presence of intermittent claudication was assessed using a validated, self administered WHO questionnaire. Further estimates of the extent of the limitation of blood flow to the lower limbs were the ankle-brachial pressure index (ABPI) and the results of reactive hyperaemia test, where systolic pressure in the ankle was measured 15 seconds after 4 minutes of arterial flow occlusion above the knee at 50 mmHg above the systolic pressure. History of myocardial infarction and angina was assessed using the WHO questionnaire.

Definite peripheral arterial disease was defined as the presence of two or more of the following: intermittent claudication, ABPI < or = 0.9 or hyperaemic reduction in ankle pressure >20%. Also, ABPI < or = 0.7 or hyperaemic reduction in ankle pressure >35% were sufficient alone for a diagnosis of peripheral vascular disease. Using these criteria, 153 of the 1592 subjects had definite peripheral vascular disease. As a part of the study, 153 age- and sex-matched controls who showed no objective or subjective evidence of peripheral vascular disease were selected to form a case–control subpopulation within the larger community study.

All 1592 participants completed two self-report personality questionnaires. The Bortner Personality Inventory, the 14-item test of Type A behaviour pattern used successfully in the small study by Cottier et al. (1983) was used to assess individuals' tendency towards pressurised behaviour. Because the measurement of negative emotions, particularly hostility, appeared to be an important consideration in research designed to discover the risk factors of atherosclerosis-based illness (Booth-Kewley and Friedman 1987), patients were asked to complete the Personality Deviance Scale (Bedford and Foulds 1978). This reliable and validated questionnaire was well-suited to the estimation of negative emotions, having three major subscales: Extrapunitiveness, Intropunitiveness and Dominance. Extrapunitiveness comprises items which estimate hostile thoughts and the tendency to be denigratory toward others. Intropunitiveness assesses individuals' tendency to be overly self-critical with separate scales estimating subjects' lack of self-confidence and degree of dependency on others. The dominance subscales assess domineering social attitudes and overt hostile acts.

Type A Behaviour Pattern (Bortner)

The following are some preliminary, unpublished results from the personality aspects of the Edinburgh Artery Study. Fig. 19.1 illustrates the results from the Bortner Personality Inventory across the four categories of peripheral vascular disease identified in the study: Intermittent claudication ($n = 67$), Major asymptomatic peripheral vascular disease ($n = 97$), Minor asymptomatic peripheral vascular disease ($n = 226$) and Normal ($n = 1038$). ANOVA testing revealed that the small difference in Bortner scores between groups was highly significant ($p<0.01$), the claudication group having the lowest mean score, the two asymptomatic clinical groups at an intermediate level and the normal group having the highest scores. The result is in the opposite direction to that which would be expected according to the hypothesis that TABP predisposes to peripheral vascular disease. Using the data from all subjects in the study, the

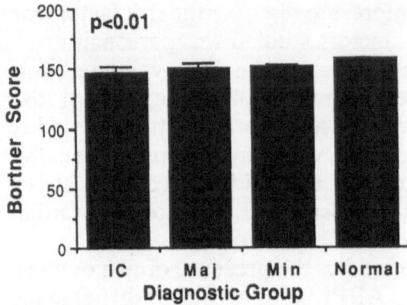

Fig. 19.1. Bortner Inventory (Type A behaviour pattern) scores and standard errors for the following groups within the Edinburgh Artery Study population: Intermittent Claudication (IC, $n = 67$), Major Asymptomatic Disease (Maj, $n = 97$), Minor Asymptomatic Disease (Min, $n = 226$), and Normal ($n = 1038$). p value refers to the one-way ANOVA result across the four groups.

responses to the Bortner Inventory were subjected to Principal Components Analysis followed by Varimax rotation. Four factors derived from the questionnaire data had eigenvalues greater than one. These were identified as Time Pressure, Competitiveness, Speed and Emotional Expression factors. Factor scores were computed and ANOVA testing was performed on the peripheral vascular disease subgroups described above. Significant differences among the groups were found for Time Pressure ($p=0.02$), Competitiveness ($p=0.007$) and Speed (<0.001), but not for Emotional Expression. These differences were in the same direction as those reported above, i.e. non-peripheral vascular disease subjects appeared more Type A prone than their diseased counterparts. Personality differences in the same direction were found for: history of myocardial infarction, but only for the Time Pressure and Speed factors (both at $p<0.05$); and for history of angina on the Bortner total score ($p<0.05$) and the Time Pressure and Speed factors (both $p<0.01$).

Personal Deviance (Foulds and Bedford)

Fig. 19.2 shows the results on the Foulds and Bedford Personal Deviance Scales for the case–control study involving 152 individuals with definite peripheral vascular disease (intermittent claudication or major asymptomatic peripheral

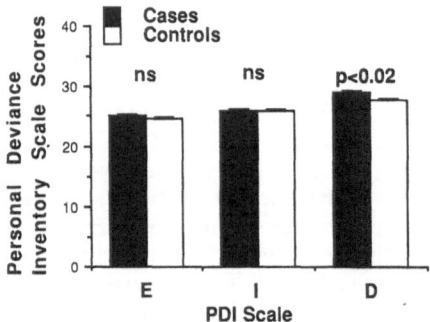

Fig. 19.2. Foulds and Bedford Personal Deviance Scale scores and standard errors for 152 definite peripheral vascular disease cases and 153 age- and sex-matched controls. Abbreviations for scales: E, Extrapunitiveness; I, Intrapunitiveness; and D, Dominance. p value and ns refer to the results of independent Student's t-tests between groups.

vascular disease) and 153 control subjects matched for age and sex. Sufferers from peripheral vascular disease scored higher on the Dominance scale ($p<0.02$) which includes two divisions; Hostile Acts where a trend toward a difference between the groups was found ($p=0.06$), and Domineering where the groups differed significantly ($p=0.04$). In the whole study population 134 individuals with a history of myocardial infarction, when compared with 1453 subjects with no such history, were significantly higher scorers on the Denigratory Toward Others subscale of the Extrapunitive scale ($p=0.03$) and on the Dominance scale ($p=0.04$). When individuals with a history of angina ($n=166$) were compared with those who had no angina (1421) there were no significant differences on any Foulds and Bedford scale.

Multiple logistic regression was performed on the results of all 1592 subjects in the study to ascertain whether putative personality risk factors for intermittent claudication lost their predictive power when smoking was taken into account. When only age, sex and the Hostile Acts score from the Foulds and Bedford Personal Deviance scales were included in the regression, the relative risk for intermittent claudication was 1.21 (95% confidence interval 1.02, 1.42; $p<0.05$). When smoking was included as a factor in the analysis the effect of the Hostile Acts score still tended toward significance. When the full Dominance scale from the Foulds and Bedford Inventory was included as a risk factor with sex and age, the odds ratio was 1.25 (95% confidence interval 0.98, 1.58; $p<0.1$) which tended toward significance, and fell to non-significance when smoking was included in the regression. Therefore, there is some suggestion of an independent effect of hostile/domineering aspects of personality among the factors which predispose to peripheral vascular disease, but smoking history variables can account for some of the overall effects of personality.

Whereas the scores from the Foulds and Bedford scales offer results in line with the original hypothesis that certain negative aspects of personality might constitute significant risk factors for atherosclerosis-based disease, the Bortner Type A scale data tend toward the opposite direction, with non-peripheral vascular disease subjects appearing to exhibit the Type A behaviour pattern more frequently. Can this be explained satisfactorily? One relevant consideration is the effect that illness might have on behaviour. Intermittent claudication limits physical activity. Moreover, the Type A hypothesis implies that a *lifetime* (or at least an adulthood) of pressurised hostile behaviour has influenced the development of cardiovascular pathology. How far are the inventories used in the present study compatible with these considerations? For each of the 14 Bortner questions the subjects were asked to rate themselves on a 40 mm line between two extreme statements (e.g. "not competitive" vs. "very competitive") in response to the following instruction, "What we would like you to do is to make a vertical line where you think you belong between the two extremes". This is likely to have focused the consideration of the subjects to the present. Therefore, it is plausible to suggest that subjects with peripheral vascular disease were indicating that they were less speedy, competitive and time-pressured in the *present*, and this might have been a consequence of peripheral vascular disease as opposed to a cause.

On the other hand, the Foulds and Bedford scales have a preface to their answering instruction which runs as follows, "This booklet contains descriptions of how you may have felt, thought or acted during *most of your life*" (emphasis in the original). This urges the subject to think about their pre-morbid personality and to respond accordingly. Perhaps, then, it is not surprising that the Foulds and

Bedford scales offered results in line with the traditional Type A hypothesis, and acted as a *trait* measure whereas the Bortner acted more as a *state* measure.

The Disease-Prone Personality?

The Edinburgh Artery Study's preliminary results offer partial corroboration for the notion that negative aspects of personality, especially hostility, may have negative consequences for physical health, especially atherosclerosis-based illness. To this extent, our results agree with the recent large-scale qualitative and quantitative reviews of the area (e.g. Booth-Kewley and Friedman 1987; Evans 1990). How should future research develop the personality–disease association? Evans (1990) urged researchers to assess the current bank of studies rather than to continue to report more "trivial" findings associated with Type A personality. He also suggests that illness states should be differentiated better when personality factors are being considered, a facet of research that has been advanced by the Edinburgh Artery Study's emphasis on intermittent claudication as a specific disease entity.

Readers with no background in personality psychology will have noted that several different scales have been mentioned above, and that even the Type A personality is a chimera. From the many available scales, which aspects of personality should be studied in future research addressing the links between temperament and disease? There seems to be some "kernel" to the notion of a Type A behaviour pattern that justifies further study as a risk factor in illness. However, it is now well recognised that Type A is multifaceted and that not all of its facets are relevant to disease risk (Booth-Kewley and Friedman 1987; Sensky 1987; Houston et al. 1986; Deary 1987; Friedman and Booth-Kewley 1987a). Therefore, further psychometric studies (e.g. May and Kline 1987; Houston et al. 1986; Eysenck and Fulker 1983) should confirm the various dimensions of the Type A conglomerate.

Some psychometric studies indicate that at least some of the Type A variance is attributable to personality dimensions long recognised in trait psychology, such as neuroticism and extraversion. For instance, Eysenck and Fulker (1983) found factors extracted from the Jenkins Activity Survey Type A scale to be closely related to these two traditional personality dimensions. Other studies have replicated the relationship between some self-report Type A scales and neuroticism in particular (Smith et al. 1989; Deary et al. 1991). When this is coupled with the recent consensus about the five major personality factors that account for the majority of the variance between individuals in temperament (McCrae and Costa 1986), there is a strong case for a new approach which takes the dimensions of neuroticism, extraversion, agreeableness, conscientiousness and openess/culture as a starting point for personality–illness relationship studies.

As was documented earlier, while there is good evidence to support the view that about 2% of the variance in coronary heart disease risk may be attributed to TABP as assessed using structured interviews, the worth of self-report Type A scales is less certain. It has emerged that some of the variance in self-report Type A scales is accounted for by the more frequent experience of anxiety and depression in some subjects (see Suls and Wan, 1989, for a meta-analysis). Taking us even further from the idea that a specific personality type might be related to a specific illness, Friedman and Booth-Kewley (1987b) provided evidence in the form of a meta-analysis to suggest that coronary heart disease, asthma and arthritis are related to anxiety, depression, anger, hostility, aggression and extraversion; and ulcer and headache are related to anxiety, depression and extraversion. In a detailed criticism of this meta-analysis Stone and Costa (1990) suggest that Friedman and Booth-Kewley (1987b) have not found a *disease*-prone personality but, rather, a *distress*-prone personality, and that their results tell us more about the relationship between neuroticism (and its attendant negative emotions) and health-related behaviour (such as somatic complaining and going to see doctors) than about personality and organic disease processes.

In summary, much of what seemed to be personality–disease relationships might turn out to have more to do with how certain personalities generate and react to somatic symptoms. Though this is interesting, it is not the original object of the exercise. However, this limitation does not implicate all attempts to find psychological factors among the predispositions to organic illness. The Edinburgh Artery Study has found a modest association between some aspects of personality and the relatively "hard" evidence for peripheral vascular disease. If replicable associations can be established between valid personality dimensions and objective indices of pathology, then the next steps will be to establish the causal mechanisms that connect temperament and disease (Suls and Rittenhouse 1990) and to attempt to alter the disease-provoking behaviours.

References

Adler R, Engel GL, MacRitchie K (1971) Psychological processes and ischaemic stroke (occlusive cerebrovascular disease). I. Observations of 32 men with 35 strokes. Psychosomatic Medicine 33:1–29

Bedford A, Foulds G (1978) Personality deviance scale (manual). NFER, Windsor

Booth-Kewley S, Friedman HS (1987) Psychological predictors of heart disease: a quantitative review. Psychol Bull 101:343–62

Bortner RW (1969) A short rating scale as a potential measure of pattern A behaviour. J Chron Dis 22:87–91

Cottier C, Adler R, Vorkauf H et al. (1983) Pressurised pattern or Type A behaviour in patients with peripheral arteriovascular disease: controlled retrospective exploratory study. Psychosom Med 45:187–93

Deary IJ (1987) Type A behaviour and coronary heart disease. Br Med J 295:606

Deary IJ, MacLullich A, Mardon J (1991) Neuroticism, extraversion and Type A behaviour: relationships with the reporting of minor physical symptoms and family history of hypertension and heart disease. Pers Ind Diff 12: in press

Evans PD (1990) Type A behaviour and coronary heart disease: when will the jury return? Br J Psychol 81:147–57

Eysenck HJ (1991) Personality as a risk factor in coronary heart disease. Eur J Pers 5:81–92

Eysenck HJ, Fulker D (1983) The components of Type A behaviour and its genetic determinants. Pers Ind Diff 4:499–505

Fowkes FGR (1988) The measurement of atherosclerotic peripheral arterial disease in epidemiological studies. Int J Epidemiol 17:201–207

Fowkes FGR, Housely E, Cawood EHH et al. (1991) Edinburgh Artery Study: prevalence of asymptomatic and symptomatic peripheral arterial disease in the general population. Int J Epidemiol 20:384–92

Friedman HS, Booth-Kewley S (1987a) Personality, type A behaviour, and coronary heart disease: the role of emotional expression. J Pers Soc Psychol 53:783–92

Friedman HS, Booth-Kewley (1987b) The "disease-prone personality". A meta-analytic view of the construct. Am Psychol 42:539–53

Friedman M, Rosenman MW (1959) Association of specific overt behaviour patterns with blood and cardiovascular findings. JAMA 169:1286–96

Gianturco OT, Breslin MS, Heyman A et al. (1974) Personality patterns and life stress in ischaemic cerebrovascular disease. I. Psychiatric findings. Stroke 5:453–60

Haynes SG, Feinleib M, Kannel WB (1980) The relationship of psychosocial factors to coronary heart disease in the Framingham study. III. Eight-year incidence of coronary heart disease. Am J Epidemiol 111:37–58

Hearn MD, Murray DM, Luepker RV (1989) Hostility, coronary heart disease, and total mortality: a 33-year follow-up study of university students. J Behav Med 12:105–21

Hecker MHL, Chesney MA, Black GW et al. (1988) Coronary-prone behaviours in the Western Collaborative Group Study. Psychosom Med 50:153–64

Helmer DC, Ragland DR, Syme SL (1991) Hostility and coronary artery disease. Am J Epidemiol 133:112–22

Houston BK, Smith TW, Zurawski RM (1986) Principal dimensions of the Framingham Type A scale: differential relationships to cardiovascular reactivity and anxiety. J Hum Stress 12:105–11

Jenkins CD, Rosenman RH, Zyzanski SJ (1974) Prediction of clinical coronary heart disease by a test for the coronary-prone behaviour pattern. N Engl J Med 290:1271–5

Koskenvuo M, Kaprio J, Rose RJ et al. (1988) Hostility as a risk factor for mortality and ischaemic heart disease in men. Psychosom Med 50:330–340

Matthews KA (1988) Coronary heart disease and Type A behaviours: update on and alternative to the Booth-Kewley and Friedman (1987) quantitative review. Psychol Bull 104:373–80

May J, Kline P (1987) Extraversion, neuroticism, obsessionality and the Type A behaviour pattern. Br J Med Psychol 60:253–9

McCrae RR, Costa PT (1986) Clinical assessment can benefit from recent advances in personality psychology. Am Psychol 42:1001–2

Menninger KA, Menninger WC (1936) Psychoanalytic observations in cardiac disorders. Am Heart J 11:10–21

Osler W (1910) The Lumleian Lectures on angina pectoris. Lancet i:839–44

Rosenman RH, Brand RJ, Jenkins D et al. (1975) Coronary heart disease in the Western Collaborative Group Study. Final follow-up experience of 8½ years. JAMA 233:872–7

Sensky T (1987) Refining thinking on Type A behaviour and coronary heart disease. Br Med J 295:69–70

Shekelle RB, Hulley SB, Neaton JD et al. (1985) The MRFIT behaviour pattern study. II. Type A behaviour and incidence of coronary heart disease. Am J Epidemiol 122:559–70

Smith TW, O'Keefe JL, Allred KD (1989) Neuroticism, symptom reports, and Type A behaviour: interpretive cautions for the Framingham scale. J Behav Med 12:1–11

Stone SV, Costa PT (1990) Disease-prone personality or distress-prone personality? The role of neuroticism in coronary heart disease. In: Friedman HS (ed) Personality and disease. Wiley, Chichester

Suls J, Rittenhouse JD (1990) Models of linkages between personality and disease. In: Friedman HS (ed) Personality and disease. Wiley, Chichester

Suls J, Wan CK (1989) The relation between Type A behaviour and chronic emotional distress: a meta-analysis. J Pers Soc Psychol 57:503–12

20 Exercise

E. HOUSLEY

It is widely believed, in the Western world at least, that exercise is of benefit in preventing cardiovascular disease in later life. This belief is based mainly on research into the role of exercise in preventing ischaemic heart disease and there has been very little research with regard to peripheral vascular disease. The reasons for this are fairly obvious, namely ischaemic heart disease is more dramatic, life threatening, and often presents at a relatively early age. It is more prevalent, posing a bigger threat to community health, and, being more prevalent, is easier to study.

I shall first review briefly the evidence linking physical inactivity to ischaemic heart disease, then discuss the work done in peripheral vascular disease and finally discuss possible pathophysiological mechanisms.

Exercise and Ischaemic Heart Disease

There have been many studies of the possible role of exercise in the prevention of ischaemic heart disease. The quality of the studies has been very variable. Powell et al. (1987) have described criteria for assessment of the quality of such studies based on the adequacy of the definition of exercise, definition of ischaemic heart disease and the soundness of the epidemiological approach. Despite the inadequacy of many studies the great majority do show that exercise is associated with a lower rate of ischaemic heart disease in later life. The small proportion failing to find a relationship tend to be the poorer studies. Morris et al. (1953) were the first to bring attention to physical inactivity as a risk factor for ischaemic heart disease with their seminal study of London bus drivers and conductors, postmen and supervisors, showing much lower mortality in men doing heavy work compared to those doing light work. Similar findings were reported by Kahn (1963) comparing letter carriers to mail clerks, Paffenbarger et al. (1970) comparing San Francisco cargo handlers to sedentary longshoremen, and many others. Studies with less positive findings such as Paul et al. (1963) and Stamler et al. (1960) generally had methodological faults such as inadequate separation of workers into high and low activity categories or failure to take difference in socioeconomic status into account.

These early studies were mostly done between 1950 and 1960 when workers in physically strenuous occupations were easy to find and define. With increasing mechanisation and consequent changes in work practices the numbers of workers engaged in hard physical work has markedly declined. It would not now, for example, be possible to repeat the San Francisco Longshoremen Study as cargo handling is done by machines. Morris et al. (1973) soon realised this and looked at leisure time activity as a protective factor against ischaemic heart disease. Also, as leisure time activity is capable of being manipulated, such studies might be more useful with a view to public health education. They looked at the reported leisure activities of London civil servants and showed that the death rate from ischaemic heart disease over an 8-year follow-up period was markedly reduced in those who had taken some vigorous exercise. Similar findings had been reported earlier by Paffenbarger et al. (1966) as part of a wider study of coronary risk factors in former college students. Further support came from Leren et al. (1975) who made the interesting observation that the most favourable coronary risk profile was in men in sedentary work but who took vigorous leisure activity. A number of studies reporting similar findings have since been published (Powell et al. 1987).

Exercise and Peripheral Vascular Disease

In contrast to the wealth of data available from studies of exercise and ischaemic heart disease there is very sparse data in the literature with regard to peripheral vascular disease. Kannel and Sorlie (1979) in a paper from the Framingham Study on physical activity and ischaemic heart disease briefly mention "in . . . peripheral vascular disease . . . the expected inverse relationship is noted, but does not reach statistical significance after adjustment for age". Karvonen (1982) followed up the Finnish cohort of the Seven Countries Study (Keys et al. 1966) and reported that, for men aged 50 to 59 years, at the 10-year follow-up "claudication was clearly associated with sedentary habits".

In the Edinburgh Artery Study we have looked at leisure time physical activity in a sample of 1592 subjects from the age–sex registers of 10 general practices in the City of Edinburgh. Equal numbers of males and females in each 5-year age group were selected and the sample matched to the socioeconomic profile of the city.

Ankle-brachial systolic blood pressure index (ABPI), measured after at least 10 minutes rest in the supine position and using a Hawksley random zero sphygmomanometer, was used as the measure of disease. The reproducibility and variability of the measurement is good (Fowkes et al. 1988). Although an ABPI of less than 0.9 is a reliable indicator of disease (Yao et al. 1969), ABPI is a continuous variable with relation to the disease, and disease may be present up to values of around 1.0. In this preliminary report I have simply taken less than 0.8 to mean "more severe disease", 0.8 to less than 0.9 "mild disease", 0.9 to less than 1.0 "possible disease" and greater than 1.0 "no disease".

Edinburgh Artery Study: Exercise

The subjects completed a structured questionnaire (Fig. 20.1) about their level of leisure time activity during the past year and during the decade when they were aged 35 to 45 years. The questionnaire was based on that used in the Welsh Heart Health Survey (1985) where it was well validated as a reliable method for quantification of leisure time activity. Recall of previous activity has also been shown to be generally reliable (Blair et al. 1991). The subjects reported on three levels of activity: light, moderate and strenuous. In this preliminary analysis and report I have looked at "no activity" compared with "some activity", however frequent, at the three levels.

Looking first at previous activity aged 35 to 45 years there was no relationship between light activity and subsequent prevalence of peripheral vascular disease. However, for moderate activity there was a significant relationship with ABPI. For those with ABPI less than 0.8, i.e. more severe disease, only 43% had taken moderate exercise whereas of those with no disease 54% had taken moderate exercise with the proportion increasing progressively with ABPI (Fig. 20.2). For strenuous activity, although the numbers are much smaller, the relationship is more striking with the proportion taking activity doubling from 6% at ABPI less than 0.8 to 12% at ABPI greater than 1.0 (Fig. 20.3).

Thus, those with disease at present had been less active, at moderate and strenuous levels, between ages 35 and 45 years, than those free of disease.

Turning to present activity we see a slightly different picture. For those taking light activity the proportion rose significantly from 75% at ABPI of less than 0.8 to 86% at ABPI greater than 1.0 (Fig. 20.4). With moderate activity there was also a significant rise from 19% to 41% (Fig. 20.5). However, in contrast to previous activity, there was no significant relationship between strenuous activity at present and ABPI.

Interpreting Associations of Exercise and Cardiovascular Disease

Correlation of physical inactivity with increased prevalence of disease does not prove causation. It is possible, indeed likely, that some or all of the lower incidence of physical activity at present in those with disease is simply a consquence of symptoms, i.e. intermittent claudication, restricting their capacity for exercise. As intermittent claudication and angina commonly coexist, angina could also be a confounding factor. Further analysis may suggest whether the correlation is causal or consequential.

The correlation of previous physical inactivity with present disease is more suggestive of a causal relationship. It seems highly unlikely that any significant amount of disease, either peripheral or coronary vascular, would have been present in the subjects between ages 35 to 45 years causing restriction of physical activity. Multivariate analysis will be necessary to show if previous physical inactivity is an independent risk factor from smoking, hypertension, hyperlipi-

PHYSICAL ACTIVITY

The following section gives examples of the sort of activities you might do or may have done REGULARLY.

LIGHT activity	MODERATE activity	STRENUOUS activity
Ballroom dancing	Badminton	Basketball
Bowling	Cricket	Competitive cycling
Light do-it-yourself	Cycling (include to and from	Competitive swimming
Light gardening	work, to shops etc)	Competitive running
Horse riding	Heavy do-it-yourself	Field sports (such as rugby,
Sailing	Golf	soccer, hockey)
Walking (including to and	Jogging	Training for strenuous
from work, to shops etc)	Swimming	sport
Yoga	Tennis	Squash

And other activities of similar intensity. Please specify others you have done.	And other activities of similar intensity. Please specify others you have done.	And other activities of similar intensity. Please specify others you have done.

....

....

1. In a typical week during the last year, on how many occasions would you take part, FOR MORE THAN 20 MINUTES EACH TIME:

    Insert 'None' if appropriate

    in LIGHT physical activity?   in summer .... times

                                         in winter .... times

    in MODERATE physical activity?   in summer .... times

                                         in winter .... times

    in STRENUOUS physical activity?   in summer .... times

                                         in winter .... times

2. In a typical week, when you were 35–45 years old, on how many occasions would you take part, FOR MORE THAN 20 MINUTES EACH TIME:

    Insert 'None' if appropriate

    in LIGHT physical activity?   in summer .... times

                                         in winter .... times

    in MODERATE physical activity?   in summer .... times

                                         in winter .... times

    in STRENUOUS physical activity?   in summer .... times

                                         in winter .... times

Fig. 20.1. Edinburgh Artery Study leisure activity questionnaire.

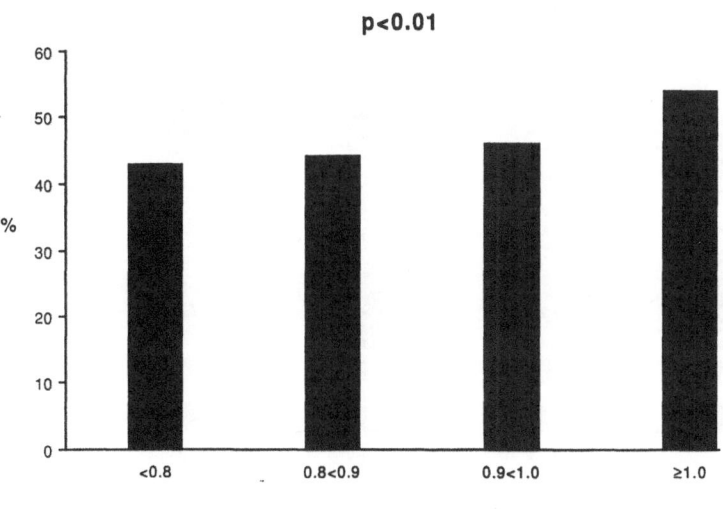

Fig. 20.2. Proportion of subjects taking moderate exercise, between ages 35 and 45 years, compared with ankle-brachial pressure index (ABPI) now (aged 55 to 74 years).

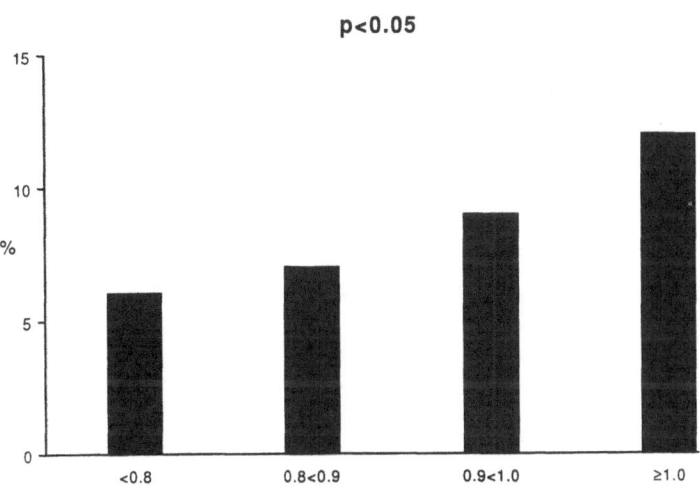

Fig. 20.3. Proportion of subjects taking strenuous exericse, between ages 35 and 45 years, compared with ankle-brachial pressure index (ABPI) now (aged 55 to 74 years).

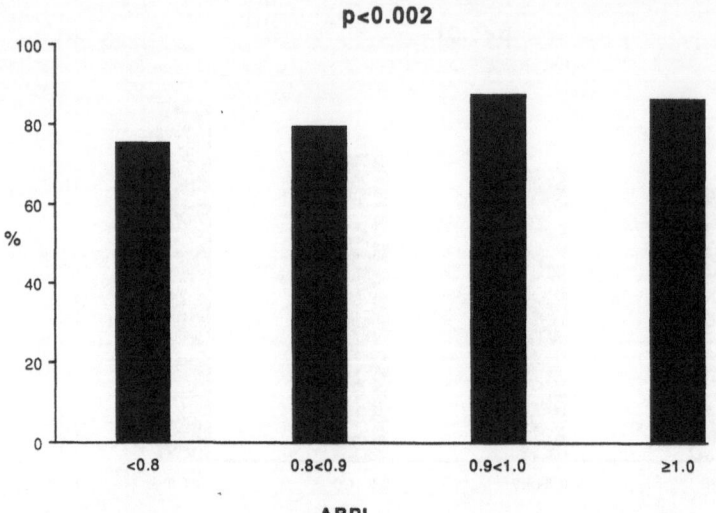

Fig. 20.4. Proportion of subjects taking light exercise now compared with ankle-brachial pressure index (ABPI).

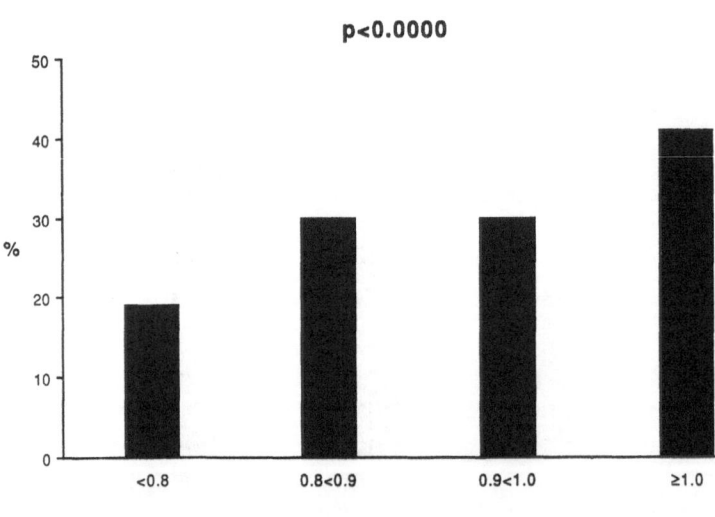

Fig. 20.5. Proportion of subjects taking moderate exercise now compared with ankle-brachial pressure index (ABPI).

daemia and diabetes. However, as similar analyses for ischaemic heart disease have shown physical inactivity to be an independent risk factor (Kannel and Sorlie 1979; Karvonen 1982; Paffenbarger and Hyde 1984) it would seem likely.

Physical inactivity appears to be a relatively weak risk factor for ischaemic heart disease with a relative risk of about 1.9 compared with two or three for smoking, hyperlipidaemia and hypertension, and probably the same holds for peripheral vascular disease (Powell et al. 1987). However it is a very common factor, e.g. in one study approximately two-thirds of a middle-aged Belgian population were physically inactive (De Backer et al. 1981). In the Edinburgh Artery Study only about 5% of subjects were taking moderate or strenuous exercise and, even between ages 35 to 45 years, less than 10% had taken moderate exercise, and only 2% strenuous exercise, at least once a day. At these levels even a relatively weak risk factor will have a considerable impact on the prevalence of a disease and be worthy of attention from public health education.

Although exercise is now generally accepted as conferring protection against cardiovascular disease there is controversy over the intensity and duration of exercise needed to confer the benefit. Physiological studies looking at the effect of exercise on other known risk factors, such as high LDL cholesterol and low HDL cholesterol, have shown that fairly intensive exercise, i.e. achieving 60% to 80% of maximum heart rate for at least 20 minutes two or three times a week, as in long distance running, soccer and speed skating, is necessary to produce significant changes but there is a dose/response relationship at all levels (Haskell 1984). Epidemiological studies (Paffenbarger et al. 1970; Karvonen 1982) also show reduced cardiovascular mortality at all levels of exercise suggesting that even low levels are worthwhile even though there is little change in other risk factors.

The mechanism of the benefit of exercise could be changes in blood lipid levels (Haskell 1984), rheological changes in the blood (Ernst and Matrai 1987), improvement of glucose intolerance and hyperinsulinaemia (Rauramaa 1984), reduction of blood pressure (Sallis et al. 1986) or some other mechanism as yet unknown. Whatever the mechanism there seems no doubt that increased levels of physical activity in earlier life are of benefit in preventing cardiovascular disease later, thus vindicating the assertion of Hippocrates over 2000 years ago that "sedentary life style is detrimental to health" (Paffenbarger and Hyde 1984).

References

Blair SN, Dowda M, Pate RR et al. (1991) Reliability of long-term recall of participation in physical activity by middle-aged men and women. Am J Epidemiol 133:266–75

De Backer G, Kornitzer M, Sobolski J et al. (1981) Physical activity and physical fitness levels of Belgian males aged 40–55 years. Cardiology 67:110–28

Ernst EEW, Matrai A (1987) Intermittent claudication, exercise, and blood rheology. Circulation 76:1110–14

Fowkes FGR, Housley E, Macintyre CCA, Prescott RJ, Ruckley CV (1988) Variability of ankle and brachial systolic pressures in the measurement of atherosclerotic peripheral arterial disease. J Epidemiol Comm Health 42:128–33

Haskell WL (1984) Exercise-induced changes in plasma lipids and lipoproteins. Prev Med 13:23–36

Kahn HA (1963) The relationship of reported coronary heart disease mortality to physical activity of work. Am J Public Health 53:1058–67

Kannel WB, Sorlie P (1979) Some health benefits of physical activity. The Framingham Study. Arch Intern Med 139:857–61

Karvonen MJ (1982) Physical activity in work and leisure time in relation to cardiovascular disease. Ann Clin Res 14 (Suppl) 34:118–23

Keys A, Aravanis C, Blackburn HW et al. (1966) Epidemiological studies related to coronary heart disease: Characteristics of men aged 40–59 in the seven countries. Acta Med Scand (Suppl) 460:169–90

Leren P, Askevold EM, Foss OP et al. (1975) The Oslo Study. Cardiovascular diseases in middle-aged and young Oslo men. Acta Med Scand (Suppl) 588:1–38

Morris JN, Heady JA, Raffle PAB, Roberts CG, Parks JW (1953) Coronary heart disease and physical activity of work. Lancet ii:1111–20

Morris JN, Chave SPW, Adam C, Sirey C, Epstein L (1973) Vigorous exercise in leisure-time and the incidence of coronary heart disease. Lancet i:333–9

Paffenbarger RS, Hyde RT (1984) Exercise in the prevention of coronary heart disease. Prev Med 13:3–22

Paffenbarger RS, Wolf PA, Notkin J, Thorne MC (1966) Chronic diseases in former college students. I. Early precursors of fatal coronary heart disease. Am J Epidemiol 83:314–28

Paffenbarger RS, Laughlin ME, Gima AS, Black RA (1970) Work activity of longshoremen as related to death from coronary heart disease and stroke. N Engl J Med 282:1109–14

Paul O, Lepper MH, Phelan WH et al. (1963) A longitudinal study of coronary heart disease. Circulation 28:20–31

Powell KE, Thompson PD, Caspersen CJ, Kendrick JS (1987) Physical activity and the incidence of coronary heart disease. Ann Rev Public Health 8:253–87

Rauramaa R (1984) Relationship of physical activity, glucose tolerance, and weight management. Prev Med 13:37–46

Sallis JF, Haskell WL, Fortmann SP, Wood PD, Vranizan KM (1986) Moderate-intensity physical activity and cardiovascular risk factors: The Stanford Five-City Project. Prev Med 15:561–8

Stamler J, Lindberg HA, Berkson DM, Shaffer A, Miller W, Poindexter A (1960) Prevalence and incidence of coronary heart disease in strata of the labor force of a Chicago industrial corporation. J Chron Dis 11:405–20

Welsh Heart Health Survey (1985) Protocol and questionnaire. Heartbeat Report No. 2, Cardiff, Wales

Yao ST, Hobbs JT, Irvine WT (1969) Ankle systolic pressure measurements in arterial disease affecting the lower extremities. Br J Surg 56:676–79

21 Economics of Prevention and Treatment

J.F. FORBES

This chapter examines a number of closely related economic aspects of peripheral vascular disease. One fundamental issue is the economic costs which peripheral vascular disease impose on patients and their families, the health service sector and society. A framework for estimating the economic cost associated with peripheral vascular disease is outlined. Economic costs in isolation offer a poor guide to assessing how resources should be allocated across the range of possible preventive and treatment options. A framework is introduced for examining the costs and benefits of preventive and treatment options and illustrates how economic evaluation is increasingly being applied to develop a more informed approach to service evaluation. The implications for peripheral vascular pro-grammes of recent fundamental changes in the pricing of health services are examined. Finally, I consider the development of explicit health service priorities and the challenge this poses to the prevention and management of peripheral vascular disease in the community.

Economic Costs of Peripheral Vascular Disease

Cost of illness studies traditionally draw a distinction between direct and indirect costs. Direct costs comprise the costs of prevention, diagnosis, treatment, rehabilitation and terminal care. Indirect costs represent the monetary value of economic losses arising from illness and disease, usually measured by the value of lost productive output as proxied by earnings. Given the increasing prevalence of peripheral vascular disease with age and limited labour force participation of persons in older age groups, indirect costs measured by losses in output and productivity are likely to be relatively small in relation to the direct costs of care and management. This narrow economic measure of indirect cost based on lost earnings or time off from work could be extended to include the impact of peripheral vascular disease on an individual's ability to sustain what they regard as normal activities or, more generally, their perceived quality of life.

Total costs or the sum of direct and indirect costs can be expressed from two contrasting perspectives. Most cost of illness estimates for peripheral vascular disease express costs on a prevalence basis, reporting costs (or more usually expenditure on selected health services) for all existing cases treated in a given calender year. Adopting a lifetime perspective, incidence-based estimates consider the expected lifetime costs associated with new cases. Despite being more difficult to implement in practice, the incidence approach is arguably more important from the perspective of estimating what likely cost savings may accrue from the effective prevention of peripheral vascular disease.

Within the cost of illness literature very little has been reported on the total economic costs of peripheral vascular disease. However, using data on the numbers of specific vascular interventions and associated hospital charges, broad expenditure estimates can be derived. These expenditure estimates, depending as they do on the frequency of interventions on treated cases, clearly ignore many of the wider costs associated with the prevalence of disease in the community. Nevertheless, they do provide some insight into the likely magnitude of expenditure devoted to the care and management of peripheral vascular disease in the treated population. For example, annual hospital expenditure on carotid endarterectomy in the United States in 1985 was approximately $1.2 billion (Maini et al. 1990) whereas total annual expenditure for surgical intervention for ruptured abdominal aortic aneurysm in the United States in 1983 was $168 million (Munoz et al. 1988).

Annual levels of expenditure based on the prevalence approach provide short-run estimates which may underestimate longer-run costs consequences. Although incidence-based estimates of the lifelong costs of peripheral vascular disease are scarce, Mackey et al. (1986) adopted the incidence approach to examine the accrual of costs over a 2-year period for limb-threatening ischaemia. Their results suggested average hospital charges in 1984 prices of $28 374 for successful revascularlisation, $56 809 for failed reconstruction and $40 563 for primary amputation. Applying these estimates to the 60 000–118 000 lower extremity amputations performed each year in the United States (Doucette et al. 1989) suggests that corresponding expenditure levels lie between $2.4 billion to $4.8 billion. These staggering expenditure figures, moreover, are best interpreted as a lower estimate of the societal costs generated over the future lifetime of individuals whose amputation was a direct consequence of the underlying pathology of peripheral vascular disease.

More comprehensive and timely estimates of the direct cost of peripheral vascular disease await improvements in the quality of routinely generated mortality data, information linking health service activities with their respective costs, and diagnostic and procedure coding of activity data. Fowkes (1991) has stressed the difficulty of conducting epidemiological investigations of peripheral vascular disease using routinely available data. Incomplete and inconsistent coding embraces not only the cause of death but also the recording of principal and secondary diagnoses and the categorisation of procedures used to treat peripheral vascular disease. Until progress is made in this area, cost of illness studies will be inevitably constrained by a fragile information base. Extending the assessment of costs beyond those falling directly on the health service is undoubtedly more challenging. Until then current estimates of the cost of peripheral vascular disease can only be interpreted as approximate, reflecting some but not all of the costs arising from this disease.

Finally, projections of direct costs over time should be interpreted with caution. Naive projections of demographic trends and future patterns of age-specific disease incidence and prevalence coupled with uncertainty regarding innovations in care and management should guarantee that forecasting errors should be the norm rather than the exception when it comes to predicting trends in the direct costs of peripheral vascular disease.

The Role of Economic Evaluation

Competing Interventions

Although the magnitude of expenditure on the care and management of peripheral vascular disease is undeniably large, cost of illness estimates on their own provide no justification for programmes or policies directed towards primary or secondary prevention. Furthermore, the cost of treatment alone cannot illuminate the relative efficiency of competing medical and surgical interventions. From the perspective of economic evaluation, the cost of specific interventions, be they preventive, curative or palliative, should be judged against their likely contribution to improved clinical outcomes. This linkage of programme cost and health outcomes lies at the heart of economic evaluation which highlights the importance of assembling robust evidence on the efficacy and effectiveness of intervening in the natural course of peripheral vascular disease.

The economic approach compares competing therapies in terms of their relative costs and relative benefits or gain in improved health outcomes (Detsky and Naglie 1990). Typically pair-wise comparisons of the costs and benefits of alternative programmes are made or the "usual" care option (which could be synonymous with no care) acts as a benchmark for comparison as is frequently the case when novel innovative interventions are evaluated. The measurement of extra benefits will be largely dictated by the stated goal of the programme, be it cases prevented, life-years gained or improvements in clinical outcomes that directly influence the patient's quality of life. Increasingly, the relationship between life-years gained and the quality of life (LaPuma and Lawlor 1990) is addressed in economic evaluations by expressing extra benefits in terms of quality adjusted life expectancy (QALE) or quality adjusted life-years (QALYs). Ultimately a judgement regarding the cost effectiveness of a particular intervention will depend on the trade-off between cost and effectiveness and the value which is placed on the extra health benefits attributed to the intervention when assessed against competing interventions (Doubilet et al. 1986).

Primary Prevention

Economic evaluation of the primary prevention of peripheral vascular disease is bedevilled by the absence of both cost and effectiveness data and by the methodological difficulty of identifying the precise contribution of life style changes to improved peripheral vascular health outcomes. More generalised preventive programmes directed towards other diseases may opportunistically alter the mix of established risk factors which may have a concomitant impact on peripheral vascular disease. The quantitative as opposed to qualitative impact of such programmes is difficult to determine with any degree of certainty and is likely to vary both over time and between populations. Indeed, a more informed understanding of how the classical risk factors such as hypertension, diabetes, smoking or high plasma cholesterol influence the development and staged progression of atherosclerosis and other types of peripheral vascular disease awaits longitudinal studies which may challenge conventional thinking based on cross-sectional investigations (Dormandy and Murray 1991).

Secondary Prevention

The secondary prevention of arterial disease in the legs is not well developed and therefore two related issues are discussed to illustrate the role of economic evaluation in secondary prevention. These are ultrasound screening for abdominal aortic aneurysm and carotid endarterectomy.

Ultrasound Screening for Abdominal Aortic Aneurysm

Ruptured abdominal aortic aneurysm accounts for about 8000 deaths in England and Wales each year. In the United States 1.2% of deaths in men over the age of 65 and 0.6% of women are attributed to aneurysm rupture. Several studies have examined the costs and effectiveness of an ultrasound screening programme for asymptomatic aneurysms followed by early elective resection, in high risk (elderly) populations. Using aortic aneurysm prevalence estimates derived from the Oxford Screening Programme for Abdominal Aortic Aneurysm (Collin et al. 1988), Russell (1990) estimated that the cost per QALY ranged from £440 for a 60-year male cohort to £1510 for a 80-year female cohort. O'Kelly and Heather (1988) reported broadly similar results for a hypothetical screening programme directed only at males aged between 65 and 74. Over the first 5 years of the programme the estimated cost per life saved was around £6400. After 10 years a significant decline in the cost per life saved (£4300) occurs as the large backlog of unscreened elderly males is largely eliminated. Crude cost per life saved estimates of between $78 000 and $108 000 have been reported using US data on costs and a series of assumptions regarding the effectiveness of a hypothetical screening programme (Quill et al. 1989).

The sensitivity of cost-effectiveness estimates to underlying assumptions is a characteristic feature of all studies which rely on hypothetical data. This is illustrated in the study of Bengtsson et al. (1989) who, using a simulation model calibrated with Swedish cost data and hybrid effectiveness estimates drawn from a wide range of published studies, estimate the relative cost-effectiveness of two screening programmes (unselected screening of all men starting at age 60 vs. selective screening of men with known intermittent claudication). Expressed in Swedish kronor (SEK) the net cost per life-year gained was 45 545 for the unselected population and 38 272 for the claudication population (US $1 = 6 SEK in 1986). A major feature of this study was the sensitivity analysis which examined the impact of a change in the assumed values of major parameters on the cost per life-year gained. For example, a doubling of the assumed elective operation mortality rate from 6% to 12% increases the cost per life-year gained by about 70% for both the unselected and claudicator populations whereas halving the elective operation mortality rate (to 3%) decreases the cost per life-year gained by around 18%. The results are also very sensitive to the assumed life expectancy for patients who survive the elective operation. If, as some studies suggest, the life expectancy of surviving aneurysmectomy patients is less than that normally attained for similarly aged cohorts in the overall population the costs per life-year gained rise considerably.

The sensitivity analysis reported by Bengtsson et al. (1989) and Quill et al. (1989) emphasise that projections of cost per life "saved" or year gained by ultrasonic screening for abdominal aortic aneurysms should be interpreted with caution. Expected costs and effects will naturally depend on the willingness of the target population to participate in the screening programme, the rescreening interval (Collin et al. 1991) and the sensitivity and specificity of the screening regime. At the beginning of a screening programme the yield (cases identified) and cost per life year gained will reflect the prevalence of abdominal aneurysms in the screened population. Over time as existing cases are identified and managed the yield will decline as it converges on the incidence of new, as opposed to existing, disease. Screening costs per case detected would consequently increase with a corresponding reduction in the total cost of both elective and emergency surgery. Finally, if screening leads to operative intervention on patients with smaller (<4.0 cm) aneurysms, the net cost per life-year gained may increase given the lower risk of small aneurysm rupture and the limited cost "savings" from the small number of averted emergency cases.

Carotid Endarterectomy

Despite the fact that experimental trials have yet to firmly establish its efficacy, the number of carotid endarterectomies performed per year in the United States increased by more than 500% between 1980 and 1990. The competing risks of invasive testing, peri-operative stroke and post-operative vascular events have yet to be clearly established. The uncertain impact on survival and quality of life in older patients with coexisting peripheral vascular disease, coronary artery disease and attendant risk factors has been a powerful incentive for prophylactic carotid endarterectomy on symptomatic as opposed to asymptomatic patients. Other outstanding issues which pose a challenge for economic evaluation of carotid endarterectomy include the relative costs and effectiveness of carotid endarterectomy vs. medical management, the place of intervention (large specialised teaching hospitals vs. smaller general or community hospitals) and the upper age limit, if any, for patients undergoing this operation (Maxwell et al. 1990).

Although it is not yet possible to integrate data on the costs of carotid endarterectomy with evidence on efficacy and effectiveness, attention has focused on the process of screening candidates for carotid endarterectomy. The costs and effectiveness of commonly used non-invasive diagnostic tests prior to carotid endarterectomy have been examined by Feussner and Matchar (1988) and Hankey and Warlow (1990). Using formal decision analysis Feussner and Matchar (1988) reviewed non-invasive screening and concluded that no diagnostic testing could be justified if surgery only led to a relative improvement of less than 0.08 QALYs per patient (or the equivalent of 1 stroke-free month). In other words, the benefit for surgery begins to offset the risk of invasive testing when the risk of stroke without surgery is greater than 3% per year in the at-risk population. If diagnostic testing is justifiable the preferred clinical strategy is non-invasive screening using duplex ultrasound or carotid Doppler ultrasound followed by carotid angiography prior to carotid endarterectomy. From a cost-effective perspective, however, the preferred testing strategy will depend on the degree of diameter stenosis which is regarded as abnormal and worthy of further investigation and subsequent management (Hankey and Warlow 1990). For example, if stenosis ⩾75% is the target the cost-effective approach is to screen using duplex ultrasound and then follow up positive cases with angiography. If, on the other hand, stenosis ⩾25% is the target it is more cost-effective to proceed directly to angiography for patients with a carotid bruit and to screen patients without a carotid bruit by duplex ultrasound.

New Technology

Technical innovation has transformed the menu of diagnostic and treatment interventions available for patients with peripheral vascular disease. Non-invasive diagnostic tools such as duplex Doppler ultrasound, combining B-mode ultrasonic imaging with Doppler frequency spectrum analysis, now complement and often supplant arteriography, the long-standing gold standard for vascular diagnosis. Duplex Doppler ultrasound can also be used in the diagnosis and management of abdominal vascular disease (intra-abdominal vascular disease, renal artery disease, mesenteric insufficiency and portal hypertension). Compared with arteriography duplex ultrasound is not only less invasive but is also less costly. These twin features may soon ensure that whenever possible arteriography is reduced to an adjunct screening test.

Following on from these innovative diagnostic tests the choice of procedures for treating peripheral vascular disease, particularly minimally invasive intervention, has increased considerably. The introduction and adoption of techniques, ranging from percutaneous transluminal angioplasty to non-thermal pulsed ultraviolet lasers, has typically preceded consideration of their relative costs and benefits. This may have unintended and possibly adverse consequences, both when less efficient techniques are adopted or when more efficient (less costly and more effective) interventions are not enthusiastically embraced by the clinical community, as Doubilet and Abrams (1984) persuasively demonstrate in the case of percutaneous transluminal angioplasty vs. surgery. Their finding that percutaneous transluminal angioplasty (whether alone or in combination with surgery) represented a more cost-effective approach compared with surgery alone was a powerful indictment of the potential inefficiency arising from delay in the introduction of (some) innovations. Whether the same conclusion emerges when percutaneous transluminal angioplasty is compared against newer techniques such as laser-thermal assisted balloon angioplasty emphasises the importance of integrating economic evaluations within well-controlled prospective randomised evaluations of novel approaches to the treatment of peripheral arterial disease. Uncritical promotion of new, or reluctant abandonment of old, approaches may inadvertently undermine the efficiency of caring for patients with peripheral vascular disease.

Paying for Peripheral Vascular Care

One of the most important factors which determines the mix and overall volume of services provided for the patient with peripheral vascular disease is the reimbursement mechanism which regulates the price of services and the financial incentives facing hospitals and physicians. The introduction of prospective payment systems based on diagnosis related groups (DRGs) heralds an important shift in the sharing of financial risk between those supplying services and those financing and purchasing services. Although originally developed and introduced in the United States, the DRG system of regulated prices has naturally attracted the interest of insurers, governments and health care financing bodies throughout the world who are searching for more effective means of controlling health service expenditure growth. Consequently, services provided for the patient with peripheral vascular disease will come under increasing scrutiny as hospitals respond to what is in effect a system of fixed and regulated prices.

Numerous studies have considered the impact of DRGs on peripheral vascular surgery (Jones 1985; Kinnison et al. 1985; Munoz et al. 1988, 1989a,b,c; Gupta and Veith, 1990). Several consistent themes emerge from this literature. First, as is the case in other surgical specialities, DRG payments generally do not fully cover the costs of older patients (≥ 70 years) undergoing peripheral vascular surgery. Munoz et al. (1989a) demonstrated that DRG payments underestimated the costs of peripheral vascular surgery performed on patients over the age of 70 by some 20%. As well as staying longer in hospital, older patients tended to make more use of clinical resources measured by emergency admission, admission to the surgical intensive care unit, blood transfusion and plasma products. Given this disparity in the cost of care and the standard level of payment, hospitals may attempt to cross-subsidise or simply restrict admission of older patients who represent an unacceptable financial risk. Second, the DRG system fails to adequately reimburse hospitals for providing care for patients with more severe peripheral vascular disease as measured by severity of illness (Jones 1985) and the presence of complications or co-morbidities (Munoz et al. 1989b). Moreover, DRG payments for patients undergoing complex arterial reconstructions for limb-threatening ischaemia significantly underestimate the costs of caring for such patients (Gupta and Veith 1990). Although reimbursement levels are close to the costs of caring for patients undergoing operations for intermittent claudication, the gap between costs and DRG payment is particularly large for patients facing limb loss. Finally, Munoz and colleagues (1988) demonstrate that hospitals that treat ruptured abdominal aortic aneurysm may be disadvantaged financially by the DRG payment system.

Like any system of administered prices, DRGs create a set of incentives which may induce a variety of responses in hospitals providing care for patients with peripheral vascular disease. Even if DRG payments are revised to take account of some of their fundamental limitations, hospitals will face increasing cost pressure that can be addressed, but probably not resolved, by altering the case-mix of patients and/or shifting the balance of care towards less costly therapeutic approaches. This may provide the most important spur yet to cost-effective innovations in peripheral vascular medicine (Kinnison et al. 1985).

Priorities for Spending on Peripheral Vascular Disease

The explicit setting of priorities for health service spending is an increasingly prominent feature of many health care systems. From an economic perspective priority setting should involve the careful assessment of programme costs and benefits. Explicit priorities have already emerged between different specialities as well as within programmes concerned with peripheral vascular disease. Throughout the United Kingdom the recent reforms of the National Health Service have focused attention on the development of more effective and informed purchasing of services which more closely reflect the perceived need, cost and expected benefit of health service interventions. Some peripheral vascular interventions (e.g., treatment for varicose veins) have already been accorded low priority by several English Health Authorities responsible for purchasing care.

In the United States, the prioritised health services list established by the Oregon Health Services Commission has attracted attention and may well serve as a model for similar exercises conducted elsewhere (Dixon and Welch 1991). The principal factors considered by the Oregon Health Services Commission were (1) how much a treatment costs, (2) what improvement in a person's quality of life it was likely to produce, and (3) how many years that improvement was likely to last. Applying these criteria to over 700 health condition/treatment pairs generated a priority ranking which will remain controversial, even when revised in light of public opinion and more robust information on the costs and benefits of treatment. For programmes directed towards the prevention and treatment of peripheral vascular disease which, like surgical treatment for peripheral athero-sclerosis and stripping/scleropathy for varicose veins, are accorded relatively low priority the future may be uncertain. If priority setting does emerge as the norm for allocating scarce resources the pay-off from clinical and economic evaluation of both preventive and treatment options in peripheral vascular disease will undoubtedly increase.

References

Bengtsson H, Bergqvist D, Jendteg S, Lindgren B, Persson U (1989) Ultrasonographic screening for abdominal aortic aneurysm: analysis of surgical decisions for cost-effectiveness. Wld J Surg 13:266–71
Collin J, Aranjo L, Walton J, Lindsell D (1988) Oxford Screening Programme for abdominal aortic aneurysm in men aged 65 to 74 years. Lancet ii:613–15

Collin J, Heather BK, Walton J (1991) Growth rates of subclinical abdominal aortic aneurysms – implications for review and rescreening programmes. Eur J Vasc Surg 5:141–4

Detsky AS, Naglie IG (1990) A clinician's guide to cost-effectiveness analysis. Ann Intern Med 113:147–54

Dixon J, Welch HG (1991) Priority setting: lessons from Oregon. Lancet 337:891–4

Dormandy JA, Murray GD (1991) The fate of the claudicant – a prospective study of 1969 claudicants. Eur J Vasc Surg 5:131–33

Doubilet P, Abrams HL (1984) The cost of underutilization. Percutaneous transluminal angioplasty for peripheral vascular disease. N Engl J Med 310:95–102

Doubilet P, Weinstein MC, McNeil BJ (1986) Use and misuse of the term 'cost effective' in medicine. N Engl J Med 314:253–6

Doucette MM, Fylling C, Knighton DR (1989) Amputation prevention in a high-risk population through comprehensive wound-healing protocol. Arch Phys Med Rehabil 70:780–5

Feussner JR, Matchar DB (1988) When and how to study the carotid arteries. Ann Intern Med 109:805–18

Fowkes FGR (1991) Peripheral vascular disease: a public health perspective. J Public Health Med 12:152–9

Gupta SK, Veith FJ (1990) Inadequacy of diagnosis related group (DRG) reimbursements of limb salvage lower extremity arterial reconstructions. J Vasc Surg 11:348–56

Hankey GJ, Warlow CP (1990) Symptomatic carotid ischaemic events: safest and most cost-effective way of selecting patients for angiography, before carotid endarterectomy. Br Med J 300:1485–91

Jones KR (1985) Predicting hospital charge and stay variation. The role of patient teaching status, controlling for diagnosis-related group, demographic characteristics, and severity of illness. Med Care 23:220–35

Kinnison ML, White RI Jr, Bowers WP, Dunlap ED (1985) Cost incentives for peripheral angioplasty. AJR 145:1241–4

LaPuma J, Lawlor EF (1990) Quality-Adjusted Life-Years: Ethical implications for physicians and policymakers. JAMA 263:2917–21

Mackey WC, McCullough JL, Conlon TP et al. (1986) The costs of surgery for limb-threatening ischemia. Surgery 99:26–35

Maini BS, Mullins TF, Catlin J, O'Mara P (1990) Carotid endarterectomy: A ten-year analysis of outcome and cost of treatment. J Vasc Surg 12:732–40

Maxwell JG, Rutherford EJ, Covington DL (1990) Community hospital carotid endarterectomy in patients over age 75. Am J Surg 160:598–603

Munoz E, Kassan MA, Chang JB (1988) Surgonomics: the costs of ruptured abdominal aortic aneurysm. Angiology 39:830–7

Munoz E, Cohen J, Chang J et al. (1989a) Socioeconomic concerns in vascular surgery: a survey of the role of age, resource consumption, and outcome in treatment cost. J Vasc Surg 9:479–86

Munoz E, Cohen JR, Goldstein J, Benacquista T, Mulloy K, Wise L (1989b) Financial risk and hospital cost in stratified, peripheral vascular surgical DRGs without complications and comorbidities. Ann Vasc Surg 3:170–6

Munoz E, Cohen J, Zelnick R, Mulloy K, Margolis I, Wise L (1989c) Hospital costs by clinical parameters for peripheral vascular surgical DRGs. J Cardiovasc Surg (Torino) 30:58–63

O'Kelly TJ, Heather BP (1988) The feasibility of screening for abdominal aortic aneurysms in a district general hospital. Ann R Coll Surg Engl 70:197–9

Quill DS, Colgan MP, Sumner DS (1989) Ultrasonic screening for the detection of abdominal aortic aneurysms. Surg Clin North Am 69:713–20

Russell JG (1990) Is screening for abdominal aortic aneurysm worthwhile? Clin Radiol 41:182–4

SECTION V

Genetics, Development and Haemostasis

---

22 Maternal and Fetal Origins of Cardiovascular Disease

D.J.P. BARKER

This chapter presents evidence that restraint of growth and development during critical periods of fetal life and infancy have an important effect on the development of cardiovascular disease.

Animal Studies

Long-term effects of interference with early growth, so-called "programming", have been demonstrated in a range of structures and functions in experimental animals.

Total somatic growth, relative organ growth, organ function and numerous metabolic activities can all be affected by transient stimuli applied in early life. Experiments showed that the adult body size of rats was more closely related to their mothers' nutrition during pregnancy and lactation than to their own genetic constitution (Dubos 1966). Maternal undernutrition during pregnancy alone permanently stunted the growth of the offspring, and was not reversed by an optimal diet after birth (Blackwell et al. 1968). The precise timing of nutritional deprivation in early life determines which organs have the greatest long-term reduction in size and DNA content. In rats, a brief period of calorie restriction immediately after birth was associated with profound reductions in the weights of the liver, spleen and thymus. There was, however, relatively little reduction in the weights of the brain and skeletal muscle. Calorie restriction immediately after weaning only reduced the weight of the thymus (Winick and Noble 1966). The metabolic activity of the rate-limiting enzymes controlling cholesterol synthesis in adult life are exquisitely sensitive to the programming effects of diet in fetal life and early infancy. Early nutrition in rats was shown to determine the response to a high fat challenge in adult life (Coates et al. 1983; Mott et al. 1991).

Geographical Studies

Early evidence that programming by an adverse environment in fetal life and infancy is a determinant of cardiovascular disease in humans came from geographical studies. The large geographical differences in death rates from

cardiovascular disease in England and Wales remain unexplained. Variations in
adult diet and cigarette smoking do not explain why the highest rates are in
industrial areas in the north and west of the country, and in some of the less
affluent rural areas such as North Wales. Rates are low throughout the south and
east, including London. It is a paradox that although the steep increase in
ischaemic heart disease during this century has been associated with rising
prosperity, the disease is now more common in poorer areas, and in lower income
groups.

One possibility is that these differences in mortality derive not from the current
environment but from the environment to which people were exposed during
childhood. The existence of detailed records of infant mortality from the
beginning of the century allows current death rates in any area of England and
Wales to be compared with infant mortality rates 60 or more years ago. This
comparison can be made with the country divided into 212 local authority
groupings. The correlations between past infant mortality and current mortality
from cardiovascular disease is remarkably strong (Barker and Osmond 1986), the
correlation coefficient being 0.73. Infant mortality is a general indicator of an
adverse environment and the conclusion drawn from these correlations is that
poor living conditions in childhood are a risk factor for cardiovascular disease – a
conclusion first put forward in 1977 by Forsdahl, who found a similar geographical relation between infant and cardiovascular mortality in the counties of
Norway.

The detailed infant mortality records in England and Wales make it possible to
distinguish neonatal mortality (deaths before 1 month of age) from post-neonatal
mortality (deaths from 1 month to 1 year). They reveal the new and surprising
clue that cardiovascular mortality in adults is closely linked to neonatal mortality
(Barker et al. 1989a). This relation is shown in Fig. 22.1. In the past neonatal
mortality was high in places where many babies had a low birthweight (Local

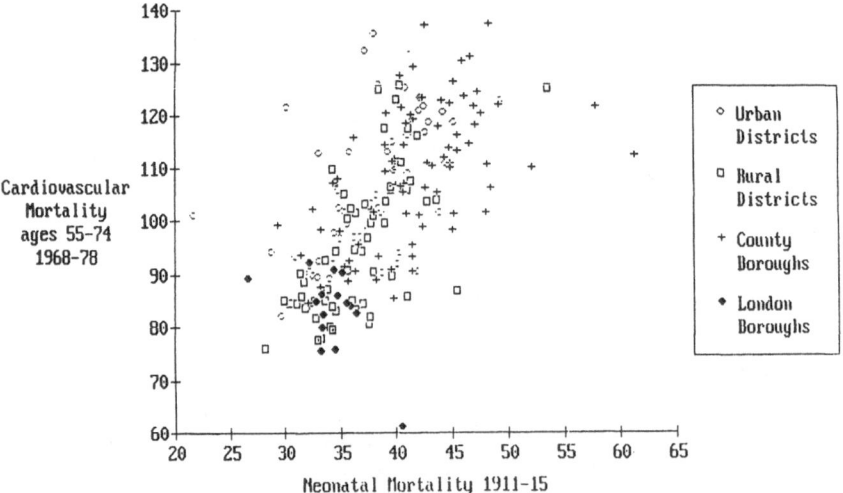

Fig. 22.1. Cardiovascular and neonatal mortality in England and Wales.

Government Board 1910). High neonatal mortality was generally associated with high maternal mortality rates, which were found in places where women had poor physique and health (Campbell et al. 1932). There is therefore a geographical association between poor maternal physique and health, poor fetal growth and high death rates from cardiovascular disease. This geographical association is reinforced by studies of migrants. A study of 2 million people who died in England and Wales during 1969 to 1972 showed that the increased risk of ischaemic heart disease and stroke among people born in northern England and in Wales persisted whether or not they moved to other parts of the country (Osmond et al. 1990).

The recent fall in stroke mortality in Britain and many other Western countries is consistent with improvement in maternal health during the past century. To explain the rise in ischaemic heart disease it seems necessary to postulate two groups of causes, one associated with poor living standards and the other associated with prosperity and linked, presumably, to the high energy Western diet.

Follow-up Studies

Further epidemiological exploration of the relation of early growth and infection to adult disease requires studies of adults in middle and old age whose early development was recorded. In Hertfordshire, England, from 1911 onwards every baby born in the county was weighed at birth, visited periodically by health visitors throughout the first year and weighed again at 1 year. From 1923 onwards all illnesses in children aged up to 5 years were recorded. The records have been preserved and it is therefore possible to trace men and women born around 60 years ago, and to relate their early development to the later occurrence of illness and death and the presence of known risk factors (Barker et al. 1989b).

In an early study 6500 men who were born in eight districts of the county between 1911 and 1930 were followed up. Table 22.1 shows standardised mortality ratios for ischaemic heart disease in the men, of whom 469 had died from the disease. The ratios fall steeply with increasing weight at 1 year, a trend not shown by deaths from non-circulatory causes. Ischaemic heart disease mortality also falls with increasing birthweight, though the relation is not as strong as with weight at 1 year. Stroke mortality shows similar trends. An interpretation of these findings is that programming of cardiovascular disease occurs both during fetal life and during infancy.

Table 22.1. Standardised mortality ratios for ischaemic heart disease according to weight at 1 year in 6500 men born during 1911–1930. Numbers of deaths in parentheses

Weight at 1 year (pounds)	Ischaemic heart disease	All non-circulatory disease
≤18	100 (36)	74 (39)
−20	84 (90)	99 (157)
−22	92 (180)	74 (215)
−24	70 (109)	67 (155)
−26	55 (44)	84 (99)
≥27	34 (10)	72 (31)
All	78 (469)	78 (696)

These findings pose the question of what pathological processes related to cardiovascular disease are programmed in early life. There is now evidence that haemostatic variables (Meade et al. unpublished observations), glucose tolerance (Hales et al. unpublished observations), blood pressure (Barker et al. 1990) and lipid metabolism (Barker 1991) are all programmed. Detailed description of the association between these variables and early growth is beyond the compass of this chapter although particular associations have been selected to illustrate certain general points.

Dr T. W. Meade and his colleagues at the MRC Epidemiology Unit at Northwick Park have measured the plasma fibrinogen levels in 591 of the Hertfordshire men who still live in the county. High plasma fibrinogen is a strong predictor of both ischaemic heart disease and stroke (Meade and North 1977; Meade et al. 1986). Levels fall with increasing weight at 1 year (Table 22.2) (Meade et al. unpublished observations). They are not, however, independently associated with birthweight. Adjustment for cigarette smoking, which is associated with raised plasma fibrinogen, does not change the trends. Cigarette smoking and failure to gain weight in infancy have an additive effect on plasma fibrinogen levels.

Table 22.2. Mean plasma fibrinogen in men aged 59 to 70 years

Weight at one year (pounds)	Number of men	Fibrinogen (g/l)[a]
≤18	37	3.21
−20	91	3.08
−22	177	3.14
−24	173	2.98
−26	80	2.95
≥27	33	2.93
All	591	3.05

[a] Geometric mean values adjusted for age and cigarette smoking.

In collaboration with Professor C. N. Hales, University of Cambridge, we have carried out glucose tolerance tests on 370 of the men who still live in Hertfordshire (Hales et al. unpublished observations). The percentage of men with impaired tolerance, defined by a 2-hour plasma glucose of 7.8 mmol/litre or more, fell progressively with increasing birthweight and weight at 1 year (Table 22.3). There were three-fold differences in the prevalence of impaired tolerance and diabetes between men with the lowest and highest early weights. Plasma 32–33 split proinsulin also fell with increasing weight at one year. Raised plasma levels of 32–33 split proinsulin are interpreted as a sign of β-cell dysfunction. These trends were independent of body mass index. The findings lead to the hypothesis that nutritional and other factors determining fetal and infant growth influence the size and function of the adult pancreatic β-cell complement. Whether and when non-insulin dependent diabetes supervenes will be determined by the rate of attrition of β-cells with ageing and the development of insulin resistance, of which the most important known determinant is obesity.

Table 22.4 shows that systolic blood pressure falls with increasing birthweight. A similar inverse relation between birthweight and adult blood pressure has been shown in a national sample of men and women aged 36 years in Britain (Barker et al. 1989c). In contrast to plasma fibrinogen and glucose intolerance, blood

Table 22.3. Impaired glucose tolerance (2-hour glucose $\geqslant$7.8 mmol/l) in men aged 59 to 70 years

Weight at one year (pounds)	Number of men	Impaired glucose tolerance		Odds ratio adjusted for body mass index (95% C.I.)
		n	%	
$\leqslant$18	23	10	43	8.2 (1.8,38)
−20	63	20	32	4.8 (1.2,19)
−22	107	32	30	4.2 (1.1,16)
−24	105	19	18	2.1 (0.5,7.9)
−26	48	9	19	2.1 (0.5,9.0)
$\geqslant$27	24	3	13	1.0
Total	370	93	25	χ^2 for trend = 14.9*

*p<0.001.

Fig. 22.4. Mean systolic pressure in men aged 59 to 70 years

Birthweight (pounds)	Number of men	Systolic pressure (mmHg)
−5.5	31	169
−6.5	95	166
−7.5	251	165
−8.5	233	163
−9.5	125	162
>9.5	56	162
All	791	164

pressure is not related to weight at 1 year independently of birthweight. This specificity of the relation of adult parameters to weight either at birth or at 1 year is a further argument in favour of long-term effects being determined during critical, perhaps brief, periods of early development.

Fetal Growth

Birthweight is a summary measure of fetal growth which includes head size, length and fatness. We now know that it greatly underestimates the relation between fetal growth and adult parameters, including blood pressure. In order to explore the association between measurements at birth and adult blood pressure we examined 449 men and women aged around 50 years who were born in one hospital in Preston (Barker et al. 1990). At that hospital, Sharoe Green, unusually detailed observations were made at birth. Table 22.5 shows the mean systolic pressures according to birthweight and placental weight. There are opposing trends such that systolic pressure falls by around 10 mmHg with increasing birthweight and rises by around 12 mmHg with increasing placental weight. Adjustment for gestation did not affect these trends. They were independent of current body mass and alcohol consumption.

An important aspect of the findings is that most people with high systolic pressure were not unusually small at birth. Rather, their birthweights were within the normal range but did not match the weight of the placenta. A feature of babies born with the heaviest placentas, weighing more than 1.5 pounds, among whom adult blood pressures were highest, was that they were disproportionate at

Table 22.5. Mean systolic pressures (mmHg) of men and women aged 46 to 54 years according to birthweight and placental weight. Numbers of people in parentheses

Birthweight (pounds)	Placental weight (pounds)				
	−1.0	−1.25	−1.5	>1.5	All
−5.5	152	154	153	206	154
	(26)	(13)	(5)	(1)	(45)
−6.5	147	151	150	166	151
	(16)	(54)	(28)	(8)	(106)
−7.5	144	148	145	160	149
	(20)	(77)	(45)	(27)	(169)
>7.5	133	148	147	154	149
	(6)	(27)	(42)	(54)	(129)
All	147	149	147	157	150
	(68)	(171)	(120)	(90)	(449)

birth, being relatively short in relation to their head circumference. This suggests that one process linking fetal growth with adult blood pressure may be diversion of fetal cardiac output away from the trunk to favour the brain.

Large placental weight is associated with clinical hypertension in later life as well as higher mean blood pressure. Among the 449 men and women the risk of having treatment for hypertension was 3.7 times greater among those with placentas weighing more than 1.5 pounds than among those whose placentas weighed 1.0 pounds or less. The causes of large placental size are largely unknown. In Preston, however, only 4 out of 56 (7%) of babies born at term to mothers in social class I and II had placentas exceeding 1.5 pounds. This compared with 62 out of 254 (24%) for mothers in the lower social classes. We suggest that one influence which links low social class with large placental weight is poor nutrition. A recent study of 8684 births in Oxford has shown that iron deficiency anaemia is associated with heavier placental weight (Godfrey et al. in press).

Conclusion

Retarded fetal and infant growth are strongly related to death from cardiovascular disease, and to risk factors for the disease. These long-term associations of retarded growth may reflect restraint of tissue growth by an adverse environment during a critical, sometimes brief, period of fetal or infant development. Which tissues are affected depends on the nature of the adverse influence and its timing. Long-term studies now in progress will extend our knowledge of the effects of earlier life programming and may give insight into the timing of critical periods.

The relation of early growth with risk factors and disease rates is continuous. The prevalence of impaired glucose tolerance (Table 22.3) and systolic blood pressure levels (Table 22.4) fall progressively up to the highest values of birthweight. It follows, therefore, that while an average birthweight is usual it may not be optimal. If the criterion of successful fetal growth is adult health and longevity, assessment of the newborn must at the least include placental weight and the proportions of the baby.

The effects of programming interact with influences in the adult environment including cigarette smoking and body weight. The existence of programming does not imply that adult influences should be discounted – though in the past we have probably overestimated their importance.

Our findings are open to the interpretation that genetic influences which are immediately manifest as growth failure in early life reveal themselves in adult life through the occurrence of degenerative disease. However, studies of the birthweights of the first born children of mothers and daughters suggest that genetic factors play only a small part in determining birthweight (Carr-Hill et al. 1987). Experiments in which newborn mice were randomly assigned to foster mothers show that individual variations in post-weaning growth rates are related more to the nutritional status of the foster mother than the origins of the offspring (Dubos 1966).

We favour an environmental explanation of our findings and suspect that natural nutrition is important. Research is needed into the maternal influences which regulate fetal and infant growth.

References

Barker DJP (1991) The intrauterine environment and adult cardiovascular disease. In: Ciba Symposium 156. The childhood environment and adult disease. John Wiley, Chichester

Barker DJP, Osmond C (1986) Infant mortality, childhood nutrition, and ischaemic heart disease in England and Wales. Lancet i:1077–81

Barker DJP, Osmond C, Law C (1989a) The intra-uterine and early postnatal origins of cardiovascular disease and chronic bronchitis. J Epidemiol Community Health 43:237–40

Barker DJP, Winter PD, Osmond C, Margetts B, Simmonds SJ (1989b) Weight in infancy and death from ischaemic heart disease. Lancet ii:577–80

Barker DJP, Osmond C, Golding J, Kuh D, Wadsworth MEJ (1989c) Growth in utero, blood pressure in childhood and adult lfe, and mortality from cardiovascular disease. Br Med J 298:564–7

Barker DJP, Bull AR, Osmond C, Simmonds SJ (1990) Fetal and placental size and risk of hypertension in adult life. Br Med J 301:259–62

Blackwell NM, Blackwell RQ, Yu TTS, Weng YS, Chow BF (1968) Further studies on growth and feed utilization in progeny of underfed mother rats. J Nutr 97:79–84

Campbell JM, Cameron D, Jones DM (1932) Ministry of Health Reports on Public Health and Medical Subjects, No. 68. High maternal mortality in certain areas. HMSO, London

Carr-Hill R, Campbell DM, Hall MH, Meredith A (1987) Is birthweight determined genetically? Br Med J 295:687–9

Coates PM, Brown SA, Sonaware BR, Koldovsky O (1983) Effect of early nutrition on serum cholesterol levels in adult rats challenged with a high fat diet. J Nutr 113:1046–50

Dubos R, Savage D, Schaedler R (1966) Biological Freudianism: lasting effects of early environmental influences. Pediatrics 38:789–800

Forsdahl A (1977) Are poor living conditions in childhood and adolescence an important risk factor for arteriosclerotic heart disease? Br J Prev Soc Med 31:91–5

Godfrey KM, Redman CWG, Barker DJP, Osmond C (1991) The effect of maternal anaemia and iron deficiency on the ratio of fetal weight to placental weight. Br J Obstet Gynaecol (in press)

Local Government Board (1910) Thirty-ninth annual report 1909–10. Supplement on infant and child mortality. HMSO, London

Meade TW, Mellows S, Brozovic M et al. (1986) Haemostatic function and ischaemic heart disease: principal results of the Northwick Park Heart Study. Lancet ii:533–37

Meade TW, North WRS (1977) Population-based distributions of haemostatic variables. Br Med Bull 33:283–88

Mott GE, Lewis DS, McGill HC (1991) Programming of cholesterol metabolism by breast or formula feeding. In: Ciba Symposium 156. The childhood environment and adult disease. John Wiley, Chichester

Osmond C, Barker DJP, Slattery JM (1990) Risk of death from cardiovascular disease and chronic
    bronchitis determined by place of birth in England and Wales. J Epidemiol Community Health
    44:139–41
Winick M, Noble A (1966) Cellular response in rats during malnutrition at various ages. J Nutr
    89:300–6

23 Genetic Analysis

J.M. CONNOR

Genetic analysis in common disorders of adulthood seeks to determine whether and which genes are involved in the predisposition to the occurrence, the prognosis and the response to various forms of therapy for a condition. Broadly these analyses are divided into complementary "old" and "new" approaches and all are hampered if onset of the condition is age-related and if exclusion of the disorder is problematical. This is clearly the case for cardiovascular disease in general where the incidence is age-related and where patients with extensive atherosclerosis may remain asymptomatic. Furthermore, reflecting the generalised nature of the disease process, symptoms may occur at diverse sites and patients presenting with intermittent claudication will have an increased risk of coronary heart disease and stroke in addition to lower limb events (Reunanen et al. 1982). Hence this review includes data from studies of coronary heart disease and stroke as well as peripheral vascular disease to reflect genetic involvement in the predisposition to atherosclerosis in general, but also indicates clues to factors favouring site-specific disease.

Background to "Old" Genetic Analyses

Genetic conditions may be subdivided, in order of frequency, into multifactorial, somatic cell genetic, single gene, chromosomal and mitochondrial disorders (Table 23.1). Multiple subtypes are apparent with a collective frequency of 33% of all births. This large number of subtypes (>5600) and the limited repertoire of disease presentation means that most diseases have a variety of different genetic origins (genetic heterogeneity). Within a family, however, a single origin is likely and classification is aided by pedigree analysis, twin studies and family correlation studies. Single gene disorders and mitochondrial disorders have diagnostic patterns of inheritance within a family whereas for the remaining disorders the distribution of patients within a family is not characteristic and a familial component may not be evident without twin or family studies. Identical twins have the same genetic makeup whereas non-identical twins have 50% of their genes in common (as do other brothers and sisters). For single gene, chromosomal and mitochondrial disorders identical twins are always either both affected or both unaffected (100% concordance). For somatic cell genetic disorders such as cancer where one or more independent mutations occur after birth the concordance rates are similar in identical and non-identical twins whereas in multifactorial traits the concordance rate in identical twins exceeds that in non-identical twins but is less than 100% due to the involvement of environmental factors. Multifactorial inheritance with the interaction of one or more genetic loci with one or more environmental factors is suspected for many common disorders of adulthood on the basis of this characteristic pattern of twin concordance. Family correlation studies are a direct extension of twin studies and seek to compare the frequency of a condition amongst groups of relatives who share the same proportions of their genes. For multifactorial conditions the frequency falls rapidly with each step further away in the pedigree whereas for single gene disorders the risk remains high in affected branches of the family. These observed frequencies in different relatives of patients with a multifactorial condition are called empiric risks and are useful for genetic counselling.

Table 23.1. Types of genetic disease

Type	Frequency/1000 births	No. of subtypes	Pedigree analysis	Twin concordance
Multifactorial disorders	60	>100	Non-diagnostic	Diagnostic
Somatic cell genetic disorders	250	>100	Non-diagnostic	Non-diagnostic
Single gene disorders	14	>5000	Diagnostic	Non-diagnostic
Chromosomal disorders	6	>400	Non-diagnostic	Non-diagnostic
Multifactorial disorders	Rare	2	Diagnostic	Non-diagnostic

Vascular Disease and "Old" Genetic Analyses

Twin studies have consistently shown an increased frequency of coronary heart disease in identical twins (ranging from 14%–90%) as compared with non-identical twins (reviewed by Goldbourt and Neufeld 1986). In each study male twins have shown a higher concordance than female twins and the concordance is higher in patients with an earlier age of onset (under 55 years of age). These studies support multifactorial inheritance of coronary heart disease and suggest that the genetic predisposition is most marked for premature disease in men.

Osler in 1897 had noted the familial aggregation of heart attacks, and multiple studies have confirmed a frequency of coronary heart disease of 2–6 times higher in patients' families than in control families (Epstein 1976; Robertson 1981; Berg 1983; Goldbourt and Neufeld 1986). This familial aggregation increases with decreasing age of affected patients (especially under 55 years of age). Hyperlipidaemia, high blood pressure and diabetes represent risk factors for coronary heart disease that have strong genetic determinants but various studies suggest that familial aggregation cannot be entirely accounted for by these and that other genes are involved. For example, Nora et al. (1980) analysed 19 independent variables in 207 patients who had a myocardial infarction before age 55 years and found that a positive family history of ischaemic heart disease was the strongest single risk factor. Furthermore in this group of patients the heritability was calculated at 0.63 (i.e. with only a 37% environmental contribution to the total predisposition). In this study 15% of their patients had hyperlipidaemia as a result of a single gene disorder but with their exclusion the heritability was still high at 0.56.

To date there have been fewer family studies in cerebrovascular disease or peripheral vascular disease. The risk of stroke is increased two- to three-fold amongst close relatives of an affected person and this risk appears to be independent of other known risk factors (Harris et al. 1989).

Such familial aggregation does not necessarily mean genetic determination since families share similar environments that may include agents causing a higher frequency of coronary heart disease in family members. There is evidence for assortative mating for coronary risk factors since spouse's families show the same extent of familial aggregation for coronary heart disease as the families of their affected husbands. Spouses of affected patients also have a higher frequency of coronary heart disease than the control population (ten Cate et al. 1984).

Background to "New" Genetic Analyses

The approaches outlined under "old" genetic analyses provide evidence for a genetic contribution to the occurrence of cardiovascular diseases but do not help to identify the nature of this contribution. Other approaches are required to identify which genes are involved and these are considered under linkage analysis, association analysis and sib-pair analysis. Linkage analysis is particularly helpful in single gene disorders and seeks to find cosegregation of the disease trait with a genetic marker within families. These studies have been greatly assisted in recent years by the development of polymorphic DNA markers. Prior to 1978 only polymorphic proteins (blood groups, enzymes, etc.) could be used as genetic markers and this approach was restricted by their limited number and uneven distribution throughout the chromosome. DNA polymorphisms (restriction fragment length polymorphisms, RFLPs; variable number of tandem repeats, VNTRs; CA repeats), in contrast, are numerous (over 2000 identified to date) and are available for all chromosomal regions. DNA from a single blood sample can be analysed for multiple markers and it has been estimated that if 150–200 are analysed then this will screen all chromosomes and allow identification of cosegregation of a particular chromosomal region with the disease (Botstein et al. 1980).

As an alternative to systematic linkage analysis using multiple markers from every chromosome a more random approach may be taken (serendipity) or candidate gene markers may be selected on the basis of observed biochemical alterations, known pathophysiology or observed linkage disequilibrium (intuitive approach). In this type of linkage study the statistical significance of observed cosegregation is measured by the lod score. This is the logarithm to the base 10 of the odds of the observed cosegregation versus that expected by chance alone and is estimated for a range of potential distances (called recombination fractions of 0–50%) between the disease and the marker (Conneally and Rivas 1980). A lod score of 3 (odds of 1000 to 1) is accepted as evidence of linkage and allows assignment of the disease to the chromosomal region of origin of the DNA marker. Lod scores have the advantage that information from different families can be combined in this type of study. Conversely failure of a DNA marker from within a particular gene to be co-inherited with a disease effectively excludes that gene as the cause of the disease.

As indicated above, linkage analysis is widely used in the investigation of discontinuous rare single gene disorders. Its application is more problematic if the condition has a continuous distribution (e.g. height or blood pressure) or has an age-related onset or difficulties in disease exclusion. A variety of statistical approaches have been suggested to assist in these situations (Risch 1990; Rotter and Landaw 1984). Further, in tomatoes a combination of selective breeding experiments and linkage analysis using multiple DNA markers has allowed identification of several quantitative trait lock for effects on the size and composition of the fruit (Paterson et al. 1988). Although a suitable series of comparable genetic markers is being developed for the human genome this approach is limited by difficulties with the structure of human pedigrees.

Association analysis compares the frequency of a particular genetic marker between patients and matched controls. To date most association analyses have been published in respect of the major histocompatibility complex (MHC). The MHC consists of a tightly linked cluster of genes, and DNA polymorphisms within such a tightly linked cluster show the phenomenon of linkage disequilibrium (Bodmer 1987). Crossing over within the cluster is infrequent and hence DNA markers in the group tend to be inherited en bloc as a haplotype. New markers or disease causing mutations arising within the cluster thus are associated with the haplotype of origin and only slowly over many generations lose this association. In the meantime linkage disequilibrium or marker association exists and an association study would reveal an excess of a particular haplotype amongst the patients as compared with the general population. Thus for example 95% of patients with insulin dependent diabetes mellitus have HLA DR3 or DR4 as compared with the combined frequency of these antigens of 50% in the normal population. Identification of controls for these studies is hampered if the disease is frequent, age-related and if disease exclusion is difficult. Selection of DNA markers may be systematic or selective on the basis of known pathophysiology, observed biochemical abnormalities or evidence from linkage analysis. Statistical significance of altered distributions of DNA markers between patients and controls is assessed by standard statistical tests and unfortunately in contrast to linkage analysis cannot be combined directly with results from other studies. Furthermore, as multiple markers are usually studied the chance of reaching statistical significance for one or more markers is high and this tends to result in a variety of conflicting potential associations.

Both linkage analysis and association analysis are hampered if a condition can be caused by different genes in different families (genetic heterogeneity) and in this respect sib-pair analysis has an advantage (Penrose 1935; Suarez and Hodge 1979). Brothers and sisters (sibs) have one half of their genes in common and hence will be genetically identical for on average one quarter of their pairs of genes. The affected sib-pair method aims to detect excessive similarity for a gene pair amongst affected sibs as such distortion suggests that this gene pair is contributing to the predisposition for a condition. Hence in sib-pairs with insulin dependent diabetes mellitus 57% have the same genes from the major histocompatibility complex (25% expected), 38% have one gene in common (50% expected) and only 5% have no gene in common for this complex (25% expected) (Field 1988). This approach needs no *a priori* information about the mode of inheritance or penetrance for a single gene disorder (both required for linkage analysis) and has been developed for affected family members other than sibs (Weeks and Lange 1988).

Vascular Disease and "New" Genetic Analyses

As indicated in the previous section of genetic markers for linkage analysis, association analysis or sib-pair analysis may be a systematic screen or a focused intuitive approach on the basis of candidate genes for the known pathophysiology, observed epidemiological associations, animal models or chance observation. To date studies have used a focused approach and there are similarities and differences for coronary heart disease, stroke and peripheral vascular disease.

Numerous epidemiological studies have confirmed cigarette smoking, elevated cholesterol, reduced high density lipoproteins, elevated blood pressure, diabetes mellitus, elevated levels of fibrinogen and a positive family history as independent risk factors for the occurrence of coronary heart disease (see review by Goldbout and Neufeld 1986). For stroke, identifiable risk factors include cigarette smoking, high blood pressure, diabetes mellitus, elevated levels of fibrinogen and a positive family history (Wilhelmson et al. 1984; Kannel et al. 1987). Similarly for peripheral vascular disease identifiable risk factors include cigarette smoking, high blood pressure, diabetes mellitus and elevated levels of fibrinogen (Fowkes 1989). Thus, whilst cigarette smoking, high blood pressure, elevated levels of fibrinogen and diabetes mellitus are risk factors for cardiovascular disease in general, abnormal lipid profiles are particularly associated with coronary heart disease. Furthermore, ethnic variation in frequency is more marked for coronary heart disease than for stroke or intermittent claudication. This variation cannot be accounted for by known risk factors and other genes appear to be involved (e.g. for coronary artery structure; Goldbourt and Neufeld 1986).

From these clues most studies have focused upon genes which are involved in the metabolism and transport of lipids in families with coronary heart disease. With premature ischaemic heart disease (onset under 55 years) about one-third of patients will have an abnormal lipid profile and of these patients one-half will have a single gene disorder. The most commonly diagnosed single gene disorders are familial hypercholesterolaemia (due to mutations in the low density lipoprotein receptor on chromosome 19), familial combined hyperlipidaemia (due to a gene locus on the long arm of chromosome 11 close to the apolipoprotein AI-

CIII-AIV gene cluster) and familial defective apo B-100. Twin and family studies have shown that normal lipid levels show a significant genetic contribution with heritabilities of 0.34 for total cholesterol, 0.66 for apolipoprotein B and 0.90 for apolipoproteins AI and AII (Berg 1983). Hence it is reasonable to expect that multifactorial elevations of lipids might arise from mutations in the genes which are normally involved in lipid metabolism. Multiple association studies have been performed in this respect and variation at the apolipoprotein E and B loci appears to account for 40% of the total genetic variance of plasma cholesterol (Sing and Davignon 1985; Law et al 1986; Talmus et al. 1987). Levels of high density lipoproteins are influenced by variation at the apolipoprotein AII locus (Scott et al. 1985) and variations in the apo AI-CIII-AIV gene cluster correlate with an increased tendency to elevated trigylcerides (Rees et al. 1985).

Increased plasma levels of fibrinogen have been shown in longitudinal studies to be associated with an increased risk of coronary heart disease and stroke (Meade et al. 1986; Wilhelmson et al. 1984; Kannel et al. 1987) and in epidemiological studies to be associated with peripheral vascular disease (Hughson et al. 1978). Plasma fibrinogen increases with age and cigarette smoking (Meade et al. 1979, 1987) but when corrected for these factors still emerged in the above studies as a risk factor on multivariate analysis. Multiple fibrinogen mutations have been characterised and these may be asymptomatic or result in a haemorrhagic tendency, recurrent episodes of thrombosis or intermittent claudication (McKusick 1990).

By analogy with lipid metabolism, genetic variation might be expected to be involved in elevated levels of plasma fibrinogen. Fibrinogen is synthesised in the liver from three polypeptide subunits, alpha, beta and gamma, which occur as a linked cluster of genes on the long arm of chromosome 4. Humphries et al. (1987) compared corrected levels of fibrinogen in 91 healthy individuals genotyped at the fibrinogen alpha and beta loci and estimated that genetic variation at these loci accounted for 15% of the total variation. This is consistent with the finding by Hamsten et al. (1987) on the basis of path analysis that 50% of the variation of plasma fibrinogen was due to genetic factors. In contrast Berg and Kierulf (1989) found low heritability of levels of fibrinogen in a Norwegian twin study and found no association between fibrinogen levels and fibrinogen alpha and beta genotypes.

Multiple association studies in cardiovascular disorders have shown an excess of the ABO blood groups A and B and a relative deficit of group O (reviewed by Mourant et al. 1971). In these studies the mean pooled relative incidence of A compared with O was 1.3. This association is not site-specific and has been confirmed by several studies in patients with intermittent claudication (Weiss 1972; Morris and Bouhoutsos 1973; Norrgard et al. 1989). In addition in the study of Norrgard et al. (1989) a statistically significant association with HLA A28 was also noted.

On a standard laboratory diet atherosclerosis is uncommon in laboratory mice but if a diet with a similar fat content to a Western man is introduced widespread atherosclerosis occurs in some strains (Bulfield 1988). Selective breeding of these strains can be combined with linkage analysis to define the contributing genetic locus (loci). So far three loci have been defined in mice using this approach which contribute to the occurrence of atherosclerosis (Stewart-Phillips et al. 1989). One locus (Ath 1) adjacent to, but distinct from, apolipoprotein AII on mouse chromosome 1 is linked to the regulation of high density lipoproteins and to the susceptibility to atherosclerosis (Paigen et al. 1987). Comparative mapping studies between mouse and man allow analogous regions of human chromosomes to be defined from which DNA markers can be identified for linkage analysis, association analysis or sib-pair analysis. This approach has the potential to identify contributing genetic loci which are unrelated to known risk factors.

References

Berg K (1983) Genetics of coronary heart disease. Progr Med Genet 5:35–90
Berk K, Kierulf P (1989) DNA polymorphism at fibrinogen loci and plasma fibrinogen concentration. Clin Genet 36:229–35
Bodmer WF (1987) The human genome sequence and the analysis of multifactorial traits. In: Ciba Foundation Symposium 130. Molecular approaches to human polygenic disease. Wiley, Chichester, pp 215–28
Botstein D, White RL, Skolnick M, Davis RW (1980) Construction of a genetic linkage map using restriction fragment length polymorphisms. Am J Hum Genet 32:314–31
Breslow JL, Deeb S, Lalouel JM et al. (1989) Genetic susceptibility to atherosclerosis. Circulation 80:724–8
Bulfield G (1988) Genetics of atherosclerosis and plasma lipoproteins in mice. TIG 4:3–4
Conneally PM, Rivas MC (1980) Linkage analysis in man. Adv Hum Genet 10:209–66
Epstein FH (1976) Genetics of ischaemic heart disease. Postgrad Med J 52:477–80
Field LI (1988) Insulin dependent diabetes mellitus: a model for the study of multifactorial disorders. Am J Hum Genet 43:793–8
Fowkes FGR (1989) Aetiology of peripheral atherosclerosis. Br Med J 298:405–6
Goldbourt U, Neufeld HN (1986) Genetic aspects of arteriosclerosis. Arteriosclerosis 6:357–77
Hamsten A, De Faire U, Iselius L, Blomback M (1987) Genetic and cultural inheritance of plasma fibrinogen concentration. Lancet ii:988–91
Harris EL, Keyl PM, Bush TL, Hale WE (1989) Family history of stroke as a risk factor for stroke in the elderly. Am J Hum Genet 45:A241
Hughson WG, Mann JI, Garrod A (1978) Intermittent claudication: prevalence and risk factors. Br Med J i:1379–81
Humphries SE, Cook M, Dubowitz M, Stirling Y, Meade TW (1987) Role of genetic variation at the fibrinogen locus in determination of plasma fibrinogen concentration. Lancet i:1452–5
Kannel WB, Wolf PA, Castelli WB et al. (1987) Fibrinogen and risk of cardiovascular disease. The Framingham Study. JAMA 258:1183–6
Law A, Wallis SC, Powell LM et al. (1986) Common DNA polymorphism within coding sequence of apolipoprotein B gene associated with altered lipid levels. Lancet i:1301–3
McKusick VA (1990) Mendelian inheritance in man, 9th edn. Johns Hopkins University Press, Baltimore
Meade TW, Chakrabarti R, Haines AP, North WRS, Stirling Y (1979) Characteristics affecting fibrinolytic activity and plasma fibrinogen concentration. Br Med J i:153–5
Meade TW, Imeson J, Stirling Y (1987) Effect of changes in smoking and other characteristics on clotting factors and the risk of ischaemic heart disease. Lancet ii:986–8
Meade TW, Mellows S, Brozovic M et al. (1986) Haemostatic function and ischaemic heart disease: principal results of the Northwick Park Heart Study. Lancet ii:533–7
Morris T, Bouhoutsos J (1973) ABO blood groups in occlusive and ectatic arterial disease. Br J Surg 60:892–3
Mourant AE, Kopec AC, Domaniewska-Sobczak K (1971) Blood groups and blood clotting. Lancet i:233–7
Nora JJ, Lortscher RH, Spangler RD, Nora AH, Kimberling WJ (1980) Genetic–epidemiologic study of early onset ischemic heart disease. Circulation 61:503–8
Norrgard O, Beckman G, Cedergren B (1989) HLA antigens, blood groups and serum protein groups in patients with intermittent claudication. Hum Hered 39:192–5

Paigen B, Mitchell D, Reue K et al. (1987) Ath-1, a gene determining the atherosclerosis
    susceptibility and high density lipoprotein levels in mice. Proc Natl Acad Sci (USA) 84:3763-7
Paterson AH, Lander ES, Hewitt JD, Peterson S, Lincoln SE, Tanksley SD (1988) Resolution of
    quantitative traits into Mendelian factors by using a complete linkage map of restriction fragment
    length polymorphisms. Nature 335:721-6
Penrose LS (1935) The detection of autosomal linkage in data which consists of pairs of brothers and
    sisters of unspecified parentage. Ann Eugen 6:133-8
Rees A, Stocks J, Sharpe CR (1985) DNA polymorphisms in the apo AI-CIII gene cluster:
    association with hypertriglyceridemia. J Clin Invest 76:1090-5
Reunanen A, Takkunen H, Aromaa A (1982) Prevalence of intermittent claudication and its effects
    on mortality. Acta Med Scand 211:249-56
Risch N (1990) Linkage analysis for genetically complex traits. Am J Hum Genet 46:222-53
Robertson FW (1981) The genetic component in coronary heart disease – review. Genet Res Cam
    37:1-16
Rotter JI, Landaw EM (1984) Measuring the genetic contribution of a single locus to a multilocus
    disease. Clin Genet 26:529-42
Scott J, Knott TJ, Priestley LM et al. (1985) High density lipoprotein composition is altered by a
    common DNA polymorphism adjacent to apolipoprotein AII in man. Lancet i:771-3
Sing CF, Davignon J (1985) Role of apolipoprotein E polymorphism in determining normal plasma
    lipid and lipoprotein variation. Am J Hum Genet 37:268-85
Stewart-Phillips JL, Lough J, Skamene E (1989) Ath-3, a new gene for atherosclerosis in the mouse.
    Clin Invest Med 12:121-6
Suarez BK, Hodge SE (1979) A simple method to detect linkage for rare recessive diseases: an
    application to juvenile diabetes. Clin Genet 15:126-36
Talmus PJ, Nazzarena B, Kessling AM et al. (1987) Apolipoprotein B gene variants are involved in
    the determination of serum cholesterol levels: a study in normo and hyperlipidemic individuals.
    Atherosclerosis 67:81-9
ten Cate LP, Boman H, Daiger SP, Motulsky AG (1984) Increased frequency of coronary heart
    disease in relatives of wives of myocardial infarction survivors: assortative mating for lifestyle and
    risk factors. Am J Cardiol 53:399-403
Weeks DE, Lange K (1988) The affected-pedigree member method of linkage analysis. Am J Hum
    Genet 42:315-26
Weiss NS (1972) ABO blood type and arteriosclerosis obliterans. Am J Hum Genet 24:65-70
Wilhelmson L, Svardsudd K, Jorsan-Bengsten et al. (1984) Fibrinogen as a risk factor for stroke and
    myocardial infarction. N Engl J Med 311:501-5

24 Fibrinogen

T.W. MEADE

There is now widespread (if fairly recent) recognition of a major thrombotic component in clinically manifest ischaemic heart disease, i.e. myocardial infarction and coronary death. Three-quarters or more of major strokes are also thrombotic or thrombo-embolic in origin. Because it is increasingly apparent that thrombogenicity is in many ways determined by lipoproteins (Miller et al. 1990), continued polarisation of the atherogenic and thrombogenic hypotheses for the pathogenesis of arterial disease is undesirable. But to the extent that thrombosis is probably the proximate intra-luminal cause of major episodes of ischaemic heart disease (Davies and Thomas 1984), it may prove that prevention and treatment with antithrombotic agents is the method of first choice, with the additional but specific treatment of other characteristics – such as hyperlipidaemia and hypertension – that contribute to the final process. This proposal may have implications for the management of peripheral vascular disease as well.

Despite the clinical importance of peripheral vascular disease and the advantages provided by an extensive and accessible system for study, aspects of the causes of peripheral vascular disease can usually only be inferred from studies on ischaemic heart disease and stroke rather than from direct evidence. Fortunately, this is only partially true in the present context.

Coagulability

There is, of course, no doubt about the central role of platelets in thrombogenesis. So far, however, there is no generally accepted test that indicates predisposition to thrombosis on account of increased platelet sensitivity. Study of the coagulation system, represented schematically in Fig. 24.1, has been more rewarding. Although this review is primarily concerned with fibrinogen, an appreciation of the steps leading to its conversion to fibrin and also of its non-coagulation effects is necessary. The function of the coagulation system is to produce thrombin, the main effect of which is traditionally considered to be the conversion of soluble fibrinogen to insoluble fibrin. As Fig. 24.1 shows, however, thrombin is also a potent platelet-aggregating agent, illustrating an important influence of the coagulation system on platelet behaviour. In addition, fibrinogen

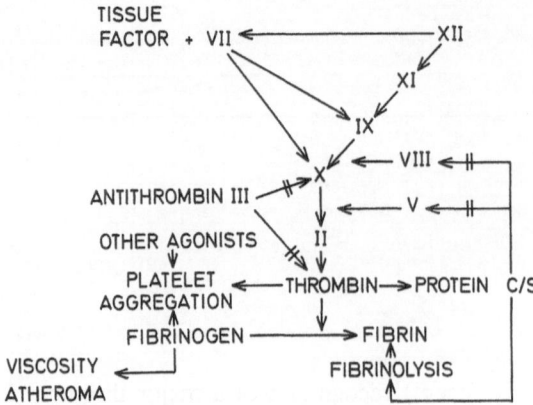

Fig. 24.1. Schematic summary of the coagulation system. Continuous lines indicate activation; barred lines indicate inhibition.

itself is also a cofactor for platelet aggregation. The activation of factor X and thus of the final pathway of coagulation is conventionally thought of as being initiated either by the contact or intrinsic system, starting with factor XII, or through the extrinsic system, so called because a contribution from outside the circulation – tissue factor – must complex with factor VII for factors IX and X to be activated. As Fig. 24.1 also shows, however, any sharp distinction between the intrinsic and extrinsic pathways is a considerable over-simplification. Through their activation of factor XII, lipoproteins such as chylomicrons and very low density lipoproteins (VLDL) contribute to the activation of factor VII, illustrating one link between the two systems. A second is the activation by factor VII and tissue factor of factor IX, a pathway the importance of which is increasingly recognised. Thrombin also stimulates the suppression of its own production through the protein C and protein S system, which reduces the activity of factor VIII and factor V and which also stimulates fibrinolytic activity. Factors II (prothrombin), VII, IX and X are vitamin K dependent clotting factors, the activity of which is reduced by warfarin. Besides being a determinant of the amount of fibrin formed when coagulation is initiated (Gurewich et al. 1976) and of platelet aggregability (Meade et al. 1985), high fibrinogen levels increase blood and plasma viscosity (Lowe 1986) and almost certainly contribute to the development of atheroma itself (Thompson and Smith 1989).

Ischaemic Heart Disease and Stroke

Two components of the coagulation system, the level of factor VII activity, VII_c, and fibrinogen appear to influence the onset of clinically manifest ischaemic heart disease and of stroke. In the case of VII_c, it is so far only the Northwick Park Heart Study (NPHS) that has shown a prospective relationship between high levels and subsequent ischaemic heart disease (Meade et al. 1980, 1986) but there is growing evidence from laboratory and clinical studies as well that they influence thrombin production and clinical events (Bauer and Rosenberg, 1987; Bauer et al. 1990). VII_c will not be considered further here except to re-emphasise its modification by warfarin and the implications this may have for the prevention and treatment of thrombosis, including thrombosis in the peripheral circulation.

A number of non-specific stimuli affect clotting factor levels. VII_c tends to be lowered by chronic or acute illness and by surgery, for example. On the other hand, these and other stimuli raise fibrinogen levels. Atheroma has many of the characteristics of an inflammatory reaction. It has therefore been suggested that the association between high fibrinogen levels and vascular disease, considered in more detail later, reflects the degree of underlying vessel wall disease rather than the direct influence of fibrinogen. There are three ways in which this important question can be elucidated to some extent. First, the pathways through which high fibrinogen levels may operate in predisposing to thrombosis (and which have already been summarised) will presumably do so whatever their origin, i.e. genetic, environmental or in response to inflammation. Thus, even if part of the explanation for high levels is a response to underlying atheroma, this contribution is likely to be functionally significant. Secondly, the extent to which high levels can be explained in this way is limited by the contribution of other influences, genetic or environmental. Several studies have now begun to explain some of the variance in fibrinogen in genetic terms (Humphries et al. 1987; Thomas et al. 1991) and in one it was suggested that the genetic contribution might be of the order of 50% (Hamsten et al. 1987). Smoking is the main environmental determinant of fibrinogen. Even here, it can theoretically be argued that the increase in fibrinogen in smokers is secondary to atherogenesis. However, fibrinogen levels begin to fall quite rapidly after smoking is discontinued (Meade et al. 1987) and to a greater extent than can be plausibly explained by the regression of atheroma, given the difficulties of achieving this. The origin of high fibrinogen levels is an important area for further study. In practical terms, however, this is a somewhat academic question in view of the several pathways through which, however they arise, high levels may predispose to thrombogenesis once they are established. The third approach is to rely much less on cross-sectional associations after the clinical event than on prospective data. Thus, the finding of high fibrinogen levels in survivors of myocardial infarction could indeed be due to the effects of the myocardial infarction itself. In a prospective setting, it is still possible that the high fibrinogen level is partly due to atheroma but at least it precedes the clinical event.

Special attention should be given to the relationship between smoking and fibrinogen, for three reasons. First, it seems increasingly likely that a major part of the relationship between smoking and ischaemic heart disease itself is mediated through the plasma fibrinogen level (Meade et al. 1987). The simultaneous inclusion of both smoking and fibrinogen in multivariate analyses may therefore be misleading unless the interpretation of their results is carefully considered. In one of the prospective studies discussed elsewhere (Meade et al. 1986), the effect of smoking "disappeared" when both variables were included, though few if any would now suggest that this means smoking is not involved in ischaemic heart disease. In another (Wilhelmsen et al. 1984), smoking appeared more important than fibrinogen. Relying too heavily on the face value results of multivariate analyses may obscure the biological message – that it is largely through fibrinogen that smoking has its effect on ischaemic heart disease. Secondly, obvious interest in the relationships between smoking, fibrinogen and peripheral vascular disease should not overlook the fact that fibrinogen levels are associated with an increased risk of ischaemic heart disease in non-smokers, as well. Thirdly, the particular importance of smoking in peripheral vascular disease emphasises the central role fibrinogen may play in its onset and progression. The very strong relationship of smoking to aortic aneurysm is not widely appreciated but of obvious importance.

Five prospective studies have now reported on the association between fibrinogen and ischaemic heart disease (Meade et al. 1980, 1986; Wilhelmsen et al. 1984; Stone and Thorp 1985; Kannel et al. 1987; Yarnell et al. 1991). In two, (Wilhelmsen et al. 1984; Kannel et al. 1987) the study population was sufficiently old to allow the relationship between fibrinogen and stroke also to be established. In summary, all have shown strong associations of high levels with both ischaemic heart disease and stroke. The relationships of fibrinogen with these major arterial events are independent of other predictive variables (subject, in one case already referred to (Wilhelmsen et al. 1984) to not including both smoking and fibrinogen in a multivariate analysis). They are about as strong as these other effects, e.g. the difference in risk of ischaemic heart disease or stroke for a standard deviation difference in fibrinogen is of the same order as for a standard deviation difference in cholesterol or blood pressure. Blood cholesterol levels are less strongly associated with the incidence of stroke than of ischaemic heart disease. At least in men, the clear associations of Fibrinogen in prospective studies indicate that it may be the best biochemical predictor of the risk of major cardiovascular disease, defined as the sum of ischaemic heart disease and stroke.

Several of the prospective studies have drawn attention to the combination of hypertension and high fibrinogen levels in its association with arterial disease, the possibility being that it is viscous blood being pumped through the circulation at high pressure that is particularly undesirable. In this context, it is worth considering an admittedly *post hoc* interpretation of a suggestive finding of the World Health Organisation (WHO) trial of clofibrate in the primary prevention of ischaemic heart disease in men (Committee of Principal Investigators, 1980). The rationale for this trial was the lipid-lowering properties of clofibrate. It is, however, also clear that clofibrate lowers fibrinogen levels. The WHO trial demonstrated a reduction of about 20% in the incidence of non-fat myocardial infarction that was attributable to clofibrate. However, this benefit was offset by an increase in mortality from a range of conditions, with the consequence that the agent has not been widely used for preventive purposes. Table 24.1 summarises

Table 24.1. WHO Clofibrate Trial: Age-standardised myocardial infarction rates per 1000 per annum

	Non-smokers		Cigarette smokers			
			<20/day		>20/day	
	Clofibrate	Placebo	Clofibrate	Placebo	Clofibrate	Placebo
Normotensive systolic BP <140 mmHg	3.4	3.5	6.2	9.6	9.6	9.1
Hypertensive systolic BP >140 mmHg	3.9	6.0	12.1	14.3	8.7*	18.3*

* $p < 0.01$

an analysis of the myocardial infarction results according to smoking status and blood pressure (Green et al. 1989). Heavy smokers had the highest fibrinogen levels. It can be seen that the protective effect against myocardial infarction was confined to the heavy smokers with the highest blood pressures. In other words, breaking the link between high fibrinogen and high blood pressure levels seemed particularly effective.

Peripheral Vascular Disease

Most of the information about fibrinogen and peripheral vascular disease is still cross-sectional, comparing fibrinogen levels in those with and without peripheral vascular disease, so that its interpretation is subject to the reservations already discussed. However, this literature is quite extensive and generally very consistent in showing strong relationships between fibrinogen and peripheral vascular disease. (The review by Leng and Fowkes (1991) also summarises the relationships of other haemostatic variables to peripheral vascular disease, which generally confirm the expectation that high levels of prothrombotic characteristics and low levels of protective mechanisms – such as fibrinolytic activity – are associated with peripheral vascular disease.) In the Scottish Heart Health Study (see Chapter 10), the odds ratios for intermittent claudication in men with plasma fibrinogen levels in the middle and high thirds of the fibrinogen distribution are 2.33 and 2.73, respectively, compared with those with levels in the low third. This association remains significant after adjusting for other variables. It is, however, interesting that the association was not observed in the smaller number of women. In the Edinburgh Artery Study, fibrinogen was significantly higher in those with low ankle-brachial systolic pressure indices (ABPI), used as a measure of arterial narrowing (F.G.R. Fowkes, personal communication).

So far, it appears to be only the Framingham Study that has related fibrinogen to the first clinical onset of peripheral vascular disease, demonstrating the expected relationship of raised levels in those who progress to peripheral vascular disease (Kannel et al. 1990). However, prospective information is also available from studies in patients with symptomatic peripheral vascular disease and concerned with the progression of disease. In a series of clinical studies first reported in 1973, Dormandy et al. (1973a) found that "a raised plasma fibrinogen was the most common single biochemical abnormality" in 126 patients with intermittent claudication, those with worsening disease showing the highest levels. This group (Dormandy et al. 1973b) also assessed the prognostic significance of several clinical and biochemical measures in 62 patients followed for between one and three years. They demonstrated a strong relationship between initially high fibrinogen levels and worsening disease, measured either as claudication distance or change in flow pattern. Other data strongly suggested that fibrinogen was exerting its deleterious effects through increased viscosity. Later, 62 patients (mostly from the previous group) were treated with clofibrate for at least 6 months. Their progress was compared with that of a matched but untreated control group of 27 patients. The most obvious effect of clofibrate was a sustained fall in whole-blood viscosity over a range of shear rates, an improvement associated with, and probably explained by, a significant fall in fibrinogen levels (Dormandy et al. 1974). Those treated with clofibrate appeared to show greater evidence of clinical improvement as well. Although this study was not based on random allocation to treatment, the results are nevertheless very suggestive of a beneficial effect due to the fibrinogen-lowering properties of clofibrate.

Wiseman et al. (1989) studied the relationships between a number of variables including fibrinogen and the patency of femoro-popliteal vein graphs in 157 patients with peripheral vascular disease. Plasma fibrinogen was the strongest predictor of graft occlusion, followed by markers of smoking.

Implications and Future Research

Particularly bearing in mind the results of the prospective studies concerned mainly with ischaemic heart disease and stroke, it seems very likely if not certain that the plasma fibrinogen concentration plays a major part in the onset and progression of peripheral vascular disease in men, though less certainly in women. This conclusion raises a number of questions and further research possibilities relevant not only to peripheral vascular disease itself but also to a more general understanding of the pathogenesis of degenerative arterial disease. First, assuming that high fibrinogen levels are causally related to onset and progression, how far do they contribute to chronic vessel wall pathology and how much to acute thrombotic and thromboembolic episodes? Secondly, do high fibrinogen levels contribute to thrombogenesis mainly through effects on viscosity, platelet aggregation or fibrin deposition? The answer to this particular question has obvious therapeutic implications. It is also worth re-emphasising the particular opportunities provided by the peripheral circulation for work on these and related topics, the results of which might well be applicable to the prevention and management of ischaemic heart disease and stroke as well.

As far as the first onset of clinical manifestations is concerned, there are of course interesting contrasts between different sites – for example, the predominance of hypertension for stroke and of smoking for peripheral vascular disease. Nevertheless, the general similarity between the characteristics associated with ischaemic heart disease, stroke and peripheral vascular disease is striking. It is, consequently, increasingly appropriate to consider the primary prevention of all types of major arterial episodes as a single, overall objective. In this case, there is little to be gained by considering the primary prevention of clinically manifest peripheral vascular disease on its own. Once disease at a particular site has occurred, however, there may be more specific considerations, e.g. preventing rhythm disturbances in ischaemic heart disease and gangrene in peripheral vascular disease. This review has concentrated almost entirely on the probable role of fibrinogen in peripheral vascular disease, but it does seem likely that other aspects of the haemostatic system are also involved (Leng and Fowkes 1991). Thus, the progression of clinical peripheral vascular disease may depend on lipid, platelet, coagulation and flow effects, the significance of which may vary from patient to patient. In turn, therefore, the appropriate pharmacological management of peripheral vascular disease may include lipid-lowering agents, aspirin and oral anticoagulants and agents with rheological properties, suggesting a future programme of randomised controlled trials leading to the increasingly rational management of peripheral vascular disease.

References

Bauer KA, Rosenberg RD (1987) The pathophysiology of the prethrombotic state in humans: Insights gained from studies using markers of hemostatic system activation. Blood 70:343–50

Bauer KA, Kass BL, ten Cate H, Bednarek MA, Hawiger JJ, Rosenberg RD (1990) Factor IX is activated in vivo by the tissue factor mechanism. Blood 76:731–6

Committee of Principal Investigators (1980) WHO co-operative trial on the primary prevention of ischaemic heart disease using clofibrate to lower serum cholesterol: mortality follow-up. Lancet ii:379–85

Davies MJ, Thomas A (1984) Thrombosis and acute coronary artery lesions in sudden cardiac ischaemic death. N Engl J Med 310:1137–40

Dormandy JA, Hoare E, Colley J, Arrowsmith DE, Dormandy TL (1973a) Clinical, haemodynamic, and biochemical findings in 126 patients with intermittent claudication. Br Med J iv:576–81

Dormandy JA, Hoare E, Khattab AH, Arrowsmith DE, Dormandy TL (1973b) Prognostic significance of theological and biochemical findings in patients with intermittent claudication. Br Med J iv: 581–3

Dormandy JA Gutteridge JMC, Hoare E, Dormandy TL (1974) Effect of clofibrate on blood viscosity in intermittent claudication. Br Med J iv:259–62

Green KG, Heady A, Oliver MF (1989) Blood pressure, cigarette smoking and heart attack in the WHO co-operative trial of clofibrate. Int J Epidemiol 18:355–60

Gurewich V, Lipinski B, Hyde E (1976) The effect of the fibrinogen concentration and the leukocyte count on intravascular fibrin deposition from soluble fibrin monomer complexes. Thrombos Haemost 36:605–14

Hamsten A, Iselius L, de Faire U, Blomback M (1987) Genetic and cultural inheritance of plasma fibrinogen concentration. Lancet ii:988–91

Humphries SE, Cook M, Dubowitz M, Stirling Y, Meade TW (1987) Role of genetic variation at the fibrinogen locus in determination of plasma fibrinogen concentrations. Lancet i:1452–5

Kannel WB, Wolf PA, Casteli WP, D'Agostino R (1987) Fibrinogen and risk of cardiovascular disease. JAMA 258:1183–6

Kannel WB, D'Agostino RB (1990) Update of fibrinogen as a major cardiovascular risk factor: the Framingham Study. J Am Coll Cardiol 15:156A (abstract)

Leng GC, Fowkes FGR (1991) Epidemiology of peripheral vascular disease. In: Forbes CD (ed) Current medical literature: Thrombosis 1:35–43

Lowe GDO (1986) Blood rheology in arterial disease. Clin Sci 71:137–46

Meade TW, North WRS, Chakrabarti R, Stirling Y, Haines AP, Thompson SG (1980) Haemostatic function and cardiovascular death: early results of a prospective study. Lancet i:1050–4

Meade TW, Vickers MV, Thompson SG, Stirling Y, Haines AP, Miller GJ (1985) Epidemiological characteristics of platelet aggregability. Br Med J 290:428–32

Meade TW, Mellows S, Brozovic M et al. (1986) Haemostatic function and ischaemic heart disease: principal results of the Northwick Park Heart Study. Lancet ii:533–7

Meade TW, Imeson JD, Stirling Y (1987) Effects of changes in smoking and other characteristics on clotting factors and the risk of ischaemic heart disease. Lancet ii:986–988

Miller GJ, Martin JC, Mitropoulos KA, Cruickshank JK (1990) Factor VII and dietary fat intake. In: Liu CY, Chien S (eds) Advances in experimental medicine and biology, vol. 281. Plenum Press, New York, pp. 145–9

Stone MC, Thorp JM (1985) Plasma fibrinogen – a major coronary risk factor. J R Coll Gen Pract 35:565–9

Thomas AE, Green FR, Kelleher CH et al. (1991) Variation in the promoter region of the B fibrinogen gene is associated with plasma fibrinogen levels in smokers and non-smokers. Thromb Haemost 65:487–90

Thompson WD, Smith EB (1989) Atherosclerosis and the coagulation system. J Pathol 159:97–106

Wilhelmsen L, Svardsudd K, Korsan-Bengtsen K, Larsson B, Welin L, Tibblin G (1984) Fibrinogen as a risk factor for stroke and myocardial infarction. N Engl J Med 311:501–5
Wiseman S, Kenchington G, Dain R et al. (1989) Influence of smoking and plasma factors on patency of femoropopliteal vein grafts. Br Med J 299:643–6
Yarnell JWG, Baker IA, Sweetnam PM et al. (1991) Fibrinogen, viscosity, and white blood cell count are major risk factors for ischaemic heart disease. Circulation 83:836–44

25 von Willebrand Factor, Beta-Thromboglobulin and Platelet Activation

F.B. SMITH

Epidemiological and clinical studies investigating haemostasis and thrombosis in the aetiology of atherosclerosis have concentrated mostly on coagulation factors, particularly fibrinogen. Recently there has been considerable interest in the role of endothelial damage and initiation of haemostasis in atherogenesis. Platelet activation plays a major part in this process and von Willebrand factor and β-thromboglobulin are two factors commonly associated with activation. In this chapter, the mechanisms by which von Willebrand factor and β-thromboglobulin may be involved in platelet activation are reviewed, and the limited clinical and epidemiological evidence relating these factors to vascular disease are examined.

Role of Platelets in Atherosclerosis and Thrombosis

Intact endothelium used to be considered to be a passive barrier preventing platelets adhering to reactive subendothelium and initiating haemostatis and clot formation (Gimbrone Jr 1987). Endothelium is now known to regulate haemostasis and thrombosis by actively influencing platelet function, coagulation and fibrinolysis. Indeed, endothelial cells have various defence mechanisms for limiting the formation of platelet aggregates to damaged sites. Not only are the endothelial cells negatively charged and repel similarly charged platelets but they can synthesise prostacyclin (PGI2) (Weksler et al. 1977) and endothelium-derived relaxing factor (EDRF) (Griffith et al. 1984) which are both potent inhibitors of platelet aggregation. Furthermore, endothelial cells synthesise ecto-adenosine diphosphatase which metabolises adenosine diphosphate (ADP), a mediator of platelet activation, and also thrombomodulin which inactivates thrombin. Thus, endothelium has several antithrombogenic effects. However, it can also promote haemostasis by releasing platelet activating factor (PAF) and von Willebrand factor (vWF) which mediates platelet adhesion to subendothelium.

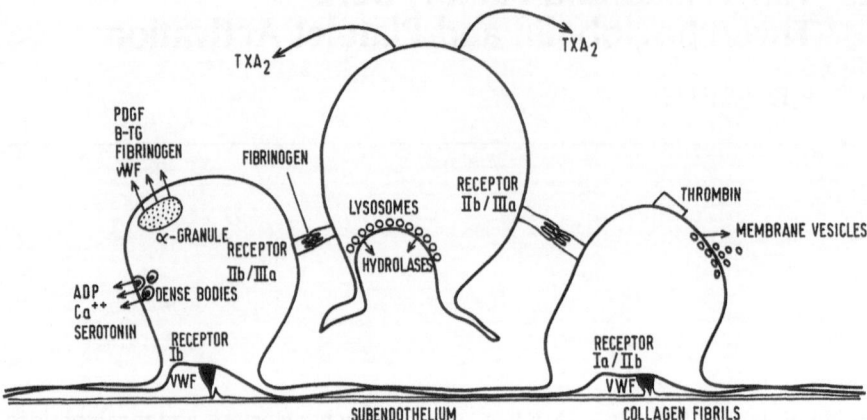

Fig. 25.1. Platelet activation showing adhesion, release and aggregation reactions.

Platelet Activation

When endothelium is injured, platelets from circulating blood quickly adhere to collagen and microfibrils exposed beneath the surface of the vessel wall. Adhesive proteins such as vWF and fibronectin present in the subendothelial matrix interact with glycoprotein receptors Ib and IIb-IIIa on the platelet surface, resulting in a monolayer of platelets spreading along the subendothelium (Sixma 1987) (Fig. 25.1).

Release of the contents of storage granules from platelets, such as ADP, serotonin, platelet derived growth factor (PDGF), fibrinogen and vWF stimulates and reinforces platelet aggregation. Fibrinogen is thought to act as a bridge between aggregating platelets via the glycoprotein receptor IIb–IIIa (Marguerie and Plow 1983). Also, activation results in the formation and release from platelets of thromboxane A2 (TXA2) which acts metabolically, together with ADP and thrombin, to reinforce aggregation.

Rearrangement of phospholipid binding sites on the surface of activated platelets and endothelial cells activates coagulation pathways leading to the release of thrombin. Thrombin has been shown to bind to human platelets at specific surface receptor sites (Shuman and Majerus 1976). It appears that a threshold level of thrombin has to build up before it can bind to platelets, thus possibly controlling the size of the fibrin clot. (Tollefsen and Majerus 1976). Thrombin also causes the conversion of fibrinogen to fibrin which stabilises aggregated platelets (Mustard et al. 1987). The resulting platelet plug and fibrin clot seal the damaged blood vessel.

Endothelial Injury

The reasons for focal damage to endothelium and the subsequent progression to thrombosis is important in implicating platelets in the pathological process. A number of factors can damage vessel walls, although it is recognised that injury may only alter the surface of endothelial cells and reduce their antithrombogenic function, rather than removing them structurally (Ross 1986). The conventional cardiovascular risk factors of hyperlipidaemia, hypertension, and smoking have each been implicated in causing damage to vascular endothelium. Chronic hyperlipidaemia may influence the endothelium by possibly increasing membrane permeability to cholesterol (Jackson and Gotto 1976). Hypertension may induce platelet aggregation by dislodging cells, particularly when there is high velocity flow over a stenosis. (Joris et al. 1982).

Smoking has been reported to cause decreased PGI2 production by endothelial cells (Reinders et al. 1986) and thus enhance the thrombotic function of vessel walls by increasing platelet aggregation. Smoke constituents have also been shown to damage endothelial cells in culture and to induce antigen and antibody complexes which are known to injure the vessel wall. Thus smoking would appear to be associated with increased platelet activation (Becker et al. 1976).

Platelets in Lesion Initiation and Thrombosis

One of the inherent mechanisms in the pathogenesis of atherosclerosis is smooth muscle cell (SMC) proliferation and migration into vessel intima. Smooth muscle cells are present in fatty streaks and predominate in advanced fibrous plaques, where they accumulate lipids (Ross 1986). The discovery in 1974 that platelets release PDGF, which is chemotactic and mitogenic for smooth muscle cells, suggested a role for platelets in atheroma formation following endothelial injury (Ross et al. 1974).

More evidence was obtained in animal studies of platelet involvement in lesion development. In one study on rabbits, repeated endothelial injury, caused by an indwelling catheter, led to a continuous deposition of platelets and fibrin leading to rapid formation of lipid rich lesions. When the catheter was removed the lesions regressed to SMC plaques (Friedman et al. 1976). A second type of injury, where the endothelium was completely removed by a balloon catheter, produced a monolayer of adherent platelets which progressed to a lesion composed of proliferating SMC and cholesterol esters. If the neointima formed after the first injury is damaged, a platelet–fibrin thrombus forms, particularly at vessel orifices, which eventually contributes to intimal thickening (Stemerman 1973). Moore et al. (1976) showed that thrombocytopenia prevented lesion development in repeated injury.

The role of platelets in arterial thrombosis is now well established. Arterial thrombi are composed of aggregated platelets within layers of fibrin and are known as "white" thrombi. Mural thrombi, usually forming at focal points in arteries such as vessel orifices and bifurcations, contribute to vessel narrowing by being incorporated into intimal lesions (Packham et al. 1967). These areas are characterised by divided blood flow, vortex formation and reduced blood velocity. This increases vessel wall interactions as platelets, rather than travelling along longitudinal axes in the centre of the vessel, are carried to the vessel wall, (Zarins et al. 1983). Vortex formation also traps platelet aggregates, and static flow permits the build-up of platelet products and coagulation factors (Karino and Goldsmith 1979).

Arterial thrombi form where atheroma causes narrowing of the blood vessel and turbulent flow. Rupture of ulcerated plaques with release of thrombogenic substances, such as collagen and necrotic cells, induces thrombosis particularly in the aorta (Friedman 1970; Davies and Thomas 1985). When narrowing of the vessel is less than 75%, the thrombi will usually remain non-occlusive. These may embolise and can cause limb, cerebral and coronary ischaemia as well as sudden death due to ventricular fibrillation (Haerem 1978). If, however, the narrowing of the blood vessel is further increased, platelets will aggregate particularly over a ruptured atherosclerotic plaque, and an occlusive thrombus will form (Falk 1983).

Occlusive thrombi have been implicated in myocardial infarction where they are usually found in coronary arteries supplying an infarcted region (Packham and Mustard 1986). Recurrent thrombosis associated with disrupted plaques is also believed to have an important part to play in unstable angina where recurrent chest pain occurs at rest. It is thought that alternating platelet aggregation and thrombosis cause vasoconstriction and reduction of blood flow by the action of TXA2 and serotonin released by platelets and hence may cause ischaemic attacks (Fuster and Chesebro 1986).

von Willebrand Factor

The production of a platelet–fibrin clot at sites of endothelial injury is crucial for normal haemostasis, thrombosis and possibly atherogenesis.

The first step in this process is platelet adhesion to various components of the subendothelium. Initial studies by Tschopp et al. (1974) using perfusion chambers developed by Baumgartner (1973) demonstrated that von Willebrand factor (vWF) was required for normal platelet adhesion to sections of rabbit aorta which were bathed in blood from patients with von Willebrand disease. Later, this bleeding disorder in patients, caused by platelet adhesion, was found to be corrected by plasma transfusions containing vWF (Weiss et al. 1978; Sakariassen et al. 1979). Therefore an essential role for vWF in platelet adhesion was established.

Synthesis and Function of vWF

von Willebrand factor is synthesised primarily in endothelial cells, but also in the megacaryocyte, the platelet producing cell contained in bone marrow. It is present in plasma, the α-granules of platelets, and in the blood vessel wall (Sixma 1987). In 1973 Jaffe et al. demonstrated that endothelial cells synthesise and release vWF in vitro, and Rand et al. (1980) later found that its secretion is followed by incorporation into the subendothelium.

von Willebrand factor has two functions in vivo. In plasma it is bound to coagulation factor VIIIC, where it may act to protect factor VIIIC from degradation (Tuddenham 1989) and to concentrate the factor at injured sites. vWF also acts as a co-factor for platelet adhesion to the subendothelium of damaged blood vessel walls (Baumgartner 1973). For optimal adhesion both endothelial and plasma vWF are required. Plasma vWF is first bound to the vessel wall and undergoes a conformational change before interaction with platelets (De Groot and Sixma 1987). Recent evidence suggests that vWF not only initiates adhesion but causes platelets to spread out as a monolayer along the subendothelium (Leytin et al. 1984). It has been demonstrated in vitro that vWF affects the rate of adhesion, but not the number of platelets adhering to non-fibrillar collagen (Aihara et al. 1984).

von Willebrand factor interacts strongly with fibrillar collagen types I, II and III but less well with non-fibrillar collagen types I and II (Sixma et al. 1987). There is some evidence that it also binds to other components of the subendothelium, such as glycolipids or heparin (Roberts et al. 1986; Fujimura et al. 1987). Two glycoprotein (gp) platelet receptors have been identified as binding sites for vWF to platelets: gp Ib, and gp IIb-IIIa. It seems that vWF must first bind to the subendothelium before platelet adhesion can take place (Bolhuis et al. 1981). Primary platelet adhesion probably occurs by subendothelial vWF binding to the gp Ib receptor, with subsequent activation and release from platelet α-granules of ADP, adrenaline and fibrinogen and the production of thrombin. The receptor gp IIb-IIIa becomes activated through the agonists ADP, thrombin and collagen and binds not only to vWF but also to fibrinogen. Fibrinogen acts as a ligand to link platelets together as aggregates. However, von Willebrand factor bound to the gp IIb-IIIa receptor may also mediate platelet aggregation because patients with afibrinogenaemia have platelet aggregates linked by vWF (Demarco et al. 1986).

The importance of shear rate of vWF–platelet interaction has been established using monoclonal antibodies. von Willebrand factor is essential for platelet adhesion to the vessel wall when shear rates are greater than 800 SEC-1 such as occurs in arteries and the microvasculature (Weiss et al. 1979). It also partly contributes to adhesion at lower shear rates but low shear adhesion is much more dependent on fibronectin (Houdijk and Sixma 1985). At high shear rates fibronectin mediated adhesion is not strong enough to bind platelets to the subendothelium. von Willebrand factor binds to the gp receptor IIb-IIIa (made available after activation) to promote maximal platelet adhesion over the matrix as well as binding to smooth muscle cells and fibroblasts (Sixma et al. 1987).

This mechanism may be important in thrombosis and atherogenesis where stenosed arteries induce high shear rates and turbulent flow with resultant formation of platelet–fibrin thrombi. von Willebrand factor has been shown to contribute to platelet thrombi build-up (Nichols et al. 1990). In animal studies, inhibition of the gp IIb-IIIa receptor by monoclonal antibodies impaired platelet adhesion and deposition as well as decreasing occlusion of arteries by platelet thrombi (Hanson et al. 1988).

Role of vWF in Thrombosis

A possible role of vWF in thrombosis was demonstrated using normal pigs and pigs with von Willebrand's disease. In a recent study conducted by Nichols et al.

(1990) hypercholesterolaemia was induced by feeding 8 normal and 6 diseased pigs with a 1%–2% cholesterol diet for 24 weeks. All the pigs but one developed coronary atherosclerosis. Occlusive thrombi were produced by superimposed stenosis and pinch injury. The phenotypically normal pigs all developed occlusive thrombi whereas the pigs with von Willebrand's disease had only platelet–fibrin microthrombi attached to the artery wall. Therefore it may be concluded that vWF is required for progression and development of an occlusive thrombosis. This may be directly due to high shear rates resulting from stenosed arteries, where increased platelet activation is produced by the interaction of receptor gp IIb-IIIa with vWF.

It should be noted however that in this study (Nichols et al. 1990) normal pigs had significantly higher levels of hypercholesterolaemia than pigs with von Willebrand's disease which may be due to different mechanisms of cholesterol metabolism between the two groups. Studies in pigs without induced atherosclerotic lesions have produced similar results. In a previous study, Nichols et al. (1986) reported occlusive thrombi in normal pigs but none in diseased pigs. Intimal thickening was produced in normal pigs fed a high lipid diet but was lacking in pigs with severe von Willebrand's disease. Attempts were made to induce atherosclerosis by transplanting normal aortic segments into pigs with von Willebrand's disease but these became resistant to atherosclerosis (Fuster et al. 1978) and no lesions developed.

vWF and Vascular Disease

In a study examining plasma vWF levels in acute myocardial infarction, peripheral vascular disease and post-operative states (Cucuianu et al. 1980), the highest level of vWF was obtained in infarction patients (213% SD 18.8% normal pool) compared with other conditions. The levels of vWF in atherosclerosis correlated with the severity of the disease. It was concluded that vWF levels may be elevated in myocardial infarction due to possible thrombi formation, but high levels in post-operative patients confirmed vWF as an acute phase reactant.

Mossard et al. (1989) estimated plasma vWF levels in patients with acute myocardial infarction, unstable angina and coronary thrombosis. von Willebrand factor was significantly increased in groups with transmural myocardial infarction and intracoronary thrombi compared with controls, but no significant differences were noted between patients with or without visible thrombi. This suggests that vWF may contribute to development of thrombosis.

In one part of the Basle Study of peripheral vascular disease (Christe et al. 1984) 89 cases and 217 controls without peripheral vascular disease were examined. von Willebrand factor levels were significantly higher in the cases than controls (254% for cases and 225% for controls, p = 0.004). Of the subjects in this study, 153 had a normal electrocardiogram and 25 had a normal electrocardiogram but previous myocardial infarction. Their vWF levels were compared and showed a higher mean level in the group with previous infarction (p<0.01) but it should be noted that there was a significant difference in mean age between the two groups which might have influenced the vWF levels.

Similar results were found in a case–control study of subjects with peripheral vascular disease selected from the cross-sectional phase of the Edinburgh Artery

Table 25.1. von Willebrand factor and β-thromboglobulin: mean levels in peripheral vascular disease and controls, and relation to smoking in Edinburgh Artery Study

	Mean levels		p	Cigarette smoking (pack-years) Spearman rank correlation coefficient	p
	Cases ($n=121$)	Controls ($n=126$)			
von Willebrand factor (% normal pool)	109.8	95.6	0.003	0.14	<0.05
β-Thromboglobulin (ng/ml)	41.1	30.9	0.002	0.15	<0.05

Study (personal observations). The mean plasma vWF was 109.8% for cases and 95.6% for controls, p<0.003, suggesting increased platelet adhesion in peripheral vascular disease (Table 25.1).

Beta-Thromboglobulin

Platelet activation stimulates release of various materials from the three types of platelet intracellular storage granules, namely the α-granules, dense bodies and lysosomes (Grette 1962) (Fig. 25.1). During the last few years, many of these products of the "platelet release reaction" have been isolated and characterised.

One of these products, beta-thromboglobulin (β-TG), is a specific protein released from the α-granules. It was first isolated, characterised and named by Moore et al. in 1975). It is thought to be a tetramer composed of four identical subunits. However, the term "β-thromboglobulin" is in fact usually used to describe a family of three almost identical proteins.

No definite biological function has yet been established for β-TG. Initial research found that β-TG is capable of binding and neutralising heparin at a ratio of 2.6 units of heparin to 1 mg of β-TG (Rucinski et al. 1979). It was also proposed that due to its high concentration in platelets (4–8 mg/10^8 platelets) β-TG might stabilise the more active proteins contained in α-granules (Begg et al. 1978). In 1979 Hope et al. reported that β-TG was found to decrease production of the platelet aggregating inhibitor, prostacyclin (PGI2), and bind to a specific receptor in bovine aortic endothelial cells.

In addition to having a possible role in thrombosis, β-TG may be involved in inflammatory reactions. This was suggested recently when the structure of β-TG was found to be similar to a group of proteins (St Charles et al. 1989), known as the small inducible gene (SIG) family of proteins, which are active in local inflammation and control mechanisms. Also, the β-TG family of proteins are now thought to be associated with cell proliferation as part of the tissue remodelling process after inflammation and activation of haemostasis (Walz et al. 1989).

Since evidence of in vivo platelet activation might be of use in indentifying or even predicting patients at risk of thrombosis and the complications of athero-sclerosis, attempts have been made in recent years to develop valid and reliable markers of this process. The development of a sensitive radioimmunoassay for β-TG in 1975 by Ludlam et al. seemed promising in detecting early platelet interaction with vessel walls. The specificity of the assay is due to the fact that β-TG is unique to platelets (Niewiarowski 1977) and the biochemical sensitivity is based upon the negligible plasma concentration of β-TG (Doyle 1980).

Age, Sex, Smoking and β-TG

In a study conducted by Zahavi et al. (1980), 219 healthy subjects were divided into age groups of 12–40 years, 41–69 years and 70–103 years. The mean plasma concentration of β-TG in each group was determined. It was found that β-TG levels increased significantly with age (see also Ludlam et al. 1975). In the same study, significant sex differences in β-TG levels were found, but only in the eldest age group, females having the higher plasma levels.

Numerous epidemiological studies have confirmed that cigarette smoking is a major risk factor for coronary heart disease and peripheral vascular disease (Doll 1984; Fowkes 1988). It is thought that one mechanism by which smoking may have an effect on pathogenesis is by inhibiting prostacyclin (PGI2) production by endothelial cells (Reinders et al. 1986). PGI2 is a potent inhibitor of platelet aggregation (Moncada et al. 1976; Czervionke et al. 1979), and so it seems probable that increased platelet activation in smokers would be detected by an increased level of β-TG. This was confirmed by a study by Duncan et al. in 1980. Smokers were found to have significantly higher levels of β-TG than non-smokers, 35.33 ng/ml compared to 21.45 ng/ml.

β-TG and Coronary Heart Disease

Measurement of β-TG has been used to examine platelet involvement in ischaemic heart disease but the results have varied. There are many reasons for the apparent inconsistencies. Although an increased plasma β-TG was reported in investigations into acute myocardial infarction by Denham et al. (1977), Matsuda (1980) and Hughes et al. (1979), the numbers of patients were too small and the results were not correlated with the severity of the disease. Complications such as renal impairment and hyperlipidaemia, which might affect the β-TG plasma levels, were also not considered.

Two studies investigating acute myocardial infarction by Rasi et al. (1982) and Van Hulsteijn et al. (1984) were consistent in reporting significantly higher plasma levels between patients and controls on hospital admission but Van Hulsteijn et al. (1984) differed from Rasi et al. (1982) by concluding that severity and location of infarct was not related to β-TG levels. This was perhaps due to inadequate clinical distinction between patients (Van Hulsteijn et al. 1984), the small numbers of patients in both studies or to lack of information on clinical complications such as renal failure (Rasi et al. 1982). It is clear too that a normal range for plasma β-TG has not been determined and varies widely between studies with considerable overlap between patients and controls.

In an interesting study by De Caterina et al. (1989), three groups of patients with myocardial ischaemia were stratified according to number of spontaneous ischaemic attacks at rest. Group I consisted of patients with no spontaneous ischaemic attacks at rest during four days of continuous electrocardiogram monitoring. Group II comprised patients with less than one ischaemic episode during four days and group III consisted of patients with more than one attack during four days. A control group was also included in the study. This grouping of patients revealed significant information which would not have been detected if analysis had been restricted to the total population because all ischaemic heart disease patients had only a slightly non-significant increase of mean plasma β-TG levels compared with controls (32 compared with 28 ng/ml). Levels were found to be no different from the controls in group I asymptomatic patients with previous myocardial infarction and in patients with stable angina pectoris. However, patients with frequent spontaneous ischaemia at rest (group III) had levels significantly higher than group II patients ($p<0.05$), group I patients ($p<0.01$), and controls ($p<0.05$). The increased β-TG levels of group III patients had been diluted by the pooling of the entire patient population. Linear regression analysis showed a positive correlation between β-TG levels and the number of spontaneous ischaemic attacks at rest ($r = 0.76$, $p<0.01$).

In this study (De Caterina et al. 1989) plasma β-TG levels were not found to be significantly raised after effort-induced ischaemia. The result suggests that induced ischaemia in itself may not cause platelet activation. However, it is more difficult to evaluate platelet activation in patients who have unpredictable ischaemic attacks as a result of thrombosis or coronary spasm. Levels of platelet markers, such as β-TG may be elevated but may be due to the preceding event rather than to the ischaemic attack.

β-TG and Peripheral and Cerebrovascular Disease

The few studies examining peripheral vascular disease are consistent in reporting elevated plasma β-TG levels. Jones et al. (1979) showed that an increase in β-TG correlated with severity of arterial disease and was significantly higher in cases compared to age- and sex-matched controls ($p<0.005$). In patients with intermittent claudication a rise in β-TG plasma levels was observed following ischaemic exercise (Forconi et al. 1987).

In the recent Edinburgh Artery Study (personal observations) 121 cases of both symptomatic and asymptomatic peripheral vascular disease had a significantly higher mean plasma β-TG than age- and sex-matched controls ($p=0.002$) (Table 25.1). The interrelationship between lifetime smoking and β-TG levels was examined by multiple regression and revealed that plasma β-TG levels were likely to be partly attributable to smoking, adding credence to the view that smoking has a possible role in platelet activation. However, β-TG was still found to be associated with peripheral vascular disease independently of smoking.

Similar associations of β-TG with disease have been found in cerebrovascular disease. Plasma β-TG levels were found to be significantly raised in patients with a cortical infarct, but normal for patients with lacunar infarct, compared with healthy controls. β-Thromboglobulin levels correlated with infarct size on computerised tomography and predicted mortality at 6 weeks (Woo et al. 1988). Similar results have been reported in another study (Taomoto et al. 1983). Stewart et al. (1983) found significantly higher levels of β-TG in transient ischaemic attacks than in age-matched controls ($p<0.05$) and suggested that β-TG may be predictive of further vascular events. Patients with acute strokes and transient ischaemic attacks were also shown in an investigation by Matsuda et al. (1980) to have raised plasma β-TG levels.

Is β-TG a good clinical marker for platelet activation? The acute sensitivity of the radioimmunoassay developed to measure β-TG levels depends upon the difference between its high concentration in platelets and minimal level in plasma. Therefore a major problem is the possibility of in vitro platelet release during blood sampling and plasma preparation. Thus it is not a feasible test to use routinely in clinical practice. Also, measurement of plasma β-TG cannot be used in patients with renal disease as the kidney is involved in its clearance and therefore an increased level in plasma may be due to impaired kidney function and not platelet activation. Similarly, plasma β-TG levels may be increased by hyperlipoproteinemia and cigarette smoking. Thus, an abnormal level of plasma β-TG cannot be regarded as an indicator of an atherosclerotic condition nor a normal level as evidence of no disease.

In conclusion, processes leading to the formation of atheroma and thrombosis are complex, involving not only platelet activation, but also coagulation and fibrinolysis. Although arterial disease is generally associated with increased plasma levels of β-thromboglobulin and von Willebrand factor, measurement of these factors is not particularly useful in identifying individuals at risk of disease, nor can the severity of disease be assessed. This is partly due to the variability in results obtained and also to the fact that no direct cause and effect relationship has been established as yet between platelet activation and disease. It seems likely that in acute episodes, a high short-lived level of these factors may occur but in chronic disease only slight increases may be detectable. However, no "normal range" for β-TG and vWF has been established and therefore it is difficult to define exactly what is an abnormal level. Consequently there may be considerable overlap of levels between diseased patients and normal controls.

References

Aihara M, Cooper HA, Wagner RH (1984) Platelet–collagen interactions: Increase in rate of adhesion of fixed washed platelets by factor VIII-related antigen. Blood 63:495–501

Baumgartner HR (1973) The role of blood flow in platelet adhesion, fibrin deposition and formation of mural thrombi. Microvasc Res 5:167–79

Becker CG, Dubin T, Wiedemann (1976) Hypersensitivity to tobacco antigen. Proc Natl Acad Sci USA 73:1712–16

Begg GS, Pepper DS, Chesterman CN, Morgan FJ (1978) Complete covalent structure of human β-thromboglobulin. Biochem 17:1739–44

Bolhuis PA, Sakariassen KS, Sander HJ, Bouma BN, Sixma JJ (1981) Binding of factor VIII-von Willebrand factor to human arterial subendothelium precedes increased platelet adherence and enhanced platelet spreading. J Lab Clin Med 97:568–76

Cella G, Zahavi J, de Haas HA, Kakkar VV (1979) β-thromboglobulin, platelet production time and platelet function in vascular disease. Br J Haematol 43:127–36

Christe M, Delley A, Marbet GA, Biland L, Duckert F (1984) Fibrinogen, factor VIII related antigen antithrombin III and $\alpha2$-antiplasmin in peripheral arterial disease. Thromb Haemostas (Stuttgart) 52(3):240–42

Cucuianu MP, Missits I, Olinic N, Roman S (1980) Increased ristocetin-cofactor in acute myocardial infarction: a component of the acute phase reaction. Thromb Haemost 43(11):41–4

Czervionke RL, Smith JB, Fry GL, Hoak JC, Haycraft DL (1979) Use of a radioimmunoassay to study thrombin induced release of PGI2 from cultured endothelium. Thromb Res 14:781–86

Davies MJ, Thomas AC (1985) Plaque fissuring – the cause of acute myocardial infarction, sudden ischaemic death, and crescendo angina. Br Heart J 53:363–73

De Caterina R, Gazzetti P, Mazzone A, Marzilli M, L'Abbate A (1989) Platelet activation in angina at rest. Evidence by paired measurement of plasma beta-thromboglobulin and platelet function 4. Eur Heart J 9:913–22

De Groot PG, Sixma JJ (1987) Role of von Willebrand factor in the vessel wall. Semin Thromb Hemost 13(4):416–24

Demarco L, Girolami A, Zimmerman TS, Ruggeri ZM (1986) von Willebrand factor interaction with the glycoprotein IIb/IIIa complex. Its role in platelet function as demonstrated in patients with congenital afibrinogenemia. J Clin Invest 77:1272–77

Denham J, Fisher M, James G, Hassan M (1977) β-thromboglobulin in venous and arterial thrombosis. Lancet i:1154

Doll R (1984) Landmark perspective: smoking and death rates. JAMA 251:2854–7

Doyle AJ, Chesterman CN, Cade JF et al. (1980) Plasma concentrations of platelet-specific proteins correlated with platelet survival. Blood 55:82–4

Duncan A, de Pratti VJ, George RR (1981) Elevated β-thromboglobulin levels associated with smoking and oral contraceptive agents in normal healthy women. Thromb Res 21:425–30

Falk E (1983) Plaque rupture with severe pre-existing stenosis precipitating coronary thrombosis. Characteristics of coronary atherosclerotic plaques underlying fatal occlusive thrombi. Br Heart J 50:127–34

Fisher M, Levine PH, Fullerton AL et al. (1982) Marker proteins of platelet activation in patients with cerebro-vascular disease. Arch Neurol 39:692–5

Forconi S, Pieragalli D, Guerrini M, Di Perri T (1987) Hemorrheology and peripheral arterial diseases. Clin Hemorheol 7:145–8

Fowkes FGR (1988) The measurement of atherosclerotic peripheral arterial disease in epidemiological surveys. Int J Epidemiol 17:248–54

Friedman M (1970) Pathogenesis of coronary thrombosis intramural and intraluminal hemorrhage. Adv Cardiol 4:20–46

Friedman RJ, Moore S, Singal DP et al. (1976) Regression of injury induced atheromatous lesions in rabbits. Arch Pathol Lab Med 100:185–95

Fujimura Y, Titani K, Holland LZ et al. (1987) A heparin binding domain of human von Willebrand factor: Characterization and localization to a tryptic fragment extending from amino acid residue VAL-449 to LYS-728. J Biol Chem 69:1734–39

Fuster V, Bowie EJW, Lewis JC et al. (1978) Resistance to arteriosclerosis in pigs with von Willebrand's disease. J Clin Invest 61:722–30

Fuster V, Chesebro JH (1986) Mechanisms of unstable angina. N Engl J Med 315:1023–25

Gimbrone MA Jr (1987) Vascular endothelium: nature's blood compatible container. Ann N Y Acad Sci 516:5–11

Green LH, Seroppian E, Handin RI (1980) Platelet activation during exercise-induced myocardial ischemia. N Engl J Med 302:193–97

Grette K (1962) Studies on the mechanism of thrombin-catalyzed hemostatic reactions in blood platelets. Acta Physiol Scand (Suppl) 56:165

Griffith TM, Edwards DH, Lewis MJ, Newby AC, Henderson AH (1984) The nature of endothelium derived vascular relaxant factor. Nature 308:645–7

Haerem JW (1978) Sudden, unexpected coronary death. The occurrence of platelet aggregates in the epicardial and myocardial vessels of man. Acta Pathol Microbiol Scand 265:1–47

Hanson SR, Pareti FI, Ruggeri ZM et al. (1988) Effects of monoclonal antibodies against the platelet glycoprotein IIb/IIIa complex on thrombosis and haemostasis in the baboon. J Clin Invest 81:149–58

Holt JC, Niewiarowski S (1979) On the relation between low affinity PF4 and β-thromboglobulin. Thromb Haemostas 42:271

Holt JC, Harris ME, Lange E et al. (1982) Platelet basic protein (PBP) is the precursor of low-affinity platelet factor 4 (LA–PF4) and β-thromboglobulin (β-TG). Fed Proc 41:528

Hope W, Martin TJ, Chesterman CN, Morgan FJ (1979) Human β-thromboglobulin inhibits PGI2 production and binds to a specific site in bovine aortic endothelial cells. Nature 282:210–12

Houdijk WPM, Sixma JJ (1985) Fibronectin in artery subendothelium is important for platelet adhesion. Blood 65:598–604

Hughes A, Daunt G, Vass G. Wickes J (1979) Beta-thromboglobulin and myocardial infarction. Seventh International Congress on Thrombosis and Haemostasis, London, p. 850 (abstract)

Jackson RL, Gotto AM Jr (eds) (1976) Hypothesis concerning membrane structure, cholesterol, and atherosclerosis. Raven Press, New York (Atherosclerosis reviews 1)

Jaffe EA, Hoyer WL, Nachman RL (1973) Synthesis of anti-hemophilic factor antigen by cultured endothelial cells. J Clin Invest 52:2757–64

Jones NAG, Zahavi J, de Haas HA et al. (1979) Platelet function in arterial vascular disease. Thromb Haemost 42:147

Joris I, Zand T, Majno G (1982) Hydrodynamic injury of the endothelium in acute aortic stenosis. Am J Pathol 106:394–408

Kaplan KC, Niewiarowski S (1985) Nomenclature of secreted platelet proteins – report of the working party on secreted platelet proteins of the subcommittee on platelets. Thromb Haemost 53:282–84

Karino T, Goldsmith HL (1979) Aggregation of human platelets in an annular vortex distal to a tubular expansion. Microvasc Res 17:217–37

Leytin VL, Garbunova NA, Misselwitz F et al. (1984) Step-by-step analysis of adhesion of human platelets to a collagen coated surface. Defect initial attachment and spreading of platelets in von Willebrand's disease. Thromb Res 34:51–63

Ludlam CA, Moore S, Bolton AE, Cash JD (1975) The release of a human platelet specific protein measured by a radio-immunoassay. Thromb Res 6:543–48

Marguerie GA, Plow EF (1983) The fibrinogen dependent pathway of platelet aggregation. Ann N Y Acad Sci 408:556–66

Matusda T, Seki T, Ogaware M et al (1980) Level of β-thromboglobulin and platelet factor 4 in various diseases. Nippon Ketsoeki Gakka Zasshi 43:871–8 (English abstract)

Moncada S, Gryglewski R, Bunting S, Vane JR (1976) An enzyme isolated from arteries transforms prostaglandin endoperoxides to an unstable substance that inhibits platelet aggregation. Nature 263:663–65

Moore S, Friedman RJ, Singal DP et al. (1976) Inhibition on injury-induced thrombo-atherosclerotic lesions by antiplatelet serum in rabbits. Thromb Haemost 35:70–81

Moore S, Pepper DS, Cash HD (1975) The isolation and characterization of platelet specific β-globulin (β-thromboglobulin) and the detection of anti-urokinase and anti-plasmin detected from thrombin-aggregated washed human platelets. Biochim Biophys Acta 379:360–9

Mossard JM, Wiesel ML, Cazenave JP et al. (1989) Increased van Willebrand factor, acute myocardial infarction, unstable angina and coronary thrombosis. Arch Mal Coeur 82:1813–18

Mustard JF, Kinlough-Rathbone RL, Packham MA (1987) The vessel wall in thrombosis. In: Colman RW, Hirsh J, Marder VJ, Salzman EW (eds) Haemostasis and thrombosis. Lippincott, Philadelphia, pp. 1073–88

Nichols TC, Bellinger DA, Johnson AJ, Lamb MA, Griggs TR (1986) Von Willebrand's disease prevents occlusive thrombosis in stenosed and injured porcine coronary arteries. Circ Res 59:15–26

Nichols TC, Bellinger DA, Tate DA et al. (1990) Von Willebrand factor and occlusive arterial thrombosis: A study in normal and von Willebrand's disease in pigs with diet-induced hypercholesterolemia and atherosclerosis. Arteriosclerosis 10:449–61

Niewiarowski S (1977) Proteins secreted by the platelet. Thromb Haemost 38:924–38

Niewiarowski S, Walz DA, James P, Rucinski B, Kueppers F (1980) Identification and separation of secreted platelet proteins by isoelectric focusing. Evidence that low-affinity platelet factor 4 is converted to β-thromboglobulin by limited proteolysis.

Packham MA, Mustard JT (1986) The role of platelets in the development and complications of atherosclerosis. Semin Hematol 23(i):8–26

Packham MA, Rowsell HC, Jorgensen L et al. (1967) Localized protein accumulation in the wall of the aorta. Exp Mol Pathol 7:214–32

Rand JH, Sussman II, Gordon RE, Chu SV, Salomon V (1980) Localization of factor VIII related antigen in human vascular subendothelium. Blood 55:752–56

Rasi V, Ikkala E, Torstila I (1982) Plasma β-thromboglobulin in acute myocardial infarction. Thromb Res 25:203–12

Reinders JH, Brinkman H-JM, Van Mourik JA, De Groot PG (1986) Cigarette smoke impairs endothelial cell prostacyclin production. Arteriosclerosis 6:15–23

Roberts DD, Williams SB, Gralnick HR, Ginsburg V (1986) Von Willebrand factor binds specifically to sulfated glycolipids. B Biol Chem 261:3306–9

Ross R (1986) The pathogenesis of atherosclerosis – an update. N Engl J Med 314:488–500

Ross R, Glomset J, Kariya B et al. (1974) A platelet-dependent serum factor that stimulates the proliferation of arterial smooth muscle cells in vitro. Proc Natl Acad Sci USA 71:1207–10

Rucinski B, Niewiarowski S, James P, Walz DA, Budzinski AZ (1979) Anti-heparin proteins secreted by human platelets. Purification, characterization, and radioimmunoassay. Blood 53:47–62

Sakariassen KS, Bolhuis PA, Sixma JJ (1979) Human blood platelet adhesion to artery subendothelium is mediated by factor VIII-von Willebrand factor bound to the subendothelium. Nature 279:636–8

Shuman MA, Majerus PN (1976) The measurement of thrombin in clotting blood by radioimmunoassay. J Clin Invest 58:1249–58

Sixma JJ (1987) Role of blood platelets, plasma proteins and the vessel wall in haemostasis. Haemost Thromb 2:283–302

Sixma JJ, Nievelstein PFEM, Houdijk PM et al. (1987) Adhesion of blood platelets to isolated components of the vessel wall. Ann N Y Acad Sci 509:103–17

St Charles R, Walz DA, Edwards BFP (1989) The three-dimensional structure of bovine platelet factor 4 at 3A resolution. J Biol Chem 264:2092–9

Stemerman MB (1973) Thrombogenesis of the rabbit arterial plaque. An electron microscope study. Am J Pathol 73:7–26

Stewart ME, Douglas JT, Lowe GDO (1983) Prognostic value of beta-thromboglobulin in patients with transient cerebral ischaemia. Lancet ii:479–82

Taomoto K, Asada M, Kawazawa Y, Matsumoto S (1983) Usefulness of the measurement of plasma beta-thromboglobulin (βTG) in cerebrovascular disease. Stroke 14:518–24

Tollefsen DM, Majerus PW (1976) Evidence for a single class of thrombin binding sites on human platelets. Biochemistry 12:282–9

Tschopp TB, Weiss HJ, Baumgartner HR (1974) Decreased adhesion of platelets to subendothelium in von Willebrand's disease. J Lab Clin Med 83:296–300

Tuddenham EGD (1989) Von Willebrand factor and its disorders: an overview of recent molecular studies. Blood Rev 3:251–62

Van Hulsteijn H, Kolff J, Briët E, Van der Laarse A, Bertina R (1984) Fibrinopeptide A and beta-thromboglobulin in patients with angina pectoris and acute myocardial infarction. Am Heart J 107:39–45

Walz A, Dewald B, Tscharner W, Baggiolini M (1989) Effects of the neutrophil activating peptide NAP-2, platelet basic protein, connecting tissue activating peptide III and platelet factor 4 on human neutrophils. J Exp Med 170:1745–50

Weiss HJ, Baumgartner HR, Tschopp TB, Turitto VT, Cohen D (1978) Correction of factor VIII of the impaired adhesion to subendothelium in von Willebrand's disease. Blood 51:267–79

Weiss HJ, Turitto VT, Baumgartner HR (1979) Effect of shear rate on platelet interaction with subendothelium in citrated and native blood. I. Shear dependent decrease of adhesion in von Willebrand's disease and in the Bernard–Soulier syndrome. J Lab Clin Med 92:750–64

Weksler BB, Marcus AS, Jaffe EA (1977) Synthesis of prostaglandin I2 (prostacyclin) by cultured human and endothelium cells. Proc Natl Acad Sci USA 74:3922–6

Woo E, Huang CY, Chan V et al. (1988) Beta thromboglobulin in cerebral infarction. J Neurol Neurosurg Psychiatr 51:557–62

Zahavi J, Jones NAG, Leyton J, Dubiel M, Kakkar VV (1980) Enhanced in vivo platelet release reactions in old healthy individuals. Thromb Res 17:329–36

Zarins CK, Giddens DP, Bharadvaj BK et al. (1983) Carotid bifurcation atherosclerosis: quantitative correlation of plaque localization with flow velocity profiles and wall shear stresses. Circ Res 53:502–14

26 Blood Rheology

G.D.O. LOWE

Blood rheology is the study of the flow behaviour of blood. Twenty years ago, John Dormandy showed that increased blood viscosity (its resistance to flow in wide vessels) was associated with decreased leg blood flow (Dormandy 1971), with intermittent claudication (Dormandy et al. 1973a), and with adverse prognosis for walking ability in claudicants (Dormandy et al. 1973b). Subsequently, the same group found that decreased blood filterability (its ability to flow in narrow vessels) was also present in patients with peripheral vascular disease, and correlated with clinical severity (Reid et al. 1976). Several other groups have since confirmed abnormal blood rheology in patients with arterial disease and its relationship to severity or outcome. This occurs not only in those with the most widespread disease and consequently the highest morbidity and mortality (peripheral vascular disease), but also in those presenting with coronary artery disease (Lowe et al. 1980a), stroke (Lowe et al. 1987a) or arterial "risk factors" (Lowe et al. 1987b). These findings have been summarised in recent reviews (Lowe 1986, 1987a, 1988, 1990a; Chien et al. 1987; Nash and Dormandy 1990).

There are two good reasons for studying blood rheology in arterial disease. The first is its role in pathophysiology (Fig. 26.1). The mechanisms by which genetic and environmental "risk factors" promote arterial disease are not clear, but presumably are those which favour atherosclerosis, thrombosis and/or ischaemia (Fig. 26.1). Atherosclerosis and thrombosis are interactive processes: arterial mural thrombi contribute to atherogenesis (Woolf 1987); while fissuring of a stenotic atherosclerotic plaque is the usual cause of the occlusive, semi-occlusive or embolising arterial thrombi which are the major causes of acute or chronic

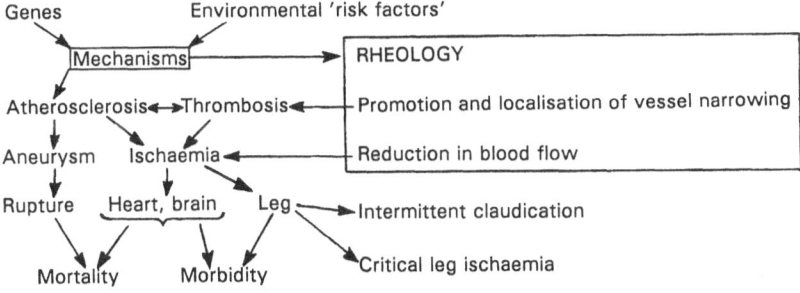

Fig. 26.1. Pathogenesis of peripheral vascular disease: the potential role of blood rheology factors.

ischaemia (Davies 1989). Rheological factors are increasingly recognised as primary and secondary risk factors for arterial events (Lowe 1986; Yarnell et al. 1991) which they may promote by both (a) localisation and promotion of atherosclerosis and thrombosis; and (b) reduction in blood flow distal to an atherothrombotic arterial stenosis (Lowe 1986; Fig. 26.1).

The second reason for studying blood rheology in arterial disease follows upon the latter concept: if ischaemia results from abnormal blood rheology as well as arterial narrowing, then modifying the blood is an alternative to dilating or bypassing narrowed arteries in treatment and prevention of ischaemia (Fig. 26.1). The great majority of patients with arterial disease are unsuitable for, unfit for, do not benefit from, or do not wish arterial surgery; furthermore our shrinking health service cannot afford it for most of them. Angioplasty or bypass grafting can also cause acute morbidity and mortality, as well as acceleration of atherosclerosis in operated arteries; and of course they do nothing for the progression of atherosclerosis and ischaemia elsewhere in the body. In contrast, "thinning the blood" is more generally applicable to arteriopaths in the population; is more generally applicable to all ischaemic organs in an arteriopath ("reaching the parts that surgeons cannot reach"); and is a rational approach to both peripheral and cerebral ischaemia, for which little medical treatment (other than aspirin and treatment of hypertension) is available (Lowe 1990b). Even in patients in whom arterial grafting is performed, there is increasing evidence that rheological factors predict graft occlusion (Bouhoutsos et al. 1974; Lowe et al. 1986; Matrai and Koller 1987; Wiseman et al. 1989), hence they should also be considered for rheological therapy to reduce the risk of graft occlusion, as well as the risk of other cardiovascular events.

Scotland now has the world's highest risk of arterial disease. In recent years, our laboratory has collaborated in several large studies of the Scottish population, measuring rheological and thrombotic variables in 30 000 blood samples. The aims of these collaborations are to establish the population ranges of these variables; their associations with genetic and environmental "risk factors" for arterial disease; and their associations with prevalent and incident arterial disease. In the Edinburgh Artery Study (Fowkes et al. 1991), blood rheology factors were measured in a random sample of over 1500 men and women aged 55–74 years, and related to both symptomatic and asymptomatic peripheral vascular disease (Lowe et al. 1991a). This is the first reported cross-sectional study of rheological variables in an older population, and the first epidemiological study of such variables in relation to both symptomatic amd asymptomatic peripheral vascular disease and its risk factors. In combination with the haemostatic factors which were also measured in this study and with studies of younger Scottish population samples (Lowe et al. 1987b, 1991b; Lee et al. 1990) our eventual aim is to clarify the roles of both rheological and thrombotic factors in arterial disease.

In the meantime, this chapter reviews the role of rheological factors in peripheral vascular disease, in the context of these recent epidemiological studies. It is first appropriate to briefly review these rheological factors and their relationships to flow in different parts of the circulation (Fig. 26.2). For more detailed accounts of blood rheology the reader is referred to recent reviews (Lowe 1986, 1987a, 1988, 1990a; Chien et al. 1987; Nash and Dormandy 1990).

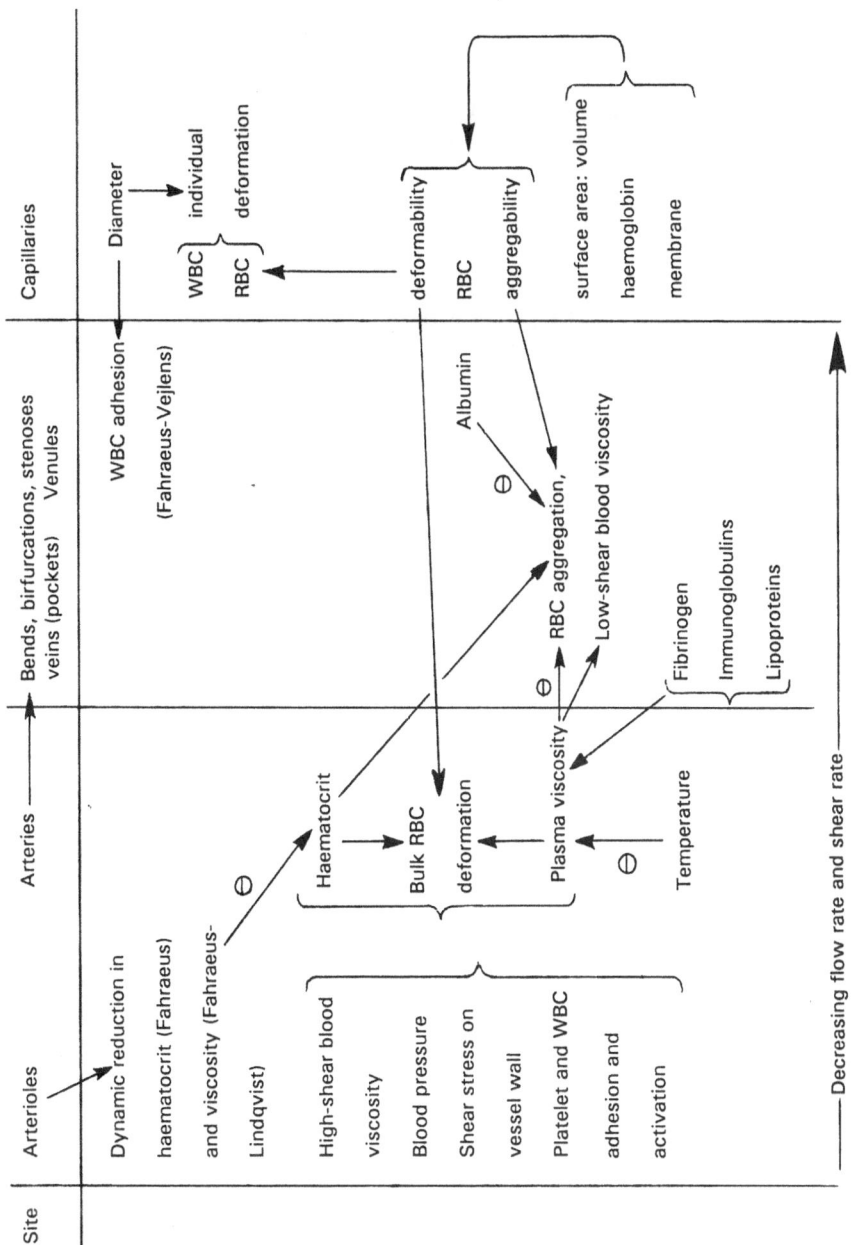

Fig. 26.2. Blood rheology factors in the circulation.

Blood Rheology Factors in the Circulation

Blood viscosity is the resistance of bulk blood to flow in wide-bore vessels. An increase in blood viscosity tends to reduce the volume rate of blood flow, in accordance with the Hagen–Poiseuille equation for the flow of simple fluids in straight rigid tubes:

$$\text{Volume flow rate} = \frac{\text{Pressure gradient} \times \text{vessel radius}^4}{8 \times \text{vessel length} \times \text{fluid viscosity}}$$

This equation can be rewritten as a first approach to blood flow in a human organ such as the leg:

$$\text{Blood flow rate} \; \alpha \; \frac{\text{Cardiac output}}{\text{Vascular hindrance} \times \text{viscosity}}$$

Hence resistance to blood flow in the leg (increases in which may cause ischaemia) is comprised of a vascular component (the combined resistance of arteries and arterial stenoses, compensatory collateral vessels, compensatory arteriolar vasodilatation, the microcirculation, and the veins) and also a rheological component (the intrinsic resistance of blood to flow in these serial parts of the circulation). Blood "viscosity" is not a constant, but varies in different parts of the circulation according to vascular geometry (vessel diameter, flow separation) which along with flow rate determines shear stresses and shear rates as well as the local composition of the blood (Fig. 26.2). Shearing is the relative movement of adjacent fluid layers, which is produced by shear stress (force per unit area, in units of milliPascals, mPa). Shear rate is the velocity gradient between adjacent fluid layers (distance per second divided by distance, resulting in units of inverse seconds, s^{-1}). Viscosity is defined as the ratio of shear stress to shear rate, in units of milliPascal.seconds (mPa.s).

Blood viscosity arises from frictional interactions between molecules, proteins and cells which are temperature-dependent. The viscosity of water at 37°C is 0.69 mPa.s; the addition of plasma proteins which disturb the flow streamlines of water almost doubles its viscosity. Mean plasma viscosity in population samples (including the Edinburgh Artery Study) is about 1.33 mPa.s, with a narrow population range (1.15–1.50 mPa.s, a 30% increase from lower to higher limit). The effect of individual plasma proteins on plasma viscosity increases with their concentration, molecular size and asymmetry: hence fibrinogen has a stronger effect than serum globulins (immunoglobulins, lipoproteins) which in turn have stronger effects than albumin. Decreases in temperature increase the viscosity of water and plasma: at 25°C (encountered in the cold extremities of persons with peripheral vascular disease) plasma viscosity increases by 30% from its value at 37°C (Lowe 1987a).

The addition of red blood cells (RBC) to plasma to form whole blood further disturbs the flow streamlines and further increases viscosity. Whole blood viscosity is determined by plasma viscosity; by the haematocrit (volume ratio of cells to whole blood); and by the deformation and aggregation of RBC in bulk flow (Fig. 26.2). Red cell deformation increases with shear rate, haematocrit (cell crowding: imagine packing balloons into boxes) and plasma viscosity (imagine pulling a balloon through a bath of treacle compared with a bath of water) and therefore partly compensates the increases in blood viscosity which arise from increases in haematocrit or plasma viscosity under high-shear conditions. High-shear viscosity is therefore a measure of blood flow resistence under normal conditions in wide-bore vessels (arteries and veins, Fig. 26.2). Mean large-vein, high-shear blood viscosity in population samples (including the Edinburgh Artery Study) is about 3.5 mPa.s (five times thicker than water) with lower values in females especially before the menopause, due to their lower haematocrit. The population range is wider than that of plasma viscosity, varying two-fold (2.2–4.5 mPa.s), corresponding to the wide population range of haematocrit (Lowe 1987a). These viscosity values are applicable to blood flow in large arteries and veins (Fig. 26.2).

In the arterioles, high flow rates and narrow diameters (under 300 μm) produce high shear rates and axial migration of the deformable RBC, which travel in the central, high-velocity fluid layers and displace platelets towards the vessel wall where they travel in the low-velocity, plasma-rich layers. The consequent reductions in dynamic haematocrit (Fåhraeus effect) and hence in dynamic whole blood viscosity (Fåhraeus–Lindqvist effect) further reduce blood flow resistance and increase blood flow (Fig. 26.2). Nevertheless, blood viscosity, plasma viscosity and haematocrit retain important influences on arterial and arteriolar flow which are relevant to atherosclerosis and arterial thrombosis. These include blood pressure (Lowe et al. 1987b, 1991b), shear stress on the vessel wall, and the adhesion and activation of platelets and leucocytes (Lowe 1987b). Abnormally high shear rates occur as blood flows through critical arterial stenoses, especially if blood viscosity is elevated: shear activation of platelets and leucocytes at these sites may be important in arterial thrombogenesis (Lowe 1987b).

Local increases in bulk blood viscosity occur with a reduction in blood flow rates and hence in shear rates, due to decreased RBC deformation and increased formation of RBC aggregates, which further disturb flow streamlines (Fig. 26.2). For example, mean blood viscosity rises from about 3.5 mPa.s at high shear rates (over 300 s^{-1}) to about 20 mPa.s at a shear rate of 1 s^{-1}. Red blood cell aggregation increases with the concentration of aggregating proteins (fibrinogen, immunoglobulins, lipoproteins), haematocrit, and RBC aggregability; and decreases with increases in albumin level and plasma viscosity (Fig. 26.2). Red blood cell aggregation occurs normally in low-flow venules, where it promotes leucocyte margination and adhesion, processes important in inflammatory reactions: the Fåhraeus–Vejlens effect (Fig. 26.2). Low-flow, low-shear conditions also occur due to flow separation at sites of atherogenesis (arterial bends and bifurcations) and thrombogenesis (distal to arterial stenoses, venous valve pockets) where local increases in viscosity may promote these processes (Lowe 1986, 1987b). Low-flow, low-shear conditions also arise in critical ischaemia, distal to critical arterial stenoses (see below).

In nutritive capillaries, whose diameters are less than those of resting RBC and leucocytes (8 μm), the concept of bulk blood viscosity does not apply, and the rheological determinants of blood flow are vessel diameter, the deformability of individual RBCs and leucocytes, and the viscosity of the plasma which lubricates their passage (Fig. 26.2). In the normal microcirculation, RBC velocity is generally high (albeit with marked spatial and temporal variations) due to high flow forces and high RBC deformability. The latter results from high ratio of surface area to volume, due to incomplete filling with haemoglobin; low internal viscosity (no nucleus); and a flexible membrane which rotates around its liquid haemoglobin content like the caterpillar tread of a military tank (Fig. 26.2). In contrast, leucocytes are 1000 times less able to traverse nutritive capillaries than RBC: their lesser deformability reflects their nucleus, organelles and spherical shape. The filterability of whole blood through micropore (5 μm diameter) filters is therefore strongly determined by leucocytes, even though there are 1000 RBC to every leucocyte in normal blood (Alderman et al. 1981).

In summary, blood viscosity and blood filterability are global measures of the resistance of blood to flow in wide-bore vessels and microvascular networks respectively. Their important determinants are plasma viscosity (an important determinant of flow in all vessels); haematocrit (much diluted in normal arterioles and in the normal microcirculation); red cell deformation (which greatly facilitates flow in the normal macro- and microcirculations); and red cell aggregation and rigid leucocytes, whose potential to increase blood flow resistance is minimised in the normal high-flow, high-shear circulation.

It will now be apparent that blood flow depends on continual interactions between cardiac output, vessel geometry and blood rheology. What are the relative roles of vessel diameter (arterial stenoses) and blood rheology in ischaemia? In the normal circulation, variations in blood rheology factors are probably easily compensated by changes in vessel diameter (note radius to the *fourth* power in the Hagen–Poiseuille equation!). Overt hyperviscosity syndromes in rare haematological disorders, can, however, cause peripheral or cerebral ischaemia despite relatively normal hearts and blood vessels, e.g. increased haematocrit and whole-blood viscosity in polycythaemias; increased plasma viscosity and red cell aggregation due to increased immunoglobulins in paraproteinaemias; and increased microvascular obstruction by rigid cells in the sickling disorders and hyperleucocytic leukaemias (Lowe 1987a, 1988).

While these disorders show the potential of extreme abnormalities of blood rheology to promote ischaemia, a much more important question is the extent to which more subtle variations in blood rheology, within population ranges, interact with arterial narrowing to produce ischaemia. In the presence of arterial stenoses, not only does systemic blood rheology (measured in arm vein blood samples) change; it also interacts with the abnormal flow conditions distal to the stenoses. The fall in driving pressure and flow rates as well as compensatory arteriolar vasodilatation reduce shear-induced haemodilution and viscosity reduction in normal arterioles, and increase low-shear red cell aggregation in venules. In the ischaemic microcirculation, RBC flow is more heterogeneous, and generally impaired due to low flow forces and reduced RBC deformability (produced by local acidosis, hyperosmolarity and calcium influx). The non-filling of nutritive skin capillaries observed in critical limb ischaemia may also be due to impaction of rigid, adhesive leucocytes or platelet aggregates; microthrombi; or reduction in capillary diameter by tissue oedema (Lowe 1987a, 1990c). Hence under the abnormal flow conditions of arterial disease, even "normal" or modestly elevated blood rheology factors have the potential to aggravate ischaemia and produce vicious cycles of reduced flow and increased blood flow resistance: a concept named "convertly abnormal blood rheology" (Lowe 1988).

Blood Rheology in Peripheral Vascular Disease

In a case–control study, Dormandy and colleagues (1973a) found a mean increase of 10% in high-shear blood viscosity in patients with intermittent claudication attending a vascular clinic. Increased plasma fibrinogen levels appeared the most important cause of the viscosity increase, which was more pronounced at low shear rates (mean increase about 25% at a shear rate of about $1 s^{-1}$), probably due to RBC aggregation by fibrinogen (Dormandy et al. 1973a; Dormandy 1975). These findings have subsequently been confirmed by other groups (e.g. Stormer et al. 1974; Blunt et al. 1980; Lowe et al. 1986), as have increased haemoglobin or haematocrit and plasma viscosity levels (Bouhoutsos et al. 1974; Lowe et al. 1986). Population studies have confirmed increased levels of fibrinogen (Hughson et al. 1978) and haematocrit (Kallero et al. 1981) in claudicants; while the Framingham prospective study has found that fibrinogen (Kannel and D'Agostino 1990) but not haematocrit (Kannel and McGee 1985) was predictive of claudication.

Several questions must be asked concerning the association of blood viscosity with claudication. Can the association be demonstrated in population studies, as distinct from selected vascular clinic referrals and controls? Is the association explained by risk factors such as cigarette-smoking (Lowe et al. 1986) which is associated with increases in blood and plasma viscosity, haematocrit and fibrinogen (Lowe et al. 1980b) and also with peripheral vascular disease? Is the association with the extent of atherosclerotic stenoses, as in coronary artery disease (Lowe et al. 1980a)? Or is the association with ischaemia in the presence of a given degree of atherosclerotic stenosis, i.e. "rheological claudication" (Dormandy 1975)?

In recent years, the rheological and chemical contributions of leucocytes to ischaemia have also been appreciated (Ernst et al. 1987; Lowe 1987a,b, 1990c; Nash and Dormandy 1990). Decreased blood filterability in claudication (Reid et al. 1976; Drummond et al. 1980) is due to leucocytosis and increased plasma viscosity rather than decreased RBC deformability (Alderman et al. 1981; Ciuffetti et al. 1988). A further decrease in blood filterability occurs after ischaemic exercise, which reflects leucocyte activation, as well as reduced RBC deformability under ischaemic conditions (Forconi et al. 1987; Shearman et al. 1988; Ciuffetti et al. 1989; Hickey et al. 1990; Neumann et al. 1990). While impaction of activated leucocytes in the nutritive microcirculation may contribute to claudication, the same questions must be asked for leucocyte activation as for blood viscosity and its determinants: what are its relationships to peripheral vascular disease, risk factors, arterial narrowing and leg ischaemia in the population?

These questions have been addressed in the Edinburgh Artery Study, in which high-shear blood viscosity and its major determinants (haematocrit, plasma viscosity, fibrinogen and an RBC deformability index) as well as plasma neutrophil elastase (a measure of neutrophil leucocyte activation) were measured in an older population sample and related to symptomatic and asymptomatic peripheral vascular disease and its conventional risk factors, including multivariate analysis (Lowe et al. 1991a). An 8.5% mean increase in whole-blood viscosity was found in persons with claudication (WHO questionnaire) compared with those with no evidence of peripheral vascular disease; those with asymptomatic disease (low ankle-brachial systolic pressure index, ABPI; and/or abnormal reactive hyperaemia tests; Fowkes et al. 1991) had intermediate values ($p<0.001$). These increases in blood viscosity are expected to be much higher under ischaemic conditions (low shear rates). The viscosity increase in claudicants was partly due to a higher haematocrit (mean increase 2.2%; $p<0.01$), but mostly due to other factors (mean increase in viscosity at standard haematocrit 5.5%; $p<0.001$). Increased plasma viscosity was largely responsible (mean increase 4.5%; $p<0.001$); a 1.5% increase in RBC rigidity index (blood viscosity standardised for haematocrit and plasma viscosity) was not statistically significant. A major cause of the increase in plasma viscosity was increased plasma fibrinogen (mean increase 14.8%; $p<0.001$). Leucocyte activation was also increased in claudicants (mean increase in elastase 21.2%; $p<0.05$) as was serum uric acid (mean increase 6.0%; $p<0.05$). Elastase levels correlated with plasma and blood viscosity, suggesting associations of leucocyte activation and increased plasma viscosity, possibly as a generalised "acute-phase reaction" or "haematological stress syndrome" in atherosclerosis (Blunt et al. 1980; Stuart 1984). Consistent with this concept, our group has recently observed increases in C-reactive protein and the cytokine, interleukin-6 (released from activated monocytes) in a case–control study of claudication (Reid 1991).

Do conventional "risk factors" such as cigarette smoking account for these associations? As in previous studies of younger populations (Lowe et al. 1987b, 1991b) blood viscosity, plasma viscosity and fibrinogen were associated with all major cardiovascular risk factors (male sex, age, smoking, blood pressure, body mass index, HDL and non-HDL cholesterol, triglycerides, social class) in the Edinburgh Artery Study (Lowe et al. 1991a), and thus may be mechanisms by which these risk factors might promote arterial disease. However, multivariate analyses including these risk factors showed that both blood viscosity ($p<0.05$) and fibrinogen ($p<0.01$) were independently associated with peripheral arterial narrowing, as measured by the ABPI. Hence blood viscosity and fibrinogen may promote narrowing of leg arteries, independently of associations with conventional risk factors: presumably via their effects on mural thrombosis and atherosclerosis (Lowe 1987b; Figs. 26.1 and 26.2). We are currently performing further studies relating rheological factors to the site and extent of atherosclerosis in the lower limb. We have previously shown that plasma viscosity and fibrinogen are associated with an imbalance of coagulation over fibrinolysis in middle-aged men, which may promote fibrin formation (Lowe et al. 1991b). Fibrinogen may also infiltrate the arterial wall (Smith et al. 1976), and we observed a significant interaction between smoking and the interaction of fibrinogen and the ABPI, possibly because smoking damages arterial endothelium (Woolf 1987) allowing fibrinogen infiltration. An interaction between smoking and fibrinogen has also been reported in the prediction of femoro-politeal graft occlusion (Wiseman et al. 1989).

On multivariate analysis including all conventional risk factors, the association between rheological factors and claudication became "non-significant" (Lowe et al. 1991a). However, this finding may be related to the small number of claudicants, and does not exclude the possibility that increases in viscosity, fibrinogen and leucocyte activation may be mechanisms whereby risk factors such as age and smoking promote the development of claudication. We are currently following up the Edinburgh Artery Study cohort to determine the relationship of rheological variables to new cases of claudication, as well as their relationships to progress of arterial disease and claudication. Rheological measurements have also been performed in over 600 new referrals for claudication to the Edinburgh Peripheral Vascular Clinic during one year; these are consistent with the abnormalities noted in the population study in claudicants (personal observations) and their prospective relationships to arterial disease are also under study in this cohort.

In the presence of a given degree of atherosclerotic narrowing (ABPI), do rheological factors predispose to leg ischaemia? In the Edinburgh Artery Study, the ABPI shows only partial correlations with claudication and reactive hyperaemia (Fowkes et al. 1991), hence other factors are implicated in ischaemia. After standardisation for the ABPI, plasma viscosity was significantly associated with the probability of claudication. The odds of claudication in the top quintile of plasma viscosity ($\geqslant$1.41 mPa.s) were 3.3 times (95%, confidence interval 1.3, 8.5) that in the bottom quintile (<1.26 mPa.s). It is interesting to compare these plasma viscosity findings with the risk of ischaemic heart disease in the South Wales prospective studies (Yarnell et al. 1991). After standardisation for major risk factors, the odds of ischaemic heart disease in the top plasma viscosity quintile were 3.2 (95% CI 1.8, 5.6) times that in the bottom quintile. In both these studies, plasma viscosity showed a stronger association with leg ischaemia (Lowe et al. 1991a) or myocardial ischaemia (Yarnell et al. 1991) than fibrinogen, suggesting that viscosity may be the more important determinant of ischaemia. In the presence of a given ABPI in the Edinburgh Artery Study, asymptomatic leg ischaemia (reactive hyperaemia test) was significantly associated with both elastase and uric acid levels (p<0.01), suggesting involvement of leucocyte activation and free radical formation (Lowe et al. 1991a).

In summary then, the Edinburgh Artery Study has confirmed the findings of previous case–control studies that blood rheology factors are associated with peripheral vascular disease. In this population study, rheological factors were related to both symptomatic and asymptomatic disease; to arterial narrowing, and to ischaemia in the presence of a standard degree of arterial narrowing; and to risk factors as well as risk factor-independent associations with arterial disease. Prospective studies of both population (n=1500) and vascular clinic referral (n=600) cohorts are in progress to determine the predictive value of rheological factors for new or progressive peripheral vascular disease, as well as for ischaemic heart disease, stroke and mortality. Further studies are also relating the blood abnormalities to the site and extent of angiographic lower limb arterial disease, in order to examine their pathophysiological role.

Blood Rheology in Critical Limb Ischaemia

As discussed previously, critical limb ischaemia (rest pain and/or pre-gangrene) is associated not only with deterioration in flow conditions (reduction in flow rates and shear rates) but also with further abnormalities in blood rheology (Dormandy 1975; Reid et al. 1976; Nash et al. 1988; and personal observations). These various abnormalities may promote non-filling of skin nutritive capillaries in critical limb ischaemia, and may also predict adverse outcome (i.e. amputation and/or mortality) Lowe 1990c). We are, therefore, currently performing prospective studies of rheological variables in prediction of critical limb ischaemia in the cohort of 600 claudicants, and in prediction of amputation or death in 200 patients with critical limb ischaemia.

Blood Rheology and Peripheral Arterial Graft Occlusion

As previously noted, high levels of haematocrit or haemoglobin (Bouhoutsos et al. 1974; Lowe et al. 1986; Matrai and Koller 1987) and fibrinogen (Wiseman et al. 1989) are predictive of arterial graft occlusion, possibly via rheological effects (Dormandy 1975). We are therefore performing a prospective study of blood rheology in relation to graft stenosis or occlusion in 200 patients with femorodistal grafts.

Rheological Therapy in Peripheral Vascular Disease

Reducing the flow resistance of blood has a much sounder rationale than use of vasodilator drugs in peripheral vascular disease (Strandness and Sumner 1975). In ischaemic limbs, vasodilator metabolites probably achieve maximal endogenous vasodilatation in both skin and muscle: exogenous vasodilator drugs may then "steal" blood from ischaemic areas by dilating other vascular beds. In contrast, rheological therapy should selectively *increase* blood flow in ischaemic areas, where low flow rates and metabolic changes locally increase the flow resistance of blood (due to haemoconcentration, red cell aggregation, decreased RBC and leucocyte deformability, and leucocyte adhesion (Lowe 1990a,c; Nash and Dormandy 1990). Indeed, rheological therapy may be the only medical way to increase nutritive blood flow when vascular reserve is exhausted.

Both exercise (Ernst and Matrai 1987a) and stopping smoking (Ernst and Matrai 1987b; Lowe et al. 1991b) reduce blood viscosity; hence there is a rheological basis for the improvement in claudication distance after patients comply with instructions to "stop smoking and keep walking". Although they may reduce arterial thrombotic events, we have recently shown that neither low-dose aspirin nor low-dose warfarin alters the abnormal blood rheology or haemostatic disturbance in patients with claudication, in a double-blind, placebo-controlled trial (Reid 1991). On the other hand, haemodilution (Ernst et al. 1987), reduction in plasma fibrinogen and viscosity by clofibrate (Dormandy et al. 1974; Postlethwaite and Dormandy 1975), reduction in leucocyte activation by pentoxifylline (Porter et al. 1982; Currie et al. 1991), or viscosity reduction by fish oils (Woodcock et al. 1984) have each shown promising results in claudicants. Further large, controlled trials are now indicated to determine whether or not such rheological therapies (in addition to advice on smoking and exercise) improve not only walking distance, but also the high morbidity and mortality from ischaemic heart disease and stroke in patients with peripheral vascular disease (Lowe 1990b). In criticial limb ischaemia, the rheological and clinical effects of intravenous prostaglandins (PGE_1 or alprostadil, PGI_2 or epoprostenol, stable prostacyclin analogues such as iloprost) are also promising (Lowe 1990c; de Gaetano et al. 1990), and the results of large, controlled studies are also awaited.

In 1597, Maister Peter Lowe of Errol, surgeon and Founder of the Royal College of Physicians and Surgeons of Glasgow, wrote that gangrene "is a mortification not altogether of the part, but tending by little and little through the great violence of the inflammation; for thereby the vaines, and arters are stopped, that the naturall heate may not passe, so the parte easily corrupteth and waxeth gangrenated. The cause is, great quantitie of blood in the member, which letteth the spirit to passe, so bindeth and intercepteth it, in such forte, that the arters cannot work their transpiration and requisite exhalation, so for want of naturall heate, the member suffocates." Almost 400 years later, the role of viscous blood and inflammatory reactions (high fibrinogen, leucocyte activation) in obstruction of the limb circulation, hinted at by Peter Lowe, appears quite plausible. By 1997, epidemiological studies and large controlled trials should provide further information on the role of blood rheology in peripheral vascular disease.

Acknowledgement. The support of the British Heart Foundation, Scottish Home and Health Department, and Medical Research Council for our rheological and haemostatic studies in peripheral vascular disease is gratefully acknowledged.

References

Alderman MJ, Ridge A, Morley AA, Ryall EG, Walsh JA (1981) Effect of total leucocyte count on whole blood filterability in patients with peripheral vascular disease. J Clin Pathol 34:163–6

Blunt RJ, George AJ, Hurlow RA, Strachan CJ, Stuart J (1980) Hyperviscosity and thrombotic changes in idiopathic and secondary Raynaud's syndrome. Br J Haematol 45:651–8

Bouhoutsos J, Morris T, Chavatzas D et al. (1974) The influence of haemoglobin and platelet levels on the results of arterial surgery. Br J Surg 51:984–6

Chien S, Dormandy J, Ernst E, Matrai A (1987) Clinical hemorheology. Martinus Nijhoff, Lancaster

Ciuffetti G, Mannarino E, Pasqualini L, Mercuri M, Lennie SE, Lowe GDO (1988) The hemorheological role of cellular factors in peripheral vascular disease. VASA 17:168–70

Ciuffetti G, Mercuri M, Mannarino E, Robinson MK, Lennie SE, Lowe GDO (1989) Peripheral vascular disease: rheologic variables during controlled ischemia. Circulation 80:348–52

Currie MS, Simel DL, Christenson RH et al. (1991) Anti-inflammatory effects of pentoxifylline in claudication. Am J Med Sci 301:85–90

Davies MJ (1989) Thrombosis and coronary atherosclerosis. In: Julian D, Kübler W, Norris RM, Swan HJC, Collen D, Verstaete M (eds) Thrombolysis in cardiovascular disease. Marcel Dekker, Basel, pp 25–43

Dormandy JA (1971) Influence of blood viscosity on blood flow and the effect of low molecular weight dextran. Br Med J ii:716–19

Dormandy JA (1975) Blood: its viscosity and circulation. In: Harcus AW, Adamson L (eds) Arteries and veins. Churchill Livingstone, Edinburgh, pp. 99–134

Dormandy JA, Hoare E, Colley J, Arrowsmith DE, Dormandy TL (1973a) Clinical, haemodynamic, rheological and biochemical findings in 126 patients with intermittent claudication. Br Med J iv:576–81

Dormandy JA, Hoare E, Khattab AH, Arrowsmith DE, Dormandy TL (1973b) Prognostic significance of rheological and biochemical findings in patients with intermittent claudication. Br Med J iv; 581–3

Dormandy JA, Gutteridge JMC, Hoare E, Dormandy TL (1974) Effect of clofibrate on blood viscosity in intermittent claudication. Br Med J iv:259–61

Drummond MM, Lowe GDO, Belch J, Barbenel JC, Forbes CD (1980) An assessment of red cell deformability using a simple filtration technique. J Clin Pathol 33:373–6

Ernst E, Matrai A (1987a) Intermittent claudication, exercise and blood rheology. Circulation 76:1110–14

Ernst E, Matrai A (1987b) Abstention from chronic smoking normalizes blood rheology. Arteriosclerosis 64:75–7

Ernst E, Matrai A, Koller L (1987) Placebo-controlled, double-blind study of haemodilution in peripheral arterial disease. Lancet i:1449–51

Ernst E, Hammerschmidt DE, Bagge U, Matrai A, Dormandy JA (1987) Leukocytes and the risk of ischemic diseases. JAMA 257:2318–24

Forconi S, Pieragalli D, Guerrini M, DiPerri I (1987) Hemorheology and peripheral arterial diseases. Clin Hemorheol 7:145–58

Fowkes FGR, Housley E, Cawood EHH et al. (1991) Edinburgh Artery Study: prevalence of asymptomatic and symptomatic peripheral arterial disease in the general population. Int J Epidemiol 20:384–92

de Gaetano G, Bertelé V, Cerletti C (1990) Mechanism of action and clinical use of prostanoids. In: Dormandy JA, Stock G (eds) Critical leg ischaemia: its pathophysiology and management. Springer-Verlag, Berlin, pp. 117–41

Hickey NC, Gosling P, Baar S, Shearman CP, Simms MH (1990) Effect of surgery on the systemic inflammatory response to intermittent claudication. Br J Surg 77:1121–4

Hughson WG, Mann JI, Garrod A (1978) Intermittent claudication: prevalence and risk factors. Br Med J i:1379–81

Kallero KS, Bergentz SE, Lindell SE, Janzon L (1981) Elevated haematocrit in patients with intermittent claudication with special regard to men below the age of 60. Bibl Haematol 47:173–84

Kannel WB, McGee DL (1985) Update on some epidemiological features of intermittent claudication: the Framingham Study. J Am Geriatr Soc 33:13–18

Kannel WB, D'Agostino RB (1990) Update on fibrinogen as a major cardiovascular risk factor: the Framingham Study. J Am Coll Cardiol 15:156A

Lee AJ, Smith WCS, Lowe GDO, Tunstall-Pedoe H (1990) Plasma fibrinogen and coronary risk factors: the Scottish Heart Health Study. J Clin Epidemiol 43:913–19

Lowe GDO (1986) Blood rheology in arterial disease. Clin Sci 71:137–46

Lowe GDO (1987a) (ed) Blood rheology and hyperviscosity syndromes. Baillière's Clinical Haematology 1, vol. 3. Baillière, London, pp. 597–867

Lowe GDO (1987b) Thrombosis and haemorheology. In: Chien S, Dormandy J, Ernst E, Matrai A (eds) Clinical hemorheology. Martinus Nijhoff, Lancaster, pp. 195–226

Lowe GDO (1988) (ed) Clinical blood rheology. CRC Press, Boca Raton

Lowe GDO (1990a) Drugs that modify red blood cell characteristics. In: Fleming JS (ed) Drugs and the delivery of oxygen to tissue. CRC Press, Boca Raton, pp. 253–64

Lowe GDO (1990b) Drugs in cerebral and peripheral arterial disease. Br Med J 300:524–8

Lowe GDO (1990c) Pathophysiology of critical leg ischaemia. In: Dormandy J, Stock G (eds) Critical leg ischaemia; its pathophysiology and management. Springer-Verlag, Berlin, pp.17–40

Lowe GDO, Drummond MM, Lorimer AR et al. (1980a) Relationship between extent of coronary artery disease and blood viscosity. Br Med J i:673–4

Lowe GDO, Drummond MM, Forbes CD, Barbenel JC (1980b) The effect of age and cigarette smoking on blood and plasma viscosity in men. Scott Med J 25:13–17

Lowe GDO, Saniabadi A, Turner A et al. (1986) Studies on haematocrit in peripheral arterial disease. Klin Wochenschr 64:969–74

Lowe GDO, Anderson J, Barbenel JC, Forbes CD (1987a) Prognostic importance of blood rheology in acute stroke. In: Hartmann A, Kuschinsky W (eds) Cerebral ischemia and haemorheology. Springer-Verlag, Berlin, pp. 496–501

Lowe GDO, Smith WCS, Tunstall-Pedoe HD et al. (1987b) Cardiovascular risk and haemorheology – results from the MONICA project, Glasgow, and the Scottish Heart Health Study. Clin Haemorheol 7:501

Lowe GDO, Donnan PT, McColl P et al. (1991a) Blood viscosity, fibrinogen and activation of coagulation and leucocytes in peripheral arterial disease: the Edinburgh Artery Study. Br J Haematol 77 (Suppl 1):27

Lowe GDO, Wood DA, Douglas JT et al. (1991b) Relationships of plasma viscosity, coagulation and fibrinolysis to coronary risk factors and angina. Thromb Haemostas 65:339–43

Lowe P (1597) The whole course of chirurgerie. Thomas Purfoot, London

Matrai A, Koller L (1987) Importance of the preoperative hemoglobin concentration in arterial surgery. Eur Surg Res 19:1–5

Nash GB, Dormandy JA (1990) The involvement of red cell aggregation and blood cell rigidity in impaired microcirculatory efficiency and oxygen delivery. In: Fleming JS (ed) Drugs and the delivery of oxygen to tissue. CRC Press, Boca Raton, pp. 227–52

Nash GB, Thomas PRS, Dormandy JA (1988) Abnormal flow properties of white cells in patients with severe ischaemia of the leg. Br Med J 296:1669–1701

Neumann F-J, Waas W, Diehm C et al. (1990) Activation and decreased deformability of neutrophils after intermittent claudication. Circulation 82:922–9

Porter JM, Cutler BS, Lee BY et al. (1982) Pentoxifylline efficacy in the treatment of intermittent claudication. Am Heart J 104:66–70

Postlethwaite JC, Dormandy JA (1975) Results of ankle systolic pressure measurements in patients with intermittent claudication being treated with clofibrate. Ann Surg 181:799–802

Reid D (1991) The clinical role of fibrinogen and fibrin in peripheral arterial disease. MD thesis, University of Glasgow

Reid HL, Dormandy JA, Barnes AJ et al. (1976) Impaired red cell deformability in peripheral arterial disease. Lancet i:666–8

Shearman CP, Gosling P, Gwynn BR, Simms MH (1988) Systemic effects associated with intermittent claudication. A model to study biochemical aspects of vascular disease? Eur J Vasc Surg 2:401–4

Smith EB, Alexander KM, Massie IB (1976) Insoluble "fibrin" in human aortic intima. Atherosclerosis 23:19–26

Stormer B, Horsch R, Kleinschmidt F et al. (1974) Blood viscosity in patients with peripheral vascular disease in the areas of low shear rates. J Cardiovasc Surg 15:577–84

Strandness DE Jr, Sumner DS (1975) Hemodynamics for surgeons. Grune and Stratton, New York

Stuart J (1984) The acute-phase reaction and haematological stress syndrome in vascular disease. Int J Microcirc: Clin Exp 3:115–29

Wiseman S, Kenchington G, Dain R et al. (1989) Influence of smoking and plasma factors on patency of femoro-popliteal vein grafts. Br Med J 299:643–6

Woodcock BE, Smith E, Lambert WH et al. (1984) Beneficial effect of fish oil on blood viscosity in peripheral vascular disease. Br Med J 288:592

Woolf N (1987) Thrombosis and atherosclerosis. In: Bloom AL, Thomas DP (eds) Haemostasis and thrombosis, 2nd edn. Churchill Livingstone, Edinburgh, pp. 651–78

Yarnell J, Baker IA, Sweetnam PM et al. (1991) Fibrinogen, viscosity and white blood cell count are major risk factors for ischemic heart disease. Circulation 83:836–44

SECTION VI

Natural History and Prevention

27 Natural History of Femoral Atheromatous Lesions

M.R. WHYMAN

"Have a chronic disease and take care of it", advised Oliver Wendell Holmes (1809–1894) as a formula for longevity. Sound advice, perhaps, but in the case of peripheral vascular disease difficult to follow. That is not to say this chronic disease is particularly hard to come by. In the Edinburgh area of Scotland around 5% of the population over 55 years of age suffer from claudication, and it is estimated that significant occlusive disease remains asymptomatic in about three times this number (Fowkes et al. 1991).

"Taking care" of the condition is an altogether more difficult proposition for both patient and vascular specialist. One reason it proves so is due to the remarkable lack of information on the natural history of the pathological lesions, and of course in an ideal world a complete understanding of the natural history of the disease would be a prerequisite for appropriate treatment.

We recognise that clinical manifestations occur at a relatively late stage in the evolution of atheroma. However, this is a stage we can readily identify and measure with reference to its clinical, radiological and haemodynamic effects. This chapter therefore reviews the natural history of advanced femoral atherosclerosis only, and proposes suggestions for improving our understanding of it.

Symptoms of the Lesion

The earliest symptom of lower limb atherosclerosis is usually calf claudication. When symptoms are confined to this site the causal lesion is most often in the superficial femoral artery, frequently the adductor (Hunter's) canal (Lindbom 1950; Mavor 1956). The typical lesion at this early stage is an atheromatous plaque which narrows the lumen sufficiently to cause a pressure drop across the stenosis. This in turn is responsible for encouraging collateral vessel formation (John and Warren 1961).

Following the usual advice to "stop smoking and keep walking" (Housley 1988), many patients experience partial or occasionally complete relief of symptoms. This might be a result of metabolic changes (Lundgren et al. 1989) and other physical adaptations as well as development of collateral vessels because clinical improvement is often not matched by any improvement in Doppler ankle pressures or radiological appearances.

Over a period of time, the stenosis progresses and complete occlusion occurs (Fig. 27.1). The final pathological event precipitating occlusion might be thrombosis or plaque dissection, but once flow has ceased in the segment, thrombosis is certain to supervene and propagation of clot along a variable distance of artery occurs.

What is poorly understood, however, is how rapidly a stenosis progresses to occlusion, what features of the stenosis predispose to rapid progression and, of great importance, how the severity or length of occlusion relates to symptoms. For an individual patient this information could be obtained by regular and accurate "lesion-specific" investigations, but this goal has not yet been achieved.

Within populations there is a poor correlation between severity of lesions and symptoms (Baker 1978). Two-thirds of patients with complete femoral artery occlusion have no symptoms at all (Widmer et al. 1964), whilst some patients with moderately severe stenoses are disabled by claudication. It is for this reason that,

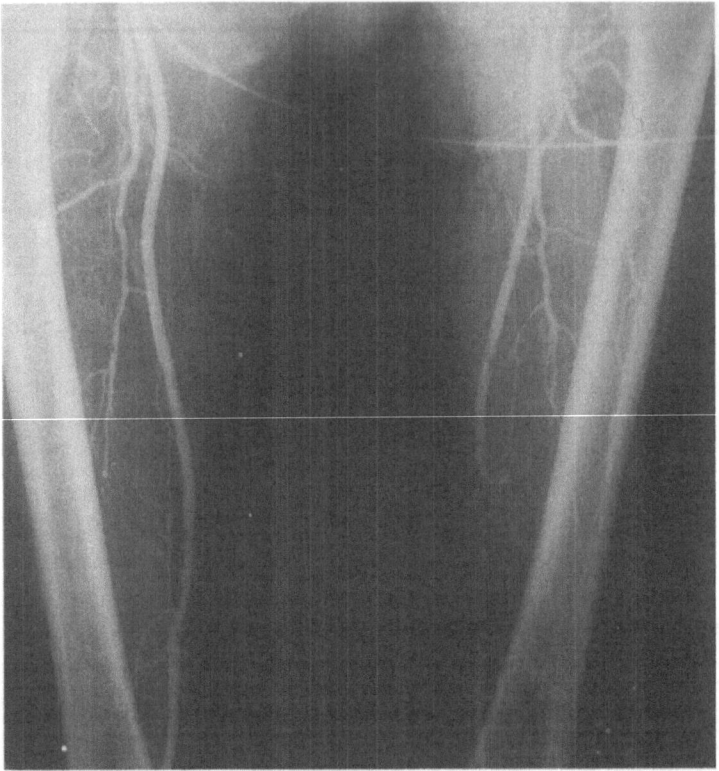

Right   **Left**

Fig. 27.1. Transfemoral arteriogram. In the left leg is a short complete occlusion of the superficial femoral artery, and in the right leg a stenosis symmetrically opposite the occlusion. How long before the stenosis becomes an occlusion and what will be the consequences?

despite there being no shortage of data on the natural history of claudication (Spaulding 1956; Bloor 1961; Begg and Richards 1962; Imperato et al. 1975; McAllister 1976; Wilson et al. 1980; Kallero 1981; Cronenwett et al. 1984; Jelnes et al. 1986), there is still a paucity of information on the natural history of the underlying lesion. Thus, although the natural history of the clinical manifestations of peripheral vascular disease and its causal lesion are inextricably linked, the two are not the same.

Claudication can cause great disability in an otherwise active person, compromise employment and make leisure pursuits unpleasant or impossible. These factors alone make it desirable to treat patients at this stage of the disease. When, in addition, there is a risk that the claudicant might deteriorate, it is even more important to know which patients to treat actively.

Actuarial data are not easy to find, and reported disease progression rates vary widely. Nevertheless, it would appear that around 25% of claudicants deteriorate within 5 years (Bloor 1961; Begg and Richards 1962; McAllister 1976; Wilson et al. 1980); deterioration being equated with progression to the next Fontaine category of disease (Fontaine 1947) or need for reconstructive surgery. Knowing that the cause of claudication is femoral stenosis or occlusion, it should be possible to monitor individuals with such lesions and determine the natural history. It might then be possible to predict how each patient will fare over time, and to advise and treat accordingly.

The Lesion "Under the Microscope"

Plaque, Stenosis and "Critical Stenosis"

I have deliberately avoided trying to define each of these terms. What is a plaque to the histopathologist may be a stenosis to the radiologist and a critical stenosis to the vascular surgeon. However, the terms are used in order to convey the idea that the atheromatous lesion goes through a series of changes, progressively encroaching on the arterial lumen until a fall in distal pressure occurs and the patient develops symptoms.

Histology

In the superficial femoral artery most primary stenoses have been shown to consist of mature atheromatous plaque (Johnson et al. 1990). "Atheroma", a term first used by Albrecht von Haller in 1755, (derivation: the Greek for "porridge") describes one of the important morphological features of a plaque. Its full development may occur gradually or in a series of step-like increments, but in any case a number of well-recognised events first occur: there is profileration of smooth muscle cells and macrophages which secrete a connective tissue matrix, plus an accumulation of intra-and extracellular lipid, most of which is cholesterol derived from plasma. Further complex events allow development of a fibrous connective tissue cap which overlies a pool of necrotic lipid-rich debris, cholesterol crystals and calcification (Ross et al. 1984), and the whole sits atop the internal elastic lamina. Exactly how and why a mature plaque liable to cause the symptoms of peripheral vascular disease develops from the earliest of atheromatous changes is not well understood.

Intra-plaque Events

Intra-plaque events are summarised in Table 27.1.

Table 27.1. Pathological events of potential importance in progression of stenosis

Intra-plaque events	Thrombosis
	Medial necrosis
	Haemorrhage
	Dissection
Extra-plaque events	Platelet accumulation
	Intravascular thrombosis
	Vessel spasm
Haemodynamic events .	Turbulence
	Haemodynamic "stress"
	Fall in distal pressure (? protective)
	Fall in limb blood flow (? pro-thrombotic)

Incorporation of Thrombus into the Plaque. This may directly promote growth of the latter (Duguid 1948), or it may effect growth of the lesion indirectly by increasing the proliferative response of the arterial wall (Woolf 1982). One-third of plaques of the superficial fermoral artery excised by atherectomy were covered by thrombus (Johnson et al. 1990), and this surface thrombus might therefore be both an integral component of a plaque in evolution and at the same time an external element which aggravates the haemodynamic effect of the lesion.

Medial Necrosis. The early plaque which barely protrudes into the lumen and offers minimal obstruction to flow may eventually thicken sufficiently to cause impaired oxygen diffusion to the inner media deeper than 1 mm. The outer media is supplied by the vasa vasorum and so a vascular watershed exists with the vulnerable central media potentially at risk of necrosis due to ischaemia. In addition, there may be a cytotoxic effect of medial oxysterols (Adams 1987) which cause further necrosis. The medial necrosis and oedema which results might add to the size of the lesion.

Intra-Plaque Haemorrhage. This is common (Woolf 1982) and may be important in the sudden transformation from plaque to significant stenosis.

Plaque Dissection. This is of potential importance during plaque growth and as the final pre-occlusive event, but very little is known about the extent to which it contributes to disease progression in the lower limb.

Platelets

The role of platelets in the formation of advanced lesions may be important, particularly in relation to thrombus formation. At the site of stenosis, where flow is altered sufficiently to cause turbulence and points of stagnation, platelets can accumulate (Fox and Hugh 1966). There may be increased platelet – vessel wall interactions which then mediate the formation of intimal microthrombi (Murphy et al. 1962; Muller-Mohnssen et al. 1978) and further stimulation of smooth muscle cell proliferation (Robertson 1960; Wesolowski et al. 1965). However, there is divergence of opinion on the exact site of platelet interaction because animal experiments have indicated that platelets accumulate mostly on the apex of a stenosis and not in the flow recirculation zone (Badimon and Badimon 1989). In any case, the process of mural thrombus formation might effect progression of an atherosclerotic narrowing (Haust 1971).

Haemodynamic Influences on the Lesion

Complex flow patterns can exist even in the healthy artery and there is much evidence suggesting that haemodynamic factors are of paramount importance not only in initiating atherogenesis but also in localising disease (Stehbens 1974). Haemodynamic stress can produce intimal tears, ulceration and mural thrombosis (Stehbens 1974), and it may therefore be responsible for some of the pathological events which are involved in evolution of a plaque and its transformation to a haemodynamically significant stenosis.

Effects of Stenosis

Haemodynamics

In the case of severe stenosis there may be a dramatic alteration in the flow patterns, with high and low flow states, turbulence, reverse flow and flow separation. Conditions of increased flow velocity which exist, e.g. within a stenosis (Barnes 1980), may exert a protective effect on atherogenesis, whereas flow separation and instability might favour atherogenesis (Zarins et al. 1981). A vortex distal to a stenosis may lead to an increase in size of the latter (Fox and Hugh 1966), thereby establishing a vicious cycle of accelerated plaque progression and further flow disturbances. Recently, it has been suggested that endothelial loss which occurs in the central region of a fibrous plaque may alter vascular reactivity (Kolodgie et al. 1990), and this in turn might accentuate a stenosis and the haemodynamic disturbances associated with it.

Velocity Changes

As blood flows through an area of narrowing there is an increase in velocity (Barnes 1980). It may therefore be possible to document change from moderate to more severe disease by measuring increases in flow velocity within a stenosis over the course of time. Thus it is possible to imagine a self-perpetuating cycle of deranged flow and acceleration of disease, which allows monitoring of the change over a relatively short period before an occlusive event occurs. A reliable instrument for screening might, therefore, allow intervention at an appropriate and timely stage in the natural history of clinical events.

Progression of Disease

The severity of the lesion may affect outcome but the direction in which it does so is unclear. Clinical deterioration is more frequent in the presence of occlusion than stenosis (Selvaag et al. 1960), possibly because the slow flow distal to an occlusion leads to the development of other occlusions via a mechanism of "thrombosis-in-situ" of stenotic segments (Humphries 1971).

On the other hand, it has been shown in animal experiments that disease distal to a stenosis advances more rapidly if there is subcritical (reduction in luminal diameter of up to 65%), rather than haemodynamically critical narrowing (Bomberger et al. 1981). In fact, it has even been suggested that a stenosis be deliberately constructed when performing a surgical bypass, in order to minimise distal disease progression (Warren et al. 1961). The hypothesis is that hypotension inhibits development of lesions (Bomberger et al. 1980).

There are, therefore, numerous mechanisms of potential importance in plaque formation, subsequent development of the lesion which encroaches on the arterial lumen, and the lesion which causes complete luminal obliteration. It is important to appreciate these possible mechanisms when looking at new means of investigating stenosis progression and when considering new types of treatment.

Occlusion

It is likely that one or more of the aforementioned processes will precipitate thrombotic occlusion, but how severe a stenosis has to become before it is a significant risk is not clear. In a series of coronary angiograms the only morphological feature of plaque which predicted total occlusion after a mean of 29 months was narrowing of greater than 75% of the luminal diameter (Halon et al. 1985). If the same could be said of femoral stenosis it would lend weight to the view that degree of narrowing is the most important measurement to make when monitoring stenosis progression.

In any case, it is probable that when the stenotic process is so advanced as to cause severely deranged flow then occlusion occurs. At this stage there is traditionally thought to be complete cessation of flow through the involved arterial segment, but it is possible that soon after thrombosis occurs there is still minimal flow through the thrombus for some time until organisation occurs and a solid obstruction to flow is present (personal observation using duplex sonography). This implies that there is still patency at the site of the stenosis, and that atherosclerosis does not bear the sole responsibility for luminal obliteration in all cases.

In some cases, of course, there may be an intra-plaque event such as haemorrhage, necrosis, dissection or thrombosis which precipitates sudden growth of the plaque to a degree sufficient to completely occlude the lumen. In this case the supervening thrombosis would not allow any flow at all, and the situation might be distinguishable from that mentioned above.

It is thought that thrombotic occlusion is initially limited to a short length of artery, but that thrombus then propagates to the next collateral vessel, usually in a proximal direction (Lindbom 1950). When the mouth of the collateral is occluded by further thrombus or atherosclerosis, the process continues until the entire length of the vessel is occluded. Proximally this is not generally beyond the common femoral bifurcation since the main collateral supply in superficial femoral artery occlusion is the large profunda femoris artery, and the medial and lateral circumflex arteries where flow is high. Distally the supreme geniculate artery provides the major distal source of re-entry into the superficial femoral artery (Humphries 1971).

After occlusion the distal circulation is dependent upon collateral vessels, and where these allow good volume flow, ischaemia is minimised. Thus the prognosis for a limb is better for occlusions of the superficial femoral than popliteal artery (Mavor 1956). It is interesting to note that it was Hunter himself who in 1785 initiated the experimental investigation of collateral circulation, but this was in the stag antler, not the femoral artery.

Once again the situation arises where it might become possible to predict, by careful screening, which patients with claudication will suffer deterioration when the occlusion lengthens. This information could then allow timely intervention by, for example, percutaneous transluminal angioplasty, which would pre-empt the worsening of symptoms.

Regression of Atherosclerosis

It must be borne in mind that regression might occur not only as the result of therapeutic intervention but also as part of the natural history of atherosclerosis.

The divergence of opinion regarding possibilities for intervention was well conveyed by Woolf (1982) quoting on the one hand Wissler and Vesselinovitch (1976): "a growing body of evidence indicates that the process of atherosclerosis is almost completely preventable and that it is substantially reversible" and on the other hand "to think of preventing atherosclerosis itself is to shut one's eyes to the nature of the condition. It is a product of ageing" (Duguid 1976).

It is likely that regression would occur more readily in an uncomplicated plaque or in early atherosclerosis than in the mature stenosis (Blankenhorn and Kramsch 1989) where haemodynamic effects might make progression inevitable. Potential for individual plaque components to contribute to regression in experimental models varies according to the component and has been summarised well recently (Armstrong et al. 1990):

Endothelium can regenerate to provide renewed cover of a denuded subendothelium and it can regain normal permeability. Intimal cells and foam cells can decrease and lipid diminish. From a functional aspect the altered vascular reactivity of atherosclerosis can be reversed during regression. The roles of fibrous connective tissue, calcification, elements of necrosis, thrombus and intraplaque haemorrhage are more controversial, and although it is possible that they might be altered in an experimental setting, there is little evidence of change in humans.

Historically, Aschoff is credited as having been the first to offer indirect evidence of regression when he noticed a reduction in aortic atherosclerosis as the First World War progressed (Aschoff 1924). This has been attributed to the imposed dietary restrictions at that time. More recently, a small number of studies of human femoral disease have been described, in the main using arteriography to measure atherosclerotic encroachment on the lumen. This method has a number of limitations, not least of which is the risk and discomfort for the patient. In addition, most of the work was conducted on patients with hyperlipidaemia after intensive drug treatment and risk factor management. Nevertheless, regression was reported in between 10% and 75% of patients (Ost and Stenson 1967; Barndt et al. 1977; Erikson et al. 1983), and in one case a femoral occlusion was reported to have actually disappeared (Ost and Stenson 1967). Individual cases of regression have also been reported (DePalma et al. 1970; Crawford et al. 1979), and one randomised trial of drug versus standard therapy in hyperlipidaemic patients showed that lesion progression was reduced by one-third in the drug-treated group compared with the control group (Duffield et al. 1983).

It is apparent that meticulous long-term studies would be required to show significant improvement in lesion size or extent in the "normolipidaemic" patient with claudication, and that the issue of regression in this context is far from solved.

Progression of Atherosclerosis in Clinical Studies

A few studies have shed light on the true natural history of femoral disease in patients with symptoms of circulatory insufficiency (Table 27.2). In these studies the arteriogram was performed in a scheduled manner regardless of symptoms rather than after deterioration of symptoms (when evidence of worsening main vessel disease was likely).

Coran and Warren (1966), in updating earlier results (Warren et al. 1964), performed annual femoral arteriography for more than 5 years in 15 patients with claudication. All patients had at least a 90% reduction in luminal diameter of the femoro-popliteal segment. It was not stated what proportion of the 19 limbs examined had occlusion as opposed to stenosis. Five limbs showed no progression of disease at all. Fourteen showed progression (increase in length or degree of narrowing or presence of new lesion), and in all the progession occurred proximal to the original lesion with only one patient experiencing worsening of symptoms. Two patients also had disease progression distal to the original lesion and one of these deteriorated symptomatically, thus lending support to the theory of Mavor (1956).

Kuthan and colleagues examined 1196 limbs by scheduled arteriography in 705 patients with peripheral vascular disease, excluding cases needing surgery (Kuthan et al. 1971). With a mean follow-up of $2\frac{1}{2}$ years, 31% of limbs had evidence of new occlusions or lengthening of an old occlusion, and 21% had worsening of a stenosis, or development of stenosis where there was none before. Again, progression rates were found to be higher proximal to the original lesion.

A small number of studies have described disease progression in patients in

Table 27.2. Progression of the disease in follow-up studies

Study	Investigation	Mean follow-up (years)	Progression[a]
Coran and Warren 1966	Scheduled angiography	5	74% of limbs
Kuthan et al. 1971	Scheduled angiography	2.5	52% of limbs
Chilvers et al. 1974	Angiography for symptoms	1–4	Increase in narrowing of superficial femoral artery by 27%–46%
Dawson and Raphael 1968	Angiography for symptoms	2	59% of limbs
Murphy et al. 1990	Angiography for symptoms	2	35% of stenoses
Tillgren et al. 1963	Angiography for symptoms	5	25% of patients
Ulrich and Siggaard-Anderson 1975	Venous occlusion plethysmography	3	35% of patients
Strandness and Stahler 1966	Segmental limb blood pressures	3	45% of limbs

[a]See text for definition.

whom arteriography was carried out for symptomatic deterioration. Chilvers et al. (1974) reported a series of patients who had more than one good quality arteriogram at intervals of between 1 and 4 years. Measuring at a fixed point in the superficial femoral artery, it was found that the average increase in narrowing (when this occurred) was 27%–46% depending on the size of the lesion, that narrowing was not inevitable and that some points even showed widening of the lumen.

Another study examined disease progression in 29 limbs and the mean observation period (time between arteriograms) was 23 months. Nineteen sets of observations were made which involved progression in 17 limbs. These observations comprised: eight extensions of occlusion, seven occlusions, three stenoses and one increase in vessel wall irregularity (Dawson and Raphael 1968). In the superficial femoral artery, six short occlusions remained unchanged in length. Again, patients requiring surgery were excluded from analysis. In a similar retrospective analysis of repeated arteriograms it was concluded that one-third of asymptomatic stenoses became symptomatic within 2 years (Murphy et al. 1990).

Where progression of stenosis is defined as development of collateral circulation, one-quarter of patients show evidence of progression within 5 years (Tillgren et al. 1963). Other methods for determining disease progression rates have been used but the diagnostic specificity of the methods must be questioned. Ulrich and Siggaard-Anderson (1975) found that of 156 claudicants demonstrated by venous occlusion plethysmography to have non-occlusive atherosclerosis, 35% developed an occlusion during a mean follow-up of 3 years.

In a study using segmental limb blood pressures (where only 60% of patients had confirmatory arteriography) to determine the extent of peripheral vascular disease, Strandness and Stahler (1966) followed up 99 limbs. Of the 80 which initially had occlusions, 34 (43%) progressed (developed further disease) in the same or other segments. The nature of this progression was not clearly definable. Of the 19 with no occlusion initially, nine (47%) developed an occlusion during a mean follow-up of 3 years.

Predicting Progression of Disease

This chapter has focused on the discrete stenotic lesion. However, when discussing the natural history of femoral lesions it must be remembered that many patients have diffuse disease rather than a discrete lesion. However, it is not easy to quantify or study diffuse disease and the best way to embark on longitudinal studies is to start with a single discrete lesion.

There are a number of risk factors which predispose to peripheral atherosclerosis of which cigarette smoking is likely to be the most important (Leng and Fowkes 1991). When trying to determine the outcome for a particular patient, it should in theory be possible to sum the relevant risk factors and obtain an overall risk of deterioration within a certain time. This pertains mainly to symptomatic deterioration and despite many risk factors having been described in clinical studies, this ideal is not yet achievable in practice. However, if one accepts that the main determinant of outcome as far as the limb is concerned is the precise nature of the offending lesion (i.e. site, length, morphology, etc.), then it would be more logical to first define the lesion accurately, next determine the risk of progression from the characteristics of that lesion, and subsequently adjust the relative risk in the light of other known risk factors. It has been said that "If factors in a particular patient permit one to reasonably say that serious worsening is imminent, a decision concerning surgery will be easier to make–and with better justification" (Humphries 1971). This statement could well apply to the patient presenting with claudication who has a lesion of the superficial femoral artery.

To date, arteriography has been the chief means of documenting disease progression. However, even when meticulously performed, it requires interpretation by the observer and there may be significant inter-observer variability (Bruins-Slot et al. 1981; Karkow 1989). Some have tried to minimise this in lower limb studies by standardising methods (Chilvers et al. 1974; Barndt et al. 1974; 1977; Crawford et al. 1979), and adapting technique (Clifford et al. 1985). The improvement in quantification of artherosclerosis by using scoring systems is an important step towards standardisation of results (Vogelberg et al. 1975, Bollinger et al. 1981), but cannot be expected to lessen observer variability. New methods of investigation are therefore to be welcomed if they are safer than, and as accurate as, arteriography. This role might soon be filled by duplex scanning. The observer variability has still to be fully evaluated for lower limb arterial duplex scanning, but it would seem the ideal tool for non-invasive, repeatable, lesion-specific documentation of femoral lesions and their physiological effects.

If either arteriography or duplex scanning is performed, the "asymptomatic" leg will often be found to have a relatively isolated stenosis, and this is usually symmetrically opposite the distal end of the occlusion in the symptomatic leg. A stenosis such as this must be one important factor helping to predict the high chance of imminent occlusion and possible clinical deterioration. It would seem the ideal choice for early intervention by, for example, ballon angioplasty. When the stenosis progresses to occlusion, this may be technically harder to treat and associated with more complications and poor results. Thus there exists a "window of opportunity" within which this treatment should be considered.

In conclusion, the natural history of peripheral vascular disease is well documented but that of the causal lesion is not. This is partly because the disease is heterogeneous, with a spectrum from diffuse to isolated lesions, but also because the methods of investigation have limitations. Either they are not lesion-specific or they involve discomfort and risk to the patient. Newer methods of examining lower limb disease such as duplex sonography offer hope of obtaining more useful information, and this might be applied to patients with specific lesions of the femoral artery in order to improve their management.

References

Adams CWM (1987) Disordered structure and function in the atherosclerotic artery. In: Olsson AG (ed) Atherosclerosis – Biology and clinical science. Churchill Livingstone, Edinburgh, pp 75–88

Armstrong ML, Heistad DD, Megan MB, Lopez JAG, Harrison DG (1990) Reversibility of atherosclerosis. Cardiovasc Clin 20:113–26

Aschoff L (1924) In: Lectures of pathology (delivered in the United States in 1924). Hoeber Medical Division, Harper and Row, New York, p 131

Badimon L, Badimon JJ (1989) Mechanisms of arterial thrombosis in nonparallel streamlines: platelet thrombi grow on the apex of stenotic severely injured vessel wall. Experimental study in the pig model. J Clin Invest 84:1134–44

Baker JD (1978) Poststress doppler ankle pressures. Arch Surg 113:1171–3

Barndt R·Jr, Blankenhorn DH, Crawford DW (1974) Prevalence of asymptomatic femoral artery atheromas in hyperlipoproteinaemic patients. Atherosclerosis 20:253–62

Barndt R, Blankenhorn DH, Crawford DW, Brooks SH (1977) Regression and progression of early femoral atherosclerosis in treated hyperlipoproteinaemic patients. Ann Intern Med 86:139

Barnes RW (1980) Haemodynamics for the vascular surgeon. Arch Surg 115:216–23

Begg TB, Richards RL (1962) The prognosis of intermittent claudication. Scott Med J 7:341–52

Blankenhorn DH, Kramsch DM (1989) Reversal of atherosis and sclerosis. The two components of atherosclerosis. Circulation 79:1–7

Bloor K (1961) Natural history of arteriosclerosis of the lower extremities. Ann R Coll Surg Eng 28:36–52

Bollinger A, Breddin K, Hess H et al. (1981) Semiquantitative assessment of lower limb atherosclerosis from routine angiographic images. Atherosclerosis 38:339–46

Bomberger RA, Zarins CK, Taylor KE, Glagov S (1980) Effects of hypotension on atherogenesis and aortic wall composition. J Surg Res 28:402–9

Bomberger RA, Zarins CK, Glagov S (1981) Resident research award. Subcritical arterial stenosis enhances distal atherosclerosis. J Surg Res 30:205–12

Bruins-Slot H, Strijbosch L, Greep JM (1981) Interobserver variability in single plane aortography. Surgery 90:497–503

Chilvers AS, Lea Thomas ML, Browse NL (1974) The progression of arteriosclerosis – A radiological study. Circulation 50:402–8

Clifford PC, Cole SEA, Rhys Davies E, Baird RN (1985) Detection of arterial stenosis: Increased accuracy using biplanar angiography and Doppler signal analysis. J Cardiovasc Surg (Torino) 26:554–7

Coran AG, Warren R (1966) Arteriographic changes in femoropopliteal arteriosclerosis obliterans. A five-year follow-up study. N Engl J Med 274:643–7

Crawford DW, Sanmarco ME, Blankenhorn DH (1979) Spatial reconstruction of human femoral atheromas showing regression. Am J Med 66:784

Cronenwett JL, Warner KG, Zelenock GB et al. (1984) Intermittent claudication. Current results of non-operative management. Arch Surg 119:430–6

Dawson JM, Raphael MJ (1968) Serial aortography in the study of peripheral vascular disease: A clinical radiological study. Br J Radiol 41:333–40

DePalma RG, Hubay CA, Insull W Jr, Robinson AV, Hartman PH (1970) Progression and regression of experimental atherosclerosis. Surg Gynecol Obstet 131:63

Duffield RGM, Lewis B, Miller NE, Jamieson CW, Brunt JNH, Colchester ACF (1983) Treatment of hyperlipidaemia retards progression of symptomatic femoral atherosclerosis. A randomised controlled trial. Lancet ii:639–42

Duguid JB (1948) Thrombosis as a factor in the pathogenesis of aortic atherosclerosis. J Pathol Bacteriol 60:57

Duguid JB (1976) Prevention of atherosclerosis. In: Duguid JB. The dynamics of atherosclerosis. Aberdeen University Press, Aberdeen, p 67

Erikson U, Helmius G, Hemmingsson A, Ruhn G, Olsson AG (1983) Measurement of atherosclerosis by arteriography and microdensitometry: Model and clinical investigations. In: Schettler FG, Gotto AM, Middelhoff F, Habenicht AJR, Jurutka KR (eds) Atherosclerosis VI. Proceedings of the Sixth International Symposium on Atherosclerosis. Springer-Verlag, Berlin, p 197

Fontaine R (1947) Les artérites oblitérantes des membres inférieurs et leur traitment. Strasbourg Med 107:303–24

Fowkes FGR, Housley E, Cawood EHH, Macintyre CCA, Ruckley CV, Prescott RJ (1991) Edinburgh Artery Study. Prevalence of asymptomatic and symptomatic peripheral arterial disease in the general population. Int J Epidemiol 20:384–92

Fox JA, Hugh AE (1966) Localisation of atheroma: a theory based on boundary layer separation. Br Heart J 23:388–99

Halon DA, Sapoznikov D, Gotsman MS, Lewis BS (1985) Can total coronary occlusions be predicted from a previous coronary arteriogram? Cathet Cardiovasc Diagn 11:455–62

Haust MD (1971) The morphogenesis and fate of potential and early atherosclerotic lesions in man. Hum Pathol 2:1–29

Housley E (1988) Treating claudication in five words. Br Med J 296:1483–4

Humphries AW (1971) The relation of the natural history of arteriosclerosis to surgical management. In: Dale WA (ed) Management of arterial occlusive disease. Year Book Medical, Chicago, pp 67–8

Imperato AM, Kim G, Davidson T et al. (1975) Intermittent claudication: its natural course. Surgery 78:795–9

Jelnes R, Gaardsting O, Hougaard Jensen K, Baekgaard N, Tonnesen KH, Schroeder T (1986) Fate in intermittent claudication: outcome and risk factors. Br Med J 293:1137–40

John HT, Warren R (1961) The stimulus to collateral circulation. Surgery 49:14–25

Johnson DE, Hinohara T, Selmon MR, Braden LJ, Simpson JB (1990) Primary peripheral arterial stenoses and restenoses excised by transluminal atherectomy: a histopathological study. J Am Coll Cardiol 15:419–25

Kallero KS (1981) Mortality and morbidity in patients with intermittent claudication as defined by venous occlusion plethysmography. A ten-year follow-up study. J Chron Dis 34:455

Karkow WS (1989) Variations in interpretation of arterial stenosis. J Cardiovasc Surg (Torino) 30:826–32

Kolodgie FD, Virmani R, Rice HE, Mergner WJ (1990) Vascular reactivity during the progression of atherosclerotic plaque. A study in WHH rabbits. Circ Res 66:1112–26

Kuthan F, Burkhalter A, Baitsch R, Ludin H, Widmer LK (1971) Development of arterial disease in lower limbs – Angiographic follow-up of 705 medical patients. Arch Surg 103:545–47

Leng GC, Fowkes FGR (1991) The epidemiology of peripheral vascular disease. Curr Med Lit: Thrombosis 1:35–43

Lindbom A (1950) Arteriosclerosis and arterial thrombosis in lower limb. Roentgenological study. Acta Radiol (Suppl) 80:1–80

Lundgren F, Dahllof A, Lundholm K, Schersten T, Volkmann R (1989) Intermittent claudication – surgical reconstruction or physical training? A prospective randomised trial of treatment efficiency. Ann Surg 209:346–55

Mavor GE (1956) Pattern of occlusion in atheroma of lower limb arteries. Br J Surg 43:352–64

McAllister FF (1976) The fate of patients with intermittent claudication managed non-operatively. Am J Surg 132:593–5

Muller-Mohnssen H, Kratzer M, Baldauf W (1978) Microthrombus formation in models of coronary arteries caused by stagnation point flow arising at the predilection site of atherosclerosis and thrombosis. In: Nerem RM, Cornhill JF (eds) The role of fluid mechanics in atherogenesis. Ohio University, Ohio, pp 12-1–12-8

Murphy EA, Rowsell HE, Downie HG, Robinson EA, Mustard JF (1962) Encrustation and atherosclerosis: The analogy between early in vivo lesions and the deposits which occur in extracorporeal circulations. Can Med Assoc J 87:259–74

Murphy P, Jeans WD, Horrocks M, Baird R (1990) A retrospective study of the fate of asymptomatic superficial femoral artery stenoses. Presented at British Institute of Radiology, Harrogate, June 1990

Ost CR, Stenson S (1967) Regression of peripheral atherosclerosis during therapy with high doses of nicotinic acid. Scand J Clin Lab Invest (Suppl) 99:241–5

Robertson JH (1960) Influence of mechanical factors on structure of peripheral arteries and the localisation of atherosclerosis. J Clin Pathol 13:199–204

Ross R, Wight TN, Strandness E, Thiele B (1984) Human atherosclerosis. I. Cell constitution and characteristics of advanced lesions of the superficial femoral artery. Am J Pathol 114:79–93

Selvaag O, Myren J, Thorsen RK, Bjornstad P (1960) Progressive tendency of arteriosclerosis obliterans of the lower extremities. Acta Chir Scand (Suppl) 253:187–95

Spaulding WB (1956) The prognosis of patients with intermittent claudication. Can Med Assoc J 75:105–111

Stehbens WE (1974) Haemodynamic production of lipid deposition, intimal tears, mural dissection and thrombosis in the blood vessel wall. Proc R Soc Lond (B) 185:357–73

Strandness DE, Stahler C (1966) Arteriosclerosis obliterans, manner and rate of progression. JAMA 196:1

Tillgren C, Stenson S, Lund F (1963) Obliterative arterial disease of the lower limbs (I) studied by means of repeated femoral arteriography. Acta Radiol 1:1161–78

Ulrich J, Siggaard-Anderson J (1975) The natural history of arteriosclerosis in the lower extremities. II. A Plethysmographic study of non-occlusive arteriosclerotic disease in the lower limbs. Dan Med Bull 22:136–140

Vogelberg KH, Berchtold P, Berger H et al. (1975) Primary hyperlipoproteinaemias as risk factors in peripheral arterial disease documented by arteriography. Atherosclerosis 22:271

Warren R, John HT, Shepherd RC, Villavicencio JL (1961) Studies on patients with arteriosclerotic obliterative disease of the femoral artery. Surgery 49:1–13

Warren R, Gomez RL, Marston JAP, Cox JST (1964) Femoropopliteal arteriosclerosis obliterans – arteriographic patterns and rates of progression. Surgery 55:135–43

Wesolowski SA, Fries CC, Sabini AM, Sawyer PN (1965) The significance of turbulence in hemic systems and in the distribution of the atherosclerotic lesion. Surgery 57:155–62

Widmer LK, Greensher A, Kannel WB (1964) Occlusion of peripheral arteries. A Study of 6400 working subjects. Circulation 30:836–41

Wilson SE, Schwartz I, Williams RA, Owens ML (1980) Occlusion of the superficial femoral artery. What happens without operation. Am J Surg 140:112–18

Wissler RW, Vesselinovitch D (1976) Studies of regression of advanced atherosclerosis in experimental animals and man. Ann NY Acad Sci 275:363–78

Woolf N (1982) Pathology of atherosclerosis. Butterworth, London

Zarins CK, Bomberger RA, Glagov S (1981) Local effects of stenoses: increased flow velocity inhibits atherogenesis. Circulation 64:221–7

28 Prognosis of Intermittent Claudication

G. DAVEY SMITH, M.J. SHIPLEY and M.G. MARMOT

Peripheral Vascular Disease: Another Manifestation of Atherosclerosis?

It is convenient to think of atherosclerosis as the common underlying pathological lesion in a range of cardiovascular conditions: coronary heart disease, thrombotic stroke, peripheral vascular disease. This cannot be the whole story, since the epidemiology of these clinical conditions is not the same. The epidemiology of stroke differs from that of coronary heart disease in a number of ways:

Rates of stroke are high in Japanese and Chinese, coronary heart disease rates are low; the high rate of stroke among Japanese and Chinese is both haemorrhagic and thrombotic

In many Western countries, mortality from stroke began to decline long before there was a decline in coronary heart disease

In the UK, immigrants from Africa and the Caribbean have high rates of stroke and low rates of coronary heart disease; South Asians have high rates of coronary heart disease but rates of stroke lower than those of people of African origin

Given the above, it is not surprising to find that risk factors for stroke are not identical with those for coronary heart disease: raised blood pressure is the dominant risk factor for stroke, while plasma lipid levels and smoking are less important than they are for coronary heart disease.

What of peripheral vascular disease? The prevalence is increased in people with evidence of angina, previous myocardial infarction, or electrocardiographic changes suggestive of ischaemia (Reid et al. 1966; Tillgren 1965; Reunanen et al. 1982), and data from Framingham show that development of intermittent claudication is strongly related to coronary risk factors (Kannel and Shurtleff 1971; Kannel and McGee 1985).

One way of examining the relation between these different clinical syndromes is to look at their prognosis. Angina is not the same as myocardial infarction, but it predicts death from coronary heart disease (Rose 1965). Intermittent claudication is related to increased death rates from cardiovascular disease (Bloor 1961; Hughson et al. 1978a; Kallero 1981; Jelnes et al. 1986; Dormandy et al. 1989). Some studies suggest, however, that this increased risk is due to the coexistence of coronary heart disease. In patients without evidence of ischaemia, the risk may not be elevated.

The Whitehall study of civil servants allows exploration of this issue. We examine the power of intermittent claudication to predict death from cardiovascular and other causes independent of the presence of ischaemic heart disease or of coronary risk factors.

Whitehall Study: Methods

In the Whitehall study, 18 403 men aged 40–64 years underwent a clinical examination and completed a questionnaire between 1967 and 1969 (Reid et al. 1974); Davey Smith et al. 1990). The electrocardiogram (ECG) was coded according to the Minnesota system, and was regarded as positive for ischaemia if Q/QS items (codes 1.1–3), ST/T items (codes 4.1–4 or 5.1–3) or left bundle branch block (code 7.1) were present. Civil service employment grade was divided into 4 levels: administrators, professionals and executives, clerical and other (mainly unskilled manual) grades. "Low work grade" refers to the lower two groups, clerical and other grades. Smoking has been categorised according to cigarette use as "current smoker", "ex-smoker" and "never smoker". In addition, adjustment for smoking habits has included a term for the number of cigarettes per day smoked by current smokers. The 640 men who smoked pipes or cigars only have been treated as a separate group in the analyses that involve smoking status.

The London School of Hygiene and Tropical Medicine chest pain and intermittent claudication questionnaire was administered by an interviewer with 938 men and was self-administered by the other subjects. The prevalence of positive response to the intermittent claudication section was similar in these two groups (Rose 1965). Data regarding angina and possible myocardial infarction have been taken from this questionnaire for the present analysis.

To be regarded as having intermittent claudication subjects had to report that they developed calf pain on walking, that they had not developed such pain when standing still or sitting, that they would stop walking or slow down when pain developed and that the pain had then usually disappeared within 10 minutes. If subjects reported that this pain had never disappeared while they were still walking they were regarded as having "probable" intermittent claudication. If they reported that the pain had disappeared while they were walking (which may have been at a slower pace) they were regarded as having "possible" intermittent claudication. Subjects reporting leg pain on walking which did not satisfy the criteria for intermittent claudication were classified as having "other leg pain".

Intermittent Claudication in Whitehall Study

Probable and possible intermittent claudication were reported by 0.8% (147) and 1% (175) of subjects respectively. Each rate more than doubled from the age groups 40–44 to 60–64 years old. "Other leg pain" was reported by 11% of the subjects.

Relation between Claudication and Ischaemic Heart Disease

Table 28.1 shows that questionnaire evidence of angina and possible myocardial infarction are more common in men with than men without claudication at baseline. The prevalence of ECG changes suggestive of ischaemia did not differ among the groups. The measure combining ECG and questionnaire evidence into "any suspect ischaemia" is more prevalent in men with claudication and is more prevalent among probable than possible cases. This latter difference is not statistically significant (p=0.12), however.

Table 28.1. Prevalence of age-adjusted risk factors in men with probable, possible, or no intermittent claudication

Risk factor	Intermittent claudication							
	No		Possible			Probable		
	%	n	%	n	p	%	n	p
Angina	4.6	837	12.9	24	<0.001	16.5	27	<0.001
Possible myocardial infarction	6.6	1181	14.0	25	<0.001	20.1	30	<0.001
Abnormal electrocardiogram	6.3	1127	5.8	13	>0.5	4.8	9	>0.5
Any suspect ischemia	15.1	2696	25.0	48	<0.001	34.0	52	<0.001

p indicates whether the proportions in the possible or probable intermittent claudication groups are different from those in the group with no intermittent claudication.

Intermittent Claudication and Coronary Risk Factors

The relationships between reported symptoms and coronary heart disease risk factors are shown in Tables 28.2 and 28.3, which show slightly elevated average levels of plasma cholesterol concentration and body mass index, and lower average ventilatory measures. There are no significant relationships with blood pressure or glucose intolerance. Smoking is more prevalent among subjects reporting intermittent claudication. Analysis of covariance suggests that the poorer ventilatory function shown by probable and possible cases is not solely due to differences in smoking behaviour.

Of the participants with intermittent claudication, 14 (8%) of the possible, and 15 (10%) of the probable cases were under treatment from their doctor at the time of the examination.

Intermittent Claudication and Subsequent Mortality

Age-adjusted mortality rates for coronary heart disease, stroke, non-cardiovascular disease, lung cancer, all neoplasms and all causes are shown in Table 28.4. Cardiovascular mortality is higher both amongst those with probable and those

Table 28.2. Mean and standard error (SEM) of age-adjusted risk factors in men with probable, possible, or no intermittent claudication

Risk factor	Intermittent claudication					
	No (n=18 066)		Possible (n=175)		Probable (n=147)	
	Mean	SEM	Mean	SEM	Mean	SEM
Systolic blood pressure (mmHg)	135.9	0.15	137.5	1.81	136.1	1.59
Diastolic blood pressure (mmHg)	84.5	0.10	83.5	1.07	84.9	1.14
Plasma cholesterol concentration (mmol/l)	5.11	0.01	5.27	0.11	5.34*	0.10
Body mass index (kg/m^2)	24.7	0.02	25.3*	0.23	25.2	0.24
FEV$_1$	3.13	0.004	3.02*	0.042	3.00**	0.044
FVC	4.03	0.005	3.84**	0.050	3.84***	0.059

FEV$_1$ forced expiratory volume in 1 second; FVC, forced vital capacity.
Test for differences between possible or probable intermittent claudication and no intermittent claudication groups: *p<0.05; **p<0.01; ***p<0.001

Table 28.3. Prevalence of age-adjusted risk factors in men with probable, possible, or no intermittent claudication

Risk factor	Intermittent claudication							
	No		Possible			Probable		
	%	n	%	n	p	%	n	p
Glucose tolerance								
Normoglycemic	93.3	16 745	89.6	154	⎫	94.2	139	⎫
Glucose intolerant	5.5	976	8.7	17	⎬ 0.09	3.1	4	⎬ 0.17
Diabetic	1.2	217	1.7	4	⎭	2.7	3	⎭
Smokers								
Never smokers	19.5	3409	16.0	25	⎫	16.0	19	⎫
Ex smokers	37.9	6595	26.3	50	⎬ 0.004	30.5	44	⎬ 0.009
Current smokers	42.6	7422	57.7	96	⎭	53.5	81	⎭
Low work grade	24.8	4278	30.4	57	0.11	29.7	50	0.07

p indicates whether the proportions in the possible or probable intermittent claudication groups are different from those in the group with no intermittent claudication.

with possible intermittent claudication. For possible, but not probable, cases, rates of non-cardiovascular causes of death are also elevated. The largest difference in mortality rates between probable and possible cases is the higher non-cardiovascular mortality rate in the latter (p=0.02).

Subjects reporting "other leg pain" had a mortality rate of 15.4 per 100 person-years, only marginally above that of 12.7 per 1000 person-years for the men reporting no leg pain. For coronary heart disease, the mortality rate for men with other leg pain was 5.8 per 100 person-years, similar to that of 4.9 per 1000 person-years for men with no leg pain.

Table 28.4 also shows that the proportions of deaths due to cardiovascular or non-cardiovascular causes amongst the possible cases differs little from that in the group without intermittent claudication (p>0.2). By contrast, among probable cases, the increase in death is confined to cardiovascular causes, hence the proportion of such deaths among all deaths is greater (p<0.001).

To examine whether the increased rates among cases reflected only a short-

Table 28.4. Age-adjusted mortality rates and number of deaths in men with probable, possible or no intermittent claudication

Cause of death	Intermittent claudication								
	No			Possible			Probable		
	Rate	n	%	Rate	n	%	Rate	n	%
Coronary heart disease	5.1	1326	39	13.3*	32	46	16.1*	35	63
Cerebrovascular disease	0.8	204	6	2.5**	6	9	2.3**	5	9
Cardiovascular disease	6.9	1788	53	17.1*	42	60	20.6*	45	80
Non-cardiovascular disease	6.1	1600	47	11.0***	28	40	4.2	11	20
Lung cancer	1.5	380	11	2.8**	8	11	0.8	2	4
All cancers	4.1	1071	31	6.6**	17	24	3.2	8	14
All causes	13.0	3401	100	28.1*	70	100	24.9*	56	100

Rates are per 1000 person-years
All causes mortality includes 13 deaths in the groups without intermittent claudication in which the specific cause of death was unknown.
Percent: percentage of deaths attributed to each cause or group of causes.
Tests for differences between possible or probable intermittent claudication and no intermittent claudication groups: * $p<0.001$; ** $p<0.05$; *** $p<0.01$.

term risk we analysed the data excluding deaths within 5 years of examination. For both the probable and the possible intermittent claudication groups all-cause and cardiovascular mortality rates were elevated by factors of approximately two and three times respectively. For the possible, but not the probable cases, non-cardiovascular mortality was elevated.

Fig 28.1 presents mortality rates, relative to those among men without claudication. These are calculated first adjusting for age alone and then for coronary risk factors, systolic blood pressure, plasma cholesterol concentration, smoking habits (including number of cigarettes smoked per day in current smokers), employment grade and degree of glucose intolerance. It is clear that the increased risk of death from cardiovascular causes cannot be accounted for by differences in coronary risk factors as measured in this study. Similarly, the increased risk of death from non-cardiovascular causes among men with possible claudication persists after adjustment for these coronary risk factors. The analysis was repeated removing men who had evidence of ischaemia at the baseline examination (Table 28.5). The same pattern of elevated mortality rates is seen.

Fig. 28.2 shows the risk factor-adjusted relative rate of coronary heart disease death among men with probable claudication compared to those without, and compares this with the relative risks for men with and without angina pectoris and with and without possible myocardial infarction. The power to discriminate men with an increased risk of subsequent death is comparable for these three different components of the London School of Hygiene Cardiovascular Questionnaire.

Implications: Intermittent Claudication

Intermittent claudication is a manifestation of atherosclerosis of large vessels. The association with other evidence of cardiovascular disease shown in the

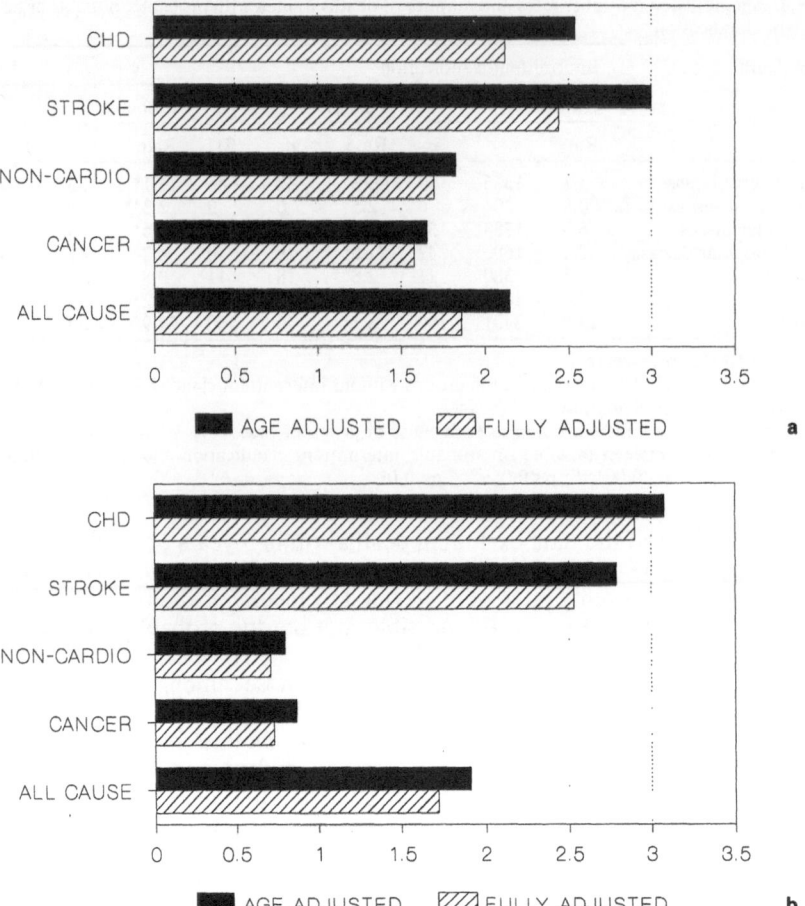

Fig. 28.1. Mortality due to coronary heart disease (CHD), stroke, non-cardiovascular causes (non-cardio), cancer, and all causes in men with intermittent claudication relative to those without, adjusted for age and "fully" adjusted for age, systolic blood pressure, plasma cholesterol, smoking habits, employment grade and degree of glucose intolerance. **a** Possible claudication; **b** probable claudication.

present study is expected and consistent with other studies (Tillgren 1965; Reid et al. 1966; Kannel and Shurtleff 1971; Reunanen et al. 1982). The absence of a relationship between intermittent claudication and ECG evidence of ischaemia is surprising in the light of other studies, but is based on relatively small numbers of subjects with both.

The relationship between coronary risk factors and the presence of intermittent claudication has been inconsistent in other prevalence studies, with the exception of smoking, which has consistently been related to intermittent claudication (Hughson et al. 1978b; Reunanen et al. 1982) to a similar degree to that seen here. Plasma cholesterol was elevated in cases of intermittent claudication in some (Schroll and Munck 1981; Reunanen et al. 1982), but not all (Hughson et al. 1978b), studies. Similarly blood pressure has been found to be elevated (Hughson

Table 28.5. Relative mortality rates and 95% confidence intervals (CI) for probable or possible intermittent claudication vs. no intermittent claudication in men with no suspected ischaemia at baseline

| Cause of death | Intermittent claudication | | | |
| | Possible | | Probable | |
	RR	95% CI	RR	95% CI
Coronary heart disease	2.82	1.8,4.5	2.84	1.8,4.5
Cerebrovascular disease	5.22	2.1,12.8	3.31	1.0,10.5
Cardiovascular disease	2.81	1.9,4.2	2.59	1.7,4.0
Noncardiovascular disease	2.06	1.3,3.2	0.56	0.2,1.3
Lung cancer	2.31	1.0,5.2	0.41	0.1,3.0
All neoplasms	1.87	1.1,3.2	0.66	0.2,1.8
All causes	2.41	1.8,3.2	1.55	1.1,2.3

RR relative rate. Relative rates are adjusted for age, systolic blood pressure, cholesterol, smoking habits, employment grade, and degree of glucose intolerance.

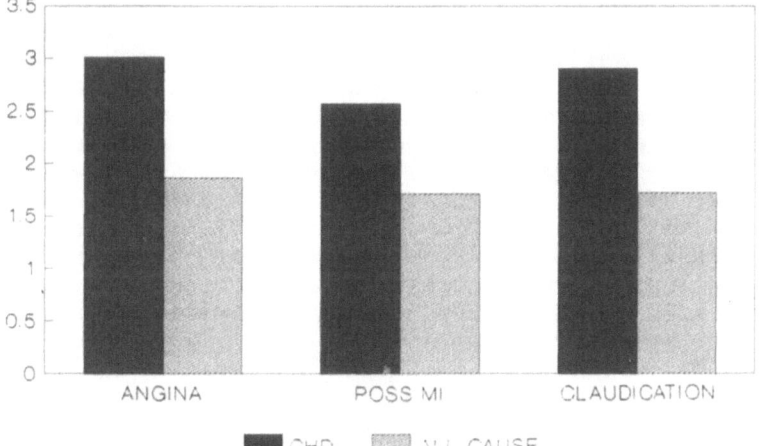

Fig. 28.2. Risk of death due to coronary heart disease (CHD) and all causes ("fully" adjusted for age, systolic blood pressure, plasma cholesterol, smoking habits, employment grade and degree of glucose intolerance) in men with a symptom compared to men without that symptom: for angina patients, those with pain of possible myocardial infarction and those with probable intermittent claudication.

et al. 1978b; Schroll and Munck 1981) or else no higher (Reunanen et al. 1982). Negative findings have been reported for body mass index (Schroll and Munck 1981) and both fasting glucose (Hughson et al. 1978b) and glucose tolerance (Reunanen et al. 1982). The present results – small elevations in plasma cholesterol and body mass index, and no significant relationship with blood pressure or glucose tolerance – add to the inconsistent set of previous findings.

The failure to detect a relationship between claudication and diabetes may be a chance finding, or could reflect selection out of employment in the civil service by participants who were both diabetic and had intermittent claudication.

In the Framingham study (Kannel and Shurtleff 1971; Kannel and McGee 1985), smoking, glucose intolerance and blood pressure were powerful independent predictors of incidence of intermittent claudication; serum cholesterol was a weak predictor, and relative weight was inversely associated with occurrence of intermittent claudication. The Framingham results may differ from those of other studies because Framingham studied the occurrence of new cases, i.e. incidence, whereas most others studied prevalence. The selective survival of individuals at lower risk may have led to underestimation of the importance of glucose tolerance and blood pressure in prevalence studies.

What Is the Nature of Possible Claudication?

The difference in criteria for possible and probable intermittent claudication is slight. In possible claudication, leg pain may have disappeared while walking; for probable claudication this could not have occurred. This small change approximately doubled the number of positive responses.

The two classes of claudication are similarly related to smoking and other risk factors cross-sectionally. Although both show higher prevalence of ischaemia than men without claudication, there was the suggestion of a greater difference in men with probable claudication. This would be as expected if weakening the criteria increased sensitivity but led to the inclusion of more false positive cases among the possibles. What is striking about the mortality follow-up of the Whitehall Study, however, is that there was only a small difference between the two intermittent claudication groups in their association with cardiovascular mortality. "Other leg pain", on the other hand, was only associated with a marginal increase in mortality risk. The rather unexpected finding regarding probable and possible intermittent claudication suggests that the category of possible intermittent claudication has acceptable validity. It is evidence that the "possible" group contains many genuine cases. In previous studies the percentages of deaths from cardiovascular causes in subjects with intermittent claudication has ranged from over 80% (Bloor 1961; Hughson et al. 1978a) to between 60%–65% (Kallero 1981). Thus the percentages of cardiovascular deaths in both the probable intermittent claudication group and in the possible intermittent claudication group lie within the previous boundaries.

It appears, however, that whereas probable and possible cases have an increased risk of cardiovascular death, only possible cases have an increased risk of dying from other causes. This non-specific mortality excess is little changed by excluding deaths within 5 years of the examination. It is unlikely therefore to be the result of more near terminal illness among the possible claudication group. If the association between possible claudication and non-cardiovascular mortality is the result of general ill-health among the possible group, it has a long persisting influence. The finding that only a small proportion of participants were under treatment from their doctor does not suggest that men with probable claudication were protected from an excess of non-cardiovascular deaths by a greater degree of medical intervention.

In conclusion, the doubling of the mortality rate in those with intermittent claudication was similar to that shown in other population-based studies using interview or questionnaire methods of diagnosis (Kannel and Shurtleff 1971; Reunanen et al. 1982). This elevation of mortality risk is less than that associated with peripheral vascular disease defined by more stringent criteria (Kallero 1981;

Criqui et al. 1985a). The elevated mortality risk seen with large vessel peripheral arterial disease appears to be independent of coexisting cardiovascular disease and risk factors (Criqui et al. 1985a). The same is true of intermittent claudication in the present study, within the limits to which it is possible to explore "independent" effects in such situations (Davey Smith and Phillips 1990). Questionnaire-defined intermittent claudication and large vessel peripheral arterial disease measured through more detailed assessments seem to overlap to a large degree (Criqui et al. 1985b) and both appear to be related to diffuse atherosclerosis, which leads to a high risk of mortality.

It appears that in population studies intermittent claudication carries with it about the same risk as angina or possible myocardial infarction assessed by questionnaire. Only a small part of this increased risk can apparently be attributed to the associated levels of coronary risk factors, and it is not dependent upon evidence of coexisting ischaemic heart disease.

References

Bloor K (1961) Natural history of arteriosclerosis of the lower extremities. Ann R Coll Surg Engl 28:36–52

Criqui MH, Coughlin SS, Fronek A (1985a) Noninvasively diagnosed peripheral arterial disease as a predictor of mortality: results from a prospective study. Circulation 72:768–73

Criqui MH, Fronek A, Klauber MR, Barrett-Connor E, Gabriel S (1985b) The sensitivity, specificity and predictive value of traditional clinical evaluation of peripheral arterial disease: results form noninvasive testing in a defined population. Circulation 71:516–22

Davey Smith G, Phillips AN (1990) Declaring independence: why we should be cautious. J Epidemiol Community Health 44:257–8

Davey Smith G, Shipley MJ, Rose G (1990) Intermittent claudication, heart disease risk factors, and mortality. Circulation 82:1925–31

Dormandy J, Mahir M, Ascady G et al. (1989) Fate of the patient with chronic leg ischaemia. J Cardiovasc Surg 30:50–7

Hughson WG, Mann JI, Tibbs DJ, Woods HF, Walton I (1978a) Intermittent claudication: factors determining outcome. Br Med J i:377–9

Hughson WG, Mann JI, Garrod A (1978b) Intermittent claudication: prevalence and risk factors. Br Med J i:379–81

Jelnes R, Gaadsting O, Hougaard JK, Baekgaard N, Tonnesen KH, Schroeder T (1986) Fate in intermittent claudication: outcome and risk factors. Br Med J 293:1137–40

Kallero KS (1981) Mortality and morbidity in patients with intermittent claudication as defined by venous occlusion plethysmography: A ten-year follow-up study. J Chronic Dis 34:455–62

Kannel WB, McGee DL (1985) Update on some epidemiologic features on intermittent claudication: The Framingham study. J Am Geriatr 33:13–18

Kannel WB, Shurtleff D (1971) The natural history of arteriosclerosis obliterans. Cardiovasc Clin 3:37–52

Reid DD, Holland WW, Humerfelt S, Rose G (1966) A cardiovascular survey of British postal workers. Lancet i:614–18

Reid DD, Brett GZ, Hamilton PJS, Jarrett RJ, Keen H, Rose G (1974) Cardiorespiratory disease and diabetes among middle-aged male civil servants. Lancet i:469–73

Reunanen A, Takkunen H, Aromaa A (1982) Prevalence of intermittent claudication and its effect on mortality. Acta Med Scand 211:249–56

Rose G (1965) Ischemic heart disease: chest pain questionnaire. Millbank Memorial Fund Q 43:32–9

Schroll M, Munck O (1981) Estimation of peripheral arteriosclerotic disease by ankle blood pressure: measurements in a population study of 60-year-old men and women. J Chronic Dis 34:261–9

Tillgren C (1965) Obliterative arterial disease of the lower limbs. III. Prognostic influence of concomitant coronary heart disease. Acta Med Scand 178:121–8

29 Factors Affecting Clinical Progression and Mortality

J.A. DORMANDY

Although much has been published on the surgical treatment of leg ischaemia, relatively little is known about claudication in general and the fate of the majority of patients with chronic leg ischaemia who never present to a specialist (usually surgical) department. In a review of the literature on the fate of the claudicant (Dormandy et al. 1989), it was concluded that on the scanty evidence available the prevalence of intermittent claudication in men below the age of 50 is approximately 1% to 1.5%, rising rapidly with age to over 5% in older men. In the few reports where men and women were studied at an early stage of the disease, their ratio was less than 2:1. However, in most studies looking at a more advanced stage, the ratio of men to women is much higher, ranging from 3:1 to 13:1. This would suggest that the prognosis of the local disease in women is much better than in men, a view supported by the recent PACK (Prevention of Atherosclerotic Complications by Ketanserin) analysis (Dormandy and Murray 1991).

Patients with claudication naturally tend to concentrate on their presenting symptom in the leg and its likely progression. When it is explained that their symptom is due to arterial disease, most patients' immediate fear is that of amputation. In fact their real and more serious problem is the coexisting arterial disease in the coronary and cerebral circulations, although these may initially be asymptomatic. The fate of the claudicant will therefore be analysed first in relation to the progression of the local disease in the legs, and what factors may influence this; then the likelihood of severe cardiovascular events and death will be considered, again looking at the possible secondary risk factors. Finally, the fate of the amputee will be briefly described. Apart from the review of the literature already mentioned (Dormandy et al. 1989), some newer publications in the last 2 years will be considered and in particular the results of the PACK Study, which have been summarised in three publications (Prevention of Atherosclerotic Complications with Ketanserin Trial Group 1989; PACK Claudication Substudy Investigators 1989; Dormandy and Murray 1991). This was a prospective, placebo-controlled study carried out in a population of 3899 intermittent claudicants (proven using Doppler pressure ratios) at all stages of the disease, recruited from 147 medical as well as surgical centres in three continents. Although the follow-up was short, this study was perhaps unique in the thoroughness with which all cardiovascular events were documented and validated; and in the exceptionally low number of cases lost to follow-up (0.2%). The first publication describes the overall primary analysis, which was essentially negative with regard to Ketanserin, although this was probably due to an unexpected interaction with potassium losing diuretics (Prevention of Atherosclerotic Complications with Ketanserin Trial Group 1989). The second publication looks more closely on the progression of the disease and symptoms on the legs (PACK Claudication Substudy Investigators 1989), while the most recent publication analyses the risk factors for progression of the local and generalised disease (Dormandy and Murray 1991).

Progression of Local Disease

The vast majority of claudicants are never referred to a specialist centre; indeed both a London and an Oxford study suggest that many claudicants do not consult any doctor at all (Hughson et al. 1978; Smith et al. 1990). Therefore conclusions about the progress of the local symptoms based on hospital series, particularly from vascular surgical units, must be suspect. Summarising the results of 10 series which were not primarily surgical, there is broad agreement that in approximately three-quarters of the cases symptoms will stabilise and may even improve, the remainder deteriorating at various rates (Dormandy et al. 1989). Although these symptoms may improve because of development of collaterals, the underlying atheromatous disease almost certainly progresses. The magnitude of the well-recognised placebo effect in terms of walking distance was shown recently when 265 established claudicants given placebo for 1 year increased their treadmill pain-free walking distance by a mean of 28.2%, with as much improvement in the first 6 months as in the second 6 months (PACK Claudication Substudy Investigators 1989). This, of course, excluded the small minority of patients whose local disease progressed to such an extent that they required an amputation or reconstructive surgery. Some of the so-called "placebo" effect could of course be due to the development of collaterals, or the effects of exercise training. The overall PACK data gives more information on a total of 3899 claudicants at various stages of their disease. 214 (5.5%) deteriorated sufficiently to require a surgical or catheter intervention in one year. Only 49 (1.2%) of the total group required a major amputation during the same period. This is close to the figures available for the overall incidence of amputation in claudicants from the only large epidemiological studies of normal populations: 1.6% in the Framingham Study (Kannel and McGee 1985) and 1.8% in the Basle study, although these are not yearly figures (Widmer et al. 1985). Again, the bias in most previous hospital-based series can be observed, the average amputation rate in six recent non-operative and three surgical series being 5% (Dormandy et al. 1989). Part of the reason for this relatively optimistic outlook, as far as progression of the original symptoms and local disease is concerned, is that most claudicants die prematurely of atheromatous disease in another territory.

Factors Affecting Progression of Local Disease

In the prospective study of 1969 PACK claudicants, the entry characteristics were analysed individually (univariate analysis) and in combination (multivariate analysis) to look for predictors of deterioration of leg ischaemia as evidenced by the relatively hard markers of either a need for arterial reconstruction, a catheter reopening procedure or amputation (Dormandy and Murray 1991). By far the most significant predictor of local progression is the ankle-brachial Doppler systolic pressure ratio. Claudicants with a pressure ratio less than 0.5 at the beginning of the period of observation (only 16% of the total study group) were 2.3 times more likely to deteriorate in the subsequent year than those with an ankle-brachial ratio above 0.5. In this study, diabetes had a hazard ratio of 1.3 and smoking during the previous 6 months 1.4. Perhaps the most surprising positive correlation was with sex, men being 1.66 times more likely to deteriorate than women. The most surprising weak correlation was with age (hazard ratio of 1.18 for increments of 10 years).

A review of the previous literature supports some of these findings. Jelnes (1986) also found that a pressure ratio above or below 0.5 was a highly significant (p=0.002) predictor of local progression of the disease. It is also generally accepted that diabetics have a more aggressive form of local disease, more likely to deteriorate rapidly presumably because of the combination of earlier atherosclerosis in the larger arteries, and diabetic microangiopathy in the arterioles and microcirculation (Hughson et al. 1978). Some national statistics on amputation rates support this view. Estimated amputation rates in the United States were 200/million/year in non-diabetics compared with 3900/million/year in diabetics (Most and Sinnock 1983).

Of the risk factors which can be influenced, there is no doubt that smoking is the most important (Hughson et al. 1978). In one study, claudicants who smoked over 40 pack-years deteriorated 3.3 times more often than those who smoked less (Cronenwett et al. 1984). There continues to be uncertainty about the relative importance of haematological and biochemical variables as secondary risk factors for progression of local disease in the legs of claudicants, although there can be little doubt about their importance as primary risk factors for the development of atherosclerotic diseases.

Overall Fate and Mortality of Claudicants

There is no doubt that the life expectancy of claudicants is very much less than that of non-claudicants of similar age; the only question is, how much less? In the PACK study, the 1969 claudicants, with a mean age of 63 years, who received placebo had a mortality of 4.3% per year (Dormandy and Murray 1991). There is a large literature with mortality data, mostly based on many fewer patients but with longer follow-up. A table of 28 publications setting out comparable mortality figures is contained in a previous review (Dormandy et al. 1989). Despite considerable disparity in the type of claudicants studied, there is a surprising agreement about the overall mortality. After 5, 10 and 15 years of follow-up, the mean mortality from all causes was approximately 30%, 50% and 70% respectively. In nine of the studies the mortality in the patients was compared with parallel age- and sex-matched general population. In all but one of these studies, the mortality of the claudicants was two to three times that of the general population at 5 years. One of the most recent epidemiological studies of claudication is the Whitehall study of over 18 000 civil servants in London (Smith et al. 1990). Its weakness is that the diagnosis of claudication was largely based on a questionnaire. The prevalence of probable claudication at the beginning of the 17-year follow-up was 0.8%. The mortality rate of these patients was 2.5% per year, compared with 1.3% for non-claudicants. (One of the most fascinating findings in this study was that only 10% of the probable claudicants were under medical care.) The suggestion, originally made by Bloor (1961), that claudication decreases life expectancy by approximately 10 years, is widely accepted (Dormandy et al. 1989; Ernst et al. 1987).

An analysis of the causes of death in claudicants is a little surprising, in that in only about half the cases was death thought to be due to myocardial ischaemia and in none of the 17 series with available information was the incidence of cardiac deaths higher than 63% (Dormandy et al. 1989). The number of claudicants dying of cerebrovascular accidents varied from 7% to 17%, but in approximately a further 10% death was due to vascular events other than stroke or cardiac causes. The majority of these were ruptured abdominal aortic aneurysms. The risk of a claudicant dying from a ruptured aortic aneurysm varied from 2% to 8% in different series and does not appear to have changed over the last few decades. Comparing this with the causes of death in the total PACK study group of 3899 patients, only 36% of these were due to definite myocardial infarction, nearly as many (28%) were due to definite strokes and again almost a quarter (26%) of all deaths were non-vascular.

Whilst there are reasonable data on the mortality of claudicants, much less is known about the incidence of serious, but non-fatal, cardiovascular events. Using clinical history and resting ECG, the prevalence of myocardial ischaemia in claudicants ranges from 40% to 60% (Dormandy et al. 1989). (The prevalence of cerebrovascular disease is so dependent on the criteria and techniques for diagnosis that it is impossible even to give an approximate range.) One of the few studies accurately recording the incidence of non-fatal cardiovascular events is the Northwick Park study of 400 claudicants (Gilliland et al. 1986). Over a 5½ year follow-up 14% of the patients had a non-fatal cardiac event and 5% a non-fatal cerebral accident. In the PACK study there was a yearly incidence of non-fatal definite myocardial infarction of 1.2% and of non-fatal major stroke of 1.0% (Dormandy and Murray 1991).

Factors Affecting Overall Fate and Mortality of Claudicants

In the analysis of the PACK data in terms of risk factors, univariate and multivariate analysis was carried out on the 1969 claudicants receiving placebo and the results were expressed as relative hazards, that is event rate with the risk factor relative to the rate without the factor (Dormandy and Murray 1991). Apart from age, the most significant risk factors for all causes of mortality (p value less than 0.01 on a univariate analysis and less than 0.05 on a multivariate analysis) were a history of coronary disease and an ankle-brachial pressure ratio below 0.5, both with a relative hazard of just over two. For all vascular deaths, plus non-fatal myocardial infarction and major stroke, in addition to the above risk factors, hypertension and diabetes were also significant at the same level. It is interesting that diabetes has not always been found to be a significant risk factor for mortality (Hughson et al. 1978).

The highly significant predictive value of the ankle-brachial pressure ratio for all causes of mortality is suprising at first glance, as this ratio only measures the severity of atherosclerosis in arteries leading to the leg. It is, however, also a finding in a smaller but longer Danish study (Jelnes et al. 1986). A possible explanation may be that, although Doppler systolic pressure measurements look at arterial disease in a non-vital circulation, it is an accurate and easy non-invasive test, for which there is no parallel in the coronary and cerebral circulations. Because atherosclerosis is a generalised disease, even the ankle-brachial pressure ratio gives a highly significant prediction of mortality, 75% of which will be caused by cardiac or cerebral accidents.

In PACK, only the smoking during the 6 months prior to entry was recorded and this was completely unrelated to mortality. If the claudicants' long-term smoking habits are considered, there can be no doubt that, as in the general population, smoking is an important risk factor for death. In the Basle study it tripled the mortality after 11 years (Biland et al. 1985). Apart from the same Basle study, which shows that diabetes is also an important risk factor for mortality, most of the evidence supporting this comes from post-reconstruction surgical series. Distal disease in the legs has also been said to be a poor prognostic factor, but this may be because of its association with diabetes. It is interesting that in the PACK analysis the total white cell count at the beginning of the period of observation was a significant risk factor for myocardial infarction, stroke and vascular mortality (p=0.05) (Dormandy and Murray 1991). There is considerable epidemiological evidence that a raised white cell count, albeit within the normal range, is both a primary and secondary risk factor for a number of other serious cardiovascular events (Ernst et al. 1987). The initial plasma cholesterol level was not a significant predictor.

Fate of the Amputee

The incidence of amputation can be derived from national statistics, epidemiological studies and data from limb-fitting centres. After appropriate adjustments, it appears that major amputation for vascular disease is required in non-diabetics in 150 to 700 patients per million population per year (Second European Consensus Document on Critical Leg Ischaemia, 1991). The risk factors for amputation have already been considered.

The fate of the vascular amputee has been recently reviewed (Dormandy and Thomas 1988). In summary, the ratio of below- to above-knee amputations varies around 1 to 1.5, although there are large individual variations. The 1-year mortality ranges from 3% to 10% for below-knee amputations and 10% to 30% for above-knee amputations. At 2 years, mortality rises to 25%–50%. Full mobility on a prosthesis is 30%–70% and 15%–30% for below-knee and above-knee amputees respectively. Two years after a successful below-knee amputation, about 30% of the patients will have died, at least 15% will have had a major amputation of the other leg and 15% will have had to be converted to an above-knee amputation.

Very recently, the HAWAII study was completed, looking prospectively at 713 below-knee amputations for ischaemic disease in 51 hospitals in 9 European countries (Dormandy et al. unpublished observations). Risk factors for primary healing and early mortality were analysed. In terms of antecedent history, the most interesting finding was that in 55% of cases the patients had no symptoms of leg ischaemia whatsoever 6 months before they required below-knee amputation; and 51% had had no attempt at a limb salvage reopening procedure. Only 42% of the stumps healed primarily and this was only related significantly to an attempt at previous revascularisation. There was a tendency for diabetics to heal better at this level. In a regression analysis, the only significant variable in terms of primary healing was again the pre-operative white cell count, higher counts being associated with a reduced chance of healing. By 3 months, 19% had been re-amputated. The mortality by 3 months was 11%. The significant risk factor associated with mortality or re-amputation within 3 months was clinically detectable myocardial disease. In the regression analysis, only a high pre-operative haematocrit was correlated with the need for re-amputation. These rather worse results in terms of primary healing, re-amputation and mortality, probably give a more accurate picture of what actually happens across Europe than studies published from individual specialised centres.

Conclusions

It is only very recently that large, prospective, epidemiological studies have begun to give accurate information on risk factors for the complications and mortality of intermittent claudication. Existing studies suggest that some of the factors traditionally accepted as primary risk factors for vascular disease in general are only weak secondary risk factors in patients who already have symptoms of claudication. Most importantly, the big risk for claudicants is their very much diminished life expectancy, due to coexistent, often asymptomatic, cardiac or cerebral vascular disease rather than deterioration of the circulation to their leg.

References

Biland L, Da Silva A, Zemp E, Widmer LK (1985) Occlusive peripheral artery disease (OPAD). Mortality and risk profile. In: Proceedings of the 13th International Congress of Angiology, Athens, 9–14th June, 1985

Bloor K (1961) Natural history of arteriosclerosis of the lower extremities. Ann R Coll Surg Engl 28:36–51

Cronenwett JL, Warner KG, Zelenock GB et al. (1984) Intermittent claudication. Arch Surg 119:430–6

Dormandy J, Mahir M, Ascady G et al. (1989) Fate of the patient with chronic leg ischaemia. J Cardiovasc Surg 30:50–7

Dormandy JA, Murray GD (1991) The fate of the claudicant – a prospective study of 1969 claudicants. Eur J Vasc Surg 5:131–3

Dormandy JA, Thomas PRS (1988) What is the natural history of a critically ischaemic patient with and without his leg? In: Greenhalgh RM, Jamieson CW, Nicolaides AN (eds) Limb salvage and amputation for vascular disease. WB Saunders, Philadelphia, pp. 11–26

Ernst E, Hammerschmidt DE, Bagge U et al. (1987) Leukocytes and the risk of ischemic diseases. JAMA 257:2318–24

Gilliland EL, Llewellyn CD, Goss DE, Lewis JD. The morbidity and mortality of stable claudicants – results of five year follow-up. Presented at 2nd International Vascular Symposium, London, September 1986

Hughson WG, Mann JI, Garrod A (1978) Intermittent claudication: prevalence and risk factors. Br Med J i:1379–81

Jelnes A, Gaardsting O, Hougaard Jensen K et al. (1986) Fate in intermittent claudication: outcome and risk factors. Br Med J 293:1137–40

Kannel WB, McGee DL (1985) Update on some epidemiological features of intermittent claudication. J Am Geriatr Soc 33:13–18

Most RS, Sinnock P (1983) The epidemiology of lower extremity amputations in diabetic individuals. Diabetes Care 6:87–91

PACK Claudication Substudy Investigators (1989) Randomized placebo-controlled, double-blind trial of ketanserin in claudicants. Circulation 80:1544–8

Prevention of Atherosclerotic Complications with Ketanserin Trial Group (1989) Prevention of atherosclerotic complications: controlled trial with ketanserin. Br Med J 298:424–31

Second European Consensus Document on Criticial Leg Ischaemia. Circulation 1991 (in press)

Smith G Davey, Shipley MJ, Rose G (1990) Intermittent claudication, heart disease risk factors, and mortality: the Whitehall Study. Circulation 82:1926–31.

Widmer LK, Biland L, Da Silva A (1985) Risk profile and occlusive peripheral artery disease (OPAD). In: Proceedings of the 13th International Congress of Angiology, Athens 9–14 June, 1985

30 Asymptomatic Carotid Stenosis in Peripheral Vascular Disease
Prevalence, Incidence and Natural History

P.J. FRANKS, M.R. ELLIS, R. CUMING, J.T. POWELL and
R.M. GREENHALGH

It is well established that patients presenting with symptomatic atherosclerotic disease in a particular region of the arterial tree may be suffering from symptomatic disease at other sites. Improvements in diagnostic techniques have now allowed for the detection of asymptomatic disease in these patients at high risk of cardiovascular events.

Population-based studies of asymptomatic carotid disease are rare because of the large numbers of patients involved, and have relied on carotid bruit as an indicator of carotid disease. The Framingham Study found an increased stroke risk in patients with carotid bruit, but this occurred more often in a territory other than that predicted by the site of bruit (Wolf et al. 1981). In Evans County a higher risk of stroke was found in men with bruit but not in women (Heyman et al. 1980). This study also showed that cervical bruits were a risk factor for death from ischaemic heart disease, indicating its role as a marker of generalised atherosclerosis.

Doppler investigations of the carotid arteries have allowed for the degree of stenosis to be estimated non-invasively. Several studies have shown the association between bruit and stenosis, with Fell et al. (1981) reporting 47% and Ellis et al. (1988) 34% of carotid arteries with bruit having greater than 50% stenosis by duplex scanning. The latter study also showed that 7.6% of the patients free from bruit had a stenosis of greater than 50%. There is little indication whether bruit or degree of stenosis is the best predictor of subsequent cerebrovascular morbidity or mortality (Barnes et al. 1980; Hennerici et al. 1982; Zhu and Norris 1990).

It is essential to understand the natural history of a condition before intervention can be advocated, yet this has not been so for asymptomatic carotid disease. In 1979 Moore and his co-workers commented that "Several alternatives to treatment have evolved, none of which is based on sound scientific evidence since the natural history of untreated pre-occlusive carotid stenosis has yet to be established". Surprisingly, from this statement they went on to describe the prophylactic use of carotid endarterectomy. Even when endarterectomy has been performed, the interpretation of results is complicated by the inclusion of both asymptomatic and symptomatic patients (Deriu et al. 1988; Forsell et al. 1988; Knudsen et al. 1990; Maini et al. 1990).

In a leading article for the *British Medical Journal* Sandercock (1987) attempted to evaluate the treatments for the disease. His conclusion was that on the available evidence the balance was against surgery because the risks of angiography and surgery outweighed the potential benefits of stroke reduction in these patients. This risk may now be reduced by the use of non-invasive duplex scanning to replace angiography. Ellis and Greenhalgh (1987) showed that there was no increased risk of stroke without antecedent transient ischaemic attack in patients with asymptomatic carotid bruit compared with age- and sex-matched arteriopaths without bruit. To achieve these results however, surgery was performed as soon as a transient ischaemic attack occurred which was more frequent in the bruit group.

Non-Invasive Assessment

The definition of significant carotid stenosis has been based largely on the ability of the method to measure the degree of luminal narrowing. It has been established that continuous wave Doppler is capable of detecting stenosis greater than 50% diameter reduction. There may be more patients with diseased arteries which go undetected because of the poor sensitivity of the machine to lower stenoses, although the more recent introduction of duplex scanning makes the detection of lower grade stenoses possible.

In assessing validity, several studies have compared the use of Doppler measurements of carotid stenosis with angiography. Hennerici et al. (1981) found a 97.3% agreement in the detection of >50% stenosis between the two techniques. Similar results were found by Rubin et al. (1987) in 32 patients undergoing both investigations using plaque morphology at operation as the gold standard. Plaque morphology agreed with duplex in 30/32 (94%) patients and 31/32 (97%) with angiography.

Prevalence of Asymptomatic Carotid Stenosis

Prevalence studies of asymptomatic disease have rarely been performed on random samples of the population. Because of the methods of assessment, studies have primarily been performed in surgical or neurological units. In 1987, Josse et al. reported a study of 528 patients referred to a neurology clinic for reasons other than peripheral or cardiac disease with no cerebral symptoms. They found that prevalence of >50% stenosis went from zero in both sexes at 45–54 years up to 6.1% in men and 6.9% in women at 75–84 years. In a study of 2009 asymptomatic patients (375 peripheral vascular disease, 264 coronary and 1370 with atherosclerotic risk factors) prevalence of >50% stenosis was 32.8%, 6.8% and 5.9% respectively (Hennerici et al. 1981). Klop et al. (1991) screened 374 patients with peripheral vascular disease and found a prevalence of 50%–75% stenosis in 10.6%, 76%–99% stenosis in 8% and complete occlusion in 8%, giving an overall prevalence of >50% stenosis of 26.6%. Faggioli et al. (1990) recorded prevalence in 539 patients undergoing coronary artery bypass and found >75% stenosis in 8.7%, whilst Colgan et al. (1988) found a prevalence of 4% with a stenosis >50% and 1% with >80% in 348 volunteers attending hospital sponsored health fairs.

Overall these results indicate that patients with peripheral vascular disease have a greater risk of carotid stenosis compared with other arteriopaths and much greater than the general population, though direct comparisons of rates are complicated by the degree of stenosis used to indicate significant disease and patient selection for different studies.

Charing Cross Asymptomatic Carotid Study

Over a 5-year period all patients attending the Vascular Surgical Service at Charing Cross Hospital for symptoms ascribed to claudication or critical ischaemia and with no history of stroke, transient ischaemic attack (TIA) or amaurosis fugax (AFx) were screened by colour coded duplex scanning (Acuson 128, Acuson Corporation) and continuous wave Doppler (Vasoflow-3, Sonicaid). The degree of internal, external and common carotid stenosis was recorded. Significant stenosis was described as a diameter reduction of more than 50% in either common or internal carotid arteries.

Of the 986 patients screened, 843 (85.6%) had no detectable stenosis (less than 50%), 112 (11.4%) had unilateral stenosis >50% and 30 (3.0%) bilateral stenoses. The prevalence of stenosis was age-specific, being 8% in those under 60 years rising to 24% in the over-80s. Prevalence in men was 14% (95% CI 11.4,16.8%) and in women 14.0% (95% CI 10.7,18.5%) (Fig. 30.1).

Incidence of Asymptomatic Carotid Stenosis

To date there have been few direct investigations of incidence of asymptomatic disease. Moll et al. (1987) used oculopneumoplethysmography to determine stenosis in 262 men and 107 women. Overall, 302 patients were normal at first visit with 17% at 2 years developing unilateral stenosis and 3% bilateral stenosis. This increased to 55% and 8% respectively at 5 years.

At Charing Cross patients with no detectable stenosis (0–49%) continued to be seen at annual follow-up clinics where carotid scans were performed. Over the period of follow-up, 27 (2.7%) developed a unilateral stenosis of greater than 50% and 3 (0.3%) developed bilateral stenoses. In all, there were 641 annual follow-up visits to the clinic for patients with no detectable stenosis at first visit.

Total annual incidence of developing >50% stenosis was 4.7% per year with no obvious difference for age. When analysed by sex there was no clear pattern, with age differences in women (zero in the under-60s rising to 9.5% at ages 60–69) probably due to the relatively small number of years of follow-up (Fig. 30.2). The overall annual incidence of >50% stenosis in men was 4.5% (95% CI 2.6,6.4) and in women 5.3% (95% CI 2.2,8.5).

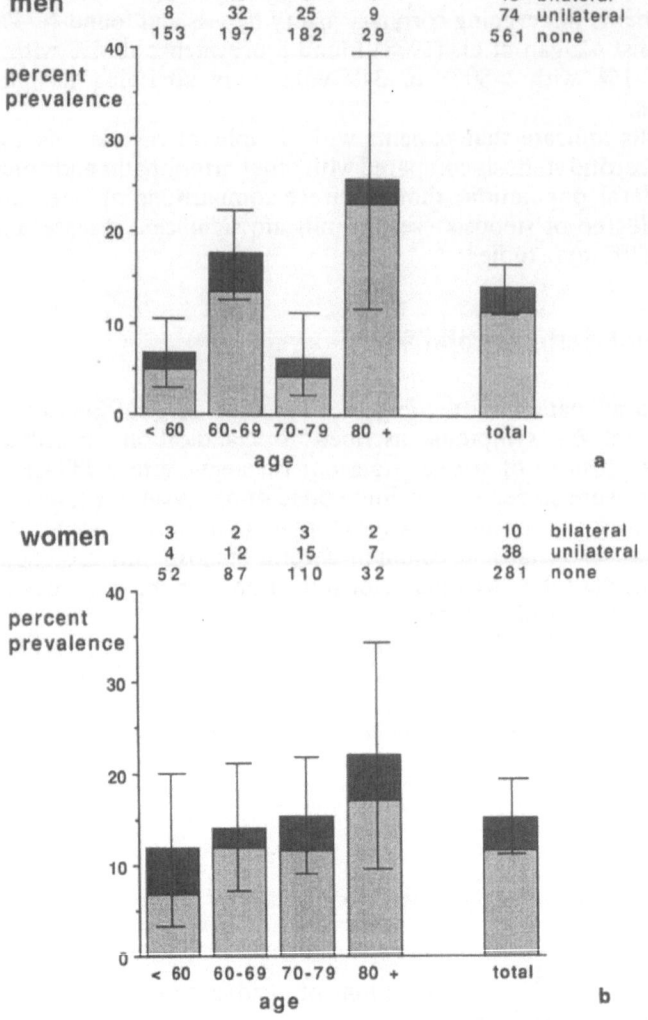

Fig. 30.1. Prevalence of asymptomatic carotid disease in patients with peripheral vascular disease broken down by age for **a** men and **b** women. Numbers of patients in each stenosis category by age group are given. 95% confidence intervals for each age band are shown by combined unilateral and bilateral stenoses. Light shaded area: unilateral stenoses; dark shaded area: bilateral stenoses.

Natural History of Patients with Asymptomatic Carotid Stenosis

Although there have been few good studies of the natural history of carotid stenosis, retrospective analysis of patients who had suffered a stroke or transient ischaemic attack has shed some light on the relative importance of carotid

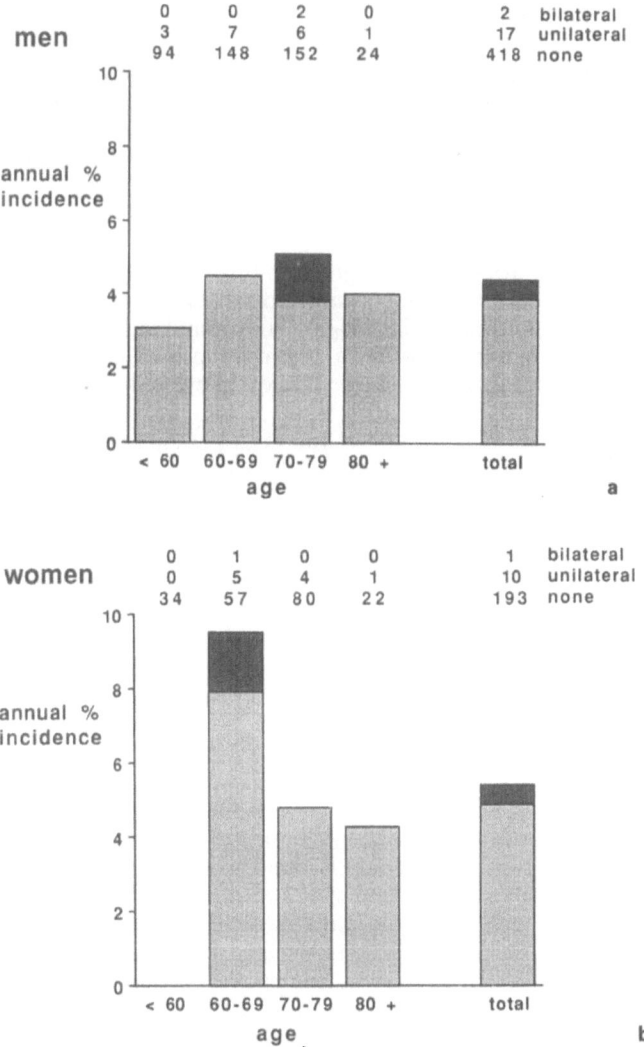

Fig. 30.2. Annual incidence of asymptomatic carotid disease in patients with peripheral vascular disease broken down by age for **a** men and **b** women. Numbers given are totals for patients who develop bilateral disease, unilateral disease or who fail to develop disease over 1 year of follow-up for each age group. Light shaded area: unilateral stenoses; dark shaded area: bilateral stenoses.

stenosis and bruit as risk factors. Zhu and Norris (1990) found that most patients with a history of stroke or transient ischaemic attack had no significant carotid stenosis (55% and 64% respectively) and that patients without neck bruits were more often without carotid disease (69% and 77% respectively). A retrospective review of 168 patients with asymptomatic stenosis >50% over 12-year follow-up (Humphries et al. 1976) found that 136 had remained asymptomatic, 2 suffered from a neurological deficit, 26 had transient ischaemic attacks leading to endarterectomy, 3 had transient ischaemic attack without endarterectomy leading to stroke, and one had stroke without warning. Faggioli et al. (1990) divided patients according to their stenosis and subsequent operation. Groups I and II had stenoses <50% (n=432) and 50%–75% (n=60) respectively, whilst group III had >75% stenosis and underwent endarterectomy (n=19), and group IV had similar stenoses but no endarterectomy (n=28). Group IV had significantly higher stroke rate when compared with all other groups (14.3% vs. 1.1%) and a higher death rate (7.1% vs. 0). No details were available on the rationale for selecting surgery or no surgery in groups III and IV.

Moll et al. (1987) found that stroke as a first sign of cerebrovascular disease was not related to the side of the stenosis. Chambers and Norris (1986) examined 500 patients with cervical bruit over 4 years of follow-up. They found that the annual incidence of stroke was 1.7% in patients with no severe stenosis compared with 5.5% in those with >75% stenosis. Although the risk of an event was higher in the group with severe stenosis, in most instances these patients did not have strokes without warning. Durward et al. (1982) followed patients with carotid stenosis contralateral to a previous operation. Of 73 patients, 3% developed symptoms in the asymptomatic side compared with 5% in the previously operated side and 4% in the vertebrobasilar territory.

At Charing Cross Hospital careful note was taken whenever a cerebrovascular symptom developed in any of the patients with or without significant asymptomatic carotid disease. These were categorised into (i) stroke (ii) transient ischaemic attack or amaurosis fugax. Records of deaths were obtained from a variety of sources, though principally through general practitioners and by flagging at the Office of Population Censuses and Surveys from where death certificates were obtained. Patients who did not attend for follow-up were contacted wherever possible to determine their reason for non-attendance.

In the 844 patients with <50% stenosis, 19 went on to develop a stroke compared with 4 of 112 with unilateral stenosis and one of 30 bilateral stenoses; of these strokes, 7 of 19, 1 of 112 and 0 of 1 were fatal. Symptoms of cerebrovascular disease (stroke, transient ischaemic attack or amaurosis fugax) developed in 37 of the 844 patients with no detectable stenosis, 10 of 112 with unilateral and 7 of 30 with bilateral stenoses. Life tables analysis illustrates the impact of carotid stenosis on the risk of stroke (Fig. 30.3a). When corrected for age and sex using Cox Proportional Hazards model, stroke risk was higher in the group with at least one stenosis with a relative risk of 1.86 but with wide confidence intervals (95% CI 0.67,5.18). There was a clear gradation for risk of all cerebrovascular symptoms according to the number of vessels involved, with more patients with carotid disease going on to develop a symptom of cerebrovascular disease (Fig. 30.3b). The risk of symptom development in the patients with at least one stenosis >50% was significantly higher than the group with no stenosis at 2.99 (95% CI 1.64,5.44).

Total mortality was similar between the three groups at 97 of 844 (11.5%), 18 of 112 (16.1%) and 5 of 30 (16.6%). When analysed by life table the death rate was 4.8% after 1 year, 10.6% after 2 and 18.6% after 3 years. As with most other studies, patients died from cardiac disease rather than stroke (Bogousslavsky et al. 1986; Moll et al. 1987; Knudsen et al. 1990).

These results have major implications for the investigation of asymptomatic carotid disease in patients with peripheral vascular disease. They show that carotid stenosis is common in these patients and that the incidence is similar for

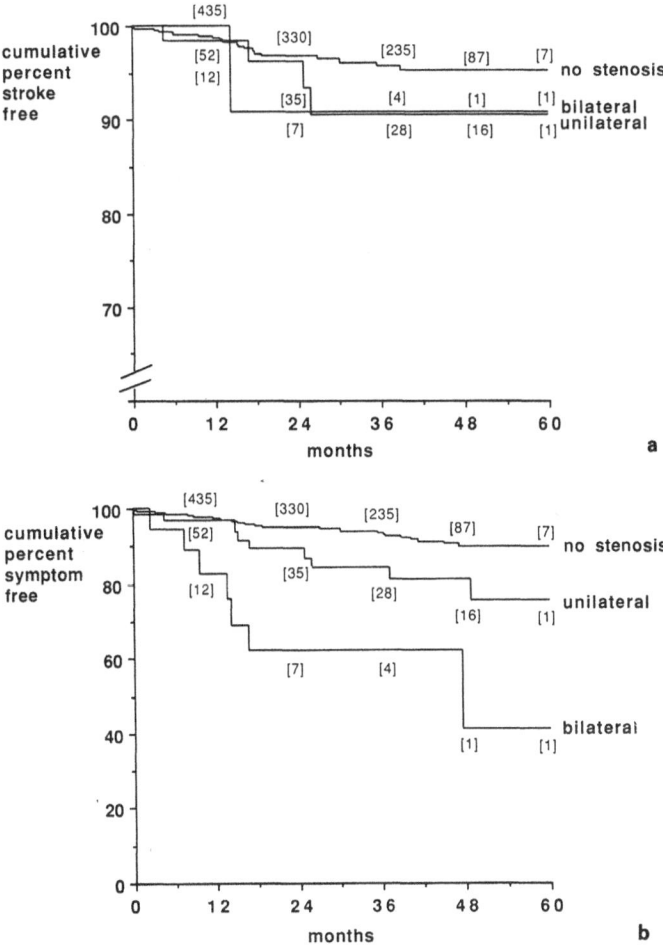

Fig. 30.3. Life table analysis for **a** stroke and **b** cerebrovascular symptoms (stroke, transient ischaemic attack or amaurosis fugax). Numbers in parentheses are patients at each annual follow-up. No statistically significant association could be found between stroke and >50% stenosis (p=0.459, log rank method). When corrected for age and sex the risk for stroke was 1.86 (95% CI 0.67,5.18) in patients with at least one stenosis compared with patients with no stenosis. For symptoms of cerebrovascular disease the relative risk was significantly higher at 2.99 with 95% confidence intervals 1.64, 5.44 (p=0.0001, log rank method).

all age groups. Patients with carotid disease have an increased risk of developing a cerebrovascular symptom. Although the risk of stroke was higher in the group with carotid stenosis the confidence intervals were wide and did not achieve statistical significance. The risk of death from cardiovascular disease was high in these patients but not significantly higher than those patients without carotid disease.

Future Research in Asymptomatic Carotid Disease

Randomised trials have been advocated to determine the potential benefits of surgery against best medical treatment (Sandmann 1987). It has recently been established that carotid endarterectomy is beneficial in reducing strokes in patients with severe stenosis (>70%) who have suffered a previous cerebrovascular symptom (Strandness 1991).

At present there are two multicentre surgical trials investigating patients with asymptomatic carotid stenosis. The Veterans Association have established a trial randomising 444 patients to carotid endarterectomy or no surgery, with all patients receiving 1300 mg aspirin per day (Towne et al. 1990). A report of the 211 patients who underwent surgery found a 30 day mortality of 1.9%, postoperative stroke in 2.4% and transient ischaemic attack in 0.9%. The results of the trial comparing morbidity and mortality with the 233 randomised to medical treatment have yet to be published.

A larger trial has been initiated by the Asymptomatic Carotid Atherosclerosis Study Group (1989). It proposes to randomise 1500 patients to surgery or follow-up with all receiving 325 mg aspirin per day.

It would be difficult to advocate screening patients with peripheral vascular disease for asymptomatic carotid stenosis for surgical intervention until the results of these trials are known. Only then will there be evidence to justify the best treatment for these patients.

References

Asymptomatic Carotid Atherosclerosis Study Group (1989) Study design for randomized prospective trial of carotid endarterectomy for asymptomatic atherosclerosis. Stroke 20:844–9

Barnes RW, Marszalek PB, Rittgeers SZ (1980) Asymptomatic carotid disease in preoperative patients. Stroke 11:136A

Bogousslavsky J, Despland P-A, Regli F (1986) Asymptomatic tight stenosis of the internal carotid artery: long-term prognosis. Neurology 36:861–3

Chambers BR, Norris JW (1986) Outcome in patients with asymptomatic neck bruits. N Engl J Med 315:860–5

Colgan MP, Strode GR, Sommer JD, Gibbs JL, Sumner DS (1988) Prevalence of asymptomatic carotid disease: results of duplex scanning in 348 unselected volunteers. J Vasc Surg 8:674–8

Deriu GP, Ballotta E, Facco E et al. (1988) Stroke risk reduction in asymptomatic and symptomatic patients treated surgically: The effectiveness of carotid endarterectomy with patch graft angioplasty. Eur J Vasc Surg 2:87–91

Durward QJ, Ferguson GG, Barr HWK (1982) The natural history of asymptomatic carotid bifurcation plaques. Stroke 13:459–64

Ellis M, Greenhalgh RM (1987) Management of asymptomatic carotid bruits. J Vasc Surg 5:869–73

Ellis MR, Meek A, Greenhalgh RM (1988) Asymptomatic carotid artery disease. In: Greenhalgh RM (ed) Indications for vascular surgery. WB Saunders, London, pp 3–14

Faggioli GL, Curl GR, Ricotta JJ (1990) The role of carotid screening before coronary artery bypass. J Vasc Surg 12:724–31

Fell G, Breslau P, Knox RA, Phillips D, Thiele BL, Strandness DE (1981) Importance of noninvasive ultrasonic-Doppler testing in the evaluation of patients with asymptomatic carotid bruits. Am Heart J 102:221–6

Forsell C, Takolander R, Bergqvist D, Bergentz S-E, Olivecrona H (1988) Long-term results after carotid artery surgery. Eur J Vasc Surg 2:93–8

Hennerici M, Aulich A, Sandeman W, Freund H-J (1981) Incidence of asymptomatic extracranial arterial disease. Stroke 12:750–8

Hennerici M, Rautenberg W, Mohr S (1982) Stroke risk from symptomless extracranial arterial lesions. Lancet ii:1180–3

Heyman A, Wilkinson WE, Heyden S et al. (1980) Risk of stroke in asymptomatic persons with cervical arterial bruits. N Engl J Med 302:838–41

Humphries AW, Young JR, Santilli PH, Beven EG, deWolfe VG (1976) Unoperated, asymptomatic significant internal carotid stenosis: a review of 182 instances. Surgery 80:695–8

Josse MO, Touboul PJ, Mas JL, Laplane D, Bousser MG (1987) Prevalence of asymptomatic internal artery stenosis. Neuroepidemiology 6:150–2

Klop RBJ, Eikelboom BC, Taks ACJM (1991) Screening of the internal carotid arteries in patients with peripheral vascular disease by colour-flow Duplex scanning. Eur J Vasc Surg 5:41–5

Knudsen L, Sillesen H, Schroeder T, Hansen HJB (1990) Eight to ten years follow-up after carotid endarterectomy: Clinical evaluation and Doppler examination of patients operated on between 1978–1980. Eur J Vasc Surg 4:259–64

Maini BS, Mullins TF, Catlin J, O'Mara P (1990) Carotid endarterectomy: a ten-year analysis of outcome and cost of treatment. J Vasc Surg 12:732–40

Moll F, Eikelboom BC, Vermeulen FEE, van Lier HJJ, Schulte BPM (1987) Risk factors in asymptomatic patients with carotid bruit. Eur J Vasc Surg 1:33–9

Moore WS, Boren C, Malone JM, Goldstone J (1979) Asymptomatic carotid stenosis. Immediate and long-term results after prophylactic endarterectomy. Am J Surg 138:228–33

Rubin JR, Bondi JA, Rhodes RS (1987) Duplex scanning versus conventional arteriography for the evaluation of carotid artery plaque morphology. Surgery 102:749–55

Sandercock P (1987) Asymptomatic carotid stenosis: spare the knife. Br Med J 294:1368–9

Sandmann W (1987) Asymptomatic carotid stenosis: Is prophylactic surgery indicated? Eur J Vasc Surg 1:147–9

Strandness (1991) The NASCET Trial. Presented to the Society for Vascular Surgery, Boston 1991

Towne JB, Weiss DG, Hobson RW (1990) First phase report of cooperative Veterans Administration asymptomatic carotid stenosis study – operative morbidity and mortality. J Vasc Surg 11:252–9

Wolf PA, Kannel WB, Sorlie P, McNamara P (1981) Asymptomatic carotid bruit and risk of stroke. The Framingham Study. JAMA 245:1442–5

Zhu CZ, Norris JW (1990) Role of carotid stenosis in ischemic stroke. Stroke 21:1131–4

31 Prevention of Vascular Events in Claudicants

L.U.CH. JANZON

Intermittent Claudication: Indicator of Generalised Atherosclerosis

Intermittent claudication is not only a symptom of leg artery occlusive disease; it is also an indicator of generalised atherosclerosis, associated with a greatly increased cardio- and cerebrovascular mortality rate. Between 30% and 50% of newly diagnosed claudicants have signs or symptoms of myocardial ischemia (Isacsson 1972; Coffman 1986). The annual mortality rate in the Framingham study among men with intermittent claudication was 39/1000 and in men free of the disease was 10/1000 (Kannel and Shurtleff 1971). Similar results have been reported by others (Reunanen et al 1982; Källerö 1981).

In the absence of surgery progression of disease is the rule. Strandness and Stahler (1966) found that 52% progressed as determined by segmental systolic pressure measurements and angiography. From the Framingham Study it has been calculated that further occlusions can be expected in 40% within 5 years. The annual amputation rate varies in published studies depending on the selection of patients from 1% to 3% (Tillgren 1965; Jelnes et al. 1986). Diabetic patients have been found to have a higher amputation rate than non-diabetic patients (Boyd 1962).

Almost all patients with intermittent claudication are or have been smokers, 30%–40% of them have hypertension, about 20% have hyperlipidemia (Hughson et al. 1978). The degree of leg blood flow impairment is related to the number of risk factors the patients have been exposed to (Janzon et al. 1981a). The risk of suffering a myocardial infarction, the most common and most serious vascular event in patients with claudication, is related to the severity of arteriosclerotic disease (Jonason and Ringqvist 1985; Lassila et al. 1986).

Value of Interventions in Claudicants

Smoking Cessations

Several observational studies have presented results on the value and importance of smoking cessation in patients with intermittent claudication. In a follow-up study of 343 claudicants Jonason and Bergström (1987) compared the clinical course in patients who stopped and continued to smoke. The 7-year incidence of rest pain in the two groups was 0 and 16%. The accumulated proportion with myocardial infarction after 10 years was 11% and 53% and the cumulated rates of cardiac death were 6% and 43%. Observed risks remained after adjustment for different confounding factors. Treatment with anti-thrombotic agents and other drugs with a potential influence on outcome was however not taken into account in the analysis.

Others have presented results showing that cessation of smoking is associated with an improvement in walking distance (Aronow et al. 1974), a lower rate of graft occlusion (Myers et al. 1978) and a lower amputation rate (Juergens et al. 1960). A common and depressing experience in many of these studies is, however, that few claudicants give up smoking. In the study by Jonason and Bergström (1987) only 1 out of 10 stopped smoking. Anti-smoking counselling can probably increase the number of quitters however (Isacsson and Janzon 1976).

Lowering Blood Pressure and Lipids

Very limited information is available on the value of blood pressure lowering and blood lipid lowering therapy in claudication. Duffield et al. (1983) showed in a small experimental study that effective treatment of hyperlipidaemia favourably reduced the angiographically detectable progression of leg arteriosclerosis during an average of 19 months of treatment. Others have reported an improved claudication distance in conjunction with clofibrate therapy (Postlethwaite and Dormandy 1975).

No studies have presented results showing that control of hypertension is associated with a lower rate of progression of leg arteriosclerosis disease. However, walking distance and degree of blood flow impairment have in one study been found to be related to the degree of blood pressure control (Janzon et al. 1981b).

Exercise

The effect of exercise on leg blood flow and walking distance has been evaluated in several controlled and non-controlled studies (Larsen and Lassen 1966; Ericsson et al. 1970; Dahlloff et al. 1976). Although results are not unequivocal it can be concluded that exercise leads to an improved walking distance. This effect does not seem to be related to an improvement of leg blood flow, however. No studies have been published evaluating long-term effects of exercise in the prevention of vascular events.

In STIMS (Janzon et al. 1990), the Swedish Ticlopidine Multicentre Study, where all smokers were given advice and help to quit smoking and all with hypertension, hyperlipidaemia and diabetes received appropriate treatment, the incidence of myocardial infarction and stroke in the control group was still about 50% higher than expected. About one out of three claudicants suffered a myocardial infarction or a stroke during the observation period of 5.6 years. The conclusion in STIMS was that a treatment programme that includes help to quit smoking and treatment of hypertension and hyperlipidemia still seems to be too little and too late for patients with peripheral vascular disease.

Anti-thrombotic Treatment

The role of anti-thrombotic treatment should be considered with these facts in mind. Based on a review of 25 trials including 29 000 patients it was the consensus at the Antiplatelet Trialists' Collaboration meeting (1988) that the vascular mortality rate in patients with symptomatic atherosclerosis can be reduced by 15% with platelet anti-aggregatory treatment. A number of studies have presented results supporting the view that anti-thrombotic treatment can prevent further deterioration of leg blood flow in patients with intermittent claudication (Hess et al. 1985; Arcan et al. 1988; Boissel et al. 1989). A more important and encouraging observation is that in patients with intermittent claudication, platelet anti-aggregatory treatment has been found to be associated with a reduced risk of myocardial infarction and stroke.

Swedish Ticlopidine Multicentre Study (STIMS)

In STIMS (Janzon et al. 1990) 687 patients with intermittent claudication were randomly allocated to treatment with Ticlopidine (Panak et al. 1983) and placebo 250 mg twice a day. All patients were given advice and help to quit smoking and all with hypertension, hyperlipidaemia and diabetes were given appropriate treatment. The mean follow-up period was 5.6 years. At termination of the trial there were 99 events (myocardial infarction, stroke and transient ischaemic attack) in the placebo group, 89 in the treatment group (p=0.24). Thirty-eight per cent (26/68) of the cases with a myocardial infarction in the placebo group versus only 25% (15/60) in the Ticlopidine group were fatal, indicating that Ticlopidine might favourably influence the course of an acute myocardial infarction in claudicants.

This hypothesis is further supported by a comparison of the number of sudden deaths in the two groups. In all, 21 patients died suddenly, 14 in the placebo group, 7 in the Ticlopidine group. At autopsy all cases were considered to be vascular deaths. Added to the primary end-points, Ticlopidine treatment was associated with a 14% lower vascular event rate (p=0.099). For the individual patient the most important question to be answered is of course if anti-thrombotic treatment will in fact improve his or her chances of survival. The 64 deaths in the Ticlopidine group and 89 in the placebo group means that treatment with Ticlopidine was associated with a 29% lower mortality rate (p=0.015). This difference was entirely explained by the lower mortality rate from ischaemic heart disease.

The outcome analysis in STIMS also included an on-treatment analysis, i.e. a comparison of the outcome in patients who actually took the study medication.

Events which occurred more than 15 days after discontinuation of treatment were not counted. The event rate, 47 events in the Ticlopidine group versus 76 events in the placebo group was 38% lower in the Ticlopidine group (p=0.017). About 30% of the effect was explained by a lower incidence of cerebrovascular events. Observed treatment effects were even more pronounced when sudden deaths were added to primary end-points and when total mortality in the two groups was compared.

At termination of the trial there were in all 89 deaths in the placebo group and 64 in the Ticlopidine group, a reduction of about 30% (p=0.015).

Ketanserin Trial (PACK)

In the so-called PACK trial (Prevention of Atherosclerotic Complications with Ketanserin), which involved 3899 claudicants, Ketanserin, an antagonist at the S_2 serotonine receptor, was evaluated against placebo in the prevention of myocardial infarction, stroke, amputation and death due to vascular disease (Prevention of Atherosclerotic Complications with Ketanserin Trial Group 1989). By the time the code was broken there were 136 and 132 events in the two treatment groups. An increased mortality in patients taking diuretics was observed early on and it was the decision by the safety committee that these patients should be withdrawn from further treatment. The higher vascular rate in the Ketanserin group at termination of the trial turned out in the further analysis to be an adverse effect which occurred when Ketanserin and potassium losing diuretics were taken together. Another adverse reaction observed early on was that Ketanserin might cause reversible disturbances of cardiac rhythm. Patients with an ECG showing a QT interval >500 ms were therefore withdrawn from further treatment.

When patients on diuretics and anti-arrhythmic drugs were excluded from the analysis there were 65 events in the Ketanserin group and 86 in the placebo group, a 23% reduction (p=0.12). When secondary events like unstable angina, transient ischaemic attack and severe hypertension were added to primary end-points it was found that Ketanserin treatment was associated with a lower event rate (p=0.02). The treatment effect was enhanced when the analysis was restricted to patients who received the trial treatment. An interesting and promising observation in PACK was the lower amputation rate in the Ketanserin group, 17 cases versus 29 in the placebo group.

Although the results in PACK indicate a beneficial effect and are in line with results from many other studies with anti-platelet treatment, the conclusion of the study group was that further trials are required to determine whether Ketanserin prevents cardiovascular complications in patients with peripheral vascular disease.

Several studies have documented the health benefits associated with smoking cessation and with treatment of hypertension and hyperlipidaemia. It is my view, although we lack data from controlled trials, that the progression of disease in intermittent claudication can be slowed down by treatment of these known arteriosclerosis risk factors. However, considering the thrombotic event underlying the most common and most serious vascular complication, i.e. acute myocardial infarction, it does not seem very rational to concentrate on a therapeutic approach which can be supposed to influence the development of atherosclerosis but not the too often life-threatening complication. Platelet anti-aggregatory agents may therefore be considered the most important part of the treatment programme for patients with peripheral vascular disease.

References

Antiplatelet Trialists' Collaboration (1988) Secondary prevention of vascular disease by prolonged antiplatelet treatment. Br Med J 296:320–31

Arcan JC, Blanchard J, Boissel JP, Destors JM, Panak E (1988) Multicenter double-blind study of Ticlopidine in the treatment of intermittent claudication and the prevention of its complications. Angiology 39:802–11

Aronow WS, Stemmer EA, Isbell MW (1974) Effect of carbon monoxide exposure on intermittent claudication. Circulation 49:415–17

Boissel JP, Peyrieux, Destors JM (1989) Is it possible to reduce the risk of cardiovascular events in subjects suffering from intèrmittent claudication of the lower limbs? Thromb Haemost 62:681–5

Boyd AM (1962) The natural course of arteriosclerosis of the lower extremities. Proc R Soc Med 55:591

Coffman JD (1986) Intermittent claudication: Not so benign. Am Heart J 112:1127–8

Dahllof AG, Holm J, Schersten T, Sivertsson R (1976) Peripheral arterial insufficiency, effect of physical training on walking tolerance, calf blood flow, and blood flow resistance. Scand J Rehabil Med 8:19–26

Duffield RGM, Miller NE, Brunt JNH, Lewis B, Jamieson CW, Colchester ACF (1983) Treatment of hyperlipidaemia retards progression of symptomatic femoral atherosclerosis. A randomised controlled trial. Lancet 1983; ii:639–42

Ericsson B, Haeger K, Lindell SE (1970) Effect of training on intermittent claudication. Angiology 21:188–92

Hess H, Deichsel G, Mietaschk A (1985) Drug-induced inhibition of platelet function delays progression of peripheral occlusive arterial disease. A prospective double-blind arteriographically controlled trial. Lancet i:415–21

Hughson WG, Mann JI, Garrod A (1978) Intermittent claudication: prevalence and risk factors. Br Med J i:1379–81

Isacsson S-O (1972) Venous occlusion plethysmography in 55-year old men. A population study in Malmö, Sweden. Acta Med Scand (suppl) 537

Isacsson SO, Janzon L (1976) Results of a quit-smoking research project in a randomly selected population. Scand J Soc Med 4:25–9

Janzon L, Bergqvist D, Boberg G et al. (1990) Prevention of myocardial infarction in stroke in patients with intermittent claudication; effects of ticlopidine. Results from STIMS, the Swedish Ticlopidine Multicentre Study. J Intern Med 227:301–8

Janzon L, Bergentz SE, Ericsson BF, Lindell SE (1981a) The arm-ankle pressure gradient in relation to cardiovascular risk factors in intermittent claudication. Circulation 63:1339–41

Janzon L, Bergentz SE, Ericsson BF, Lindell SE (1981b) Intermittent claudication and hypertension. Ankle pressure and walking distance in patients with well-treated and non-treated hypertension. Angiology 32:175–9

Jelnes R, Gaardsting O, Hougaard Jensen K, Bækgaard N, Tønnesen KH, Schroeder T (1986) Fate in intermittent claudication: outcome and risk factors. Br Med J 293:1137–40

Jonason T, Bergström R (1987) Cessation of smoking in patients with intermittent claudication. Effects on the risk of peripheral vascular complications, myocardial infarction and mortality. Acta Med Scand 221:253–60

Jonason T, Ringqvist I (1985) Mortality and morbidity in patients with intermittent claudication in relation to the location of the occlusive atherosclerosis in the leg. Angiology 36:310–14

Juergens JL, Barker NW, Hines Jr EA (1960) Arteriosclerosis obliterans: Review of 520 cases with special reference to pathogenic and prognostic factors. Circulation xxi:188–95

Källerö KS (1981) Mortality and morbidity in patients with intermittent claudication as defined by venous occlusion plethysmography. A ten-year follow-up study. J Chron Dis 34:455–672

Kannel WB, Shurtleff D (1971) The natural history of arteriosclerosis obliterans. Cardiovasc Clin 3:37–52

Larsen OA, Lassen NA (1966) Effect of daily muscular exercise in patients with intermittent claudication. Lancet ii:1093–6

Lassila R, Lepäntalo M, Lindfors O (1986) Peripheral arterial disease – natural outcome. Acta Med Scand 220:295–301

Myers KA, King RB, Scott DF, Johnson N, Morris PJ (1978) The effect of smoking on the late patency of arterial reconstructions in the legs. Br J Surg 65:267–71

Panak E, Maffrand JP, Picard-Fraire C, Vallee E, Blanchard J, Ronucci R (1983) Ticlopidine, a promise for the prevention and treatment of thrombosis and its complications. Haemostasis 13 (Suppl):1–54

Postlethwaite JC, Dormandy JA (1975) Results of ankle systolic pressure measurements in patients with intermittent claudication being treated with clofibrate. Ann Surg 181:799–802

Prevention of Atherosclerotic Complications with Ketanserin Trial Group (1989) Prevention of atherosclerotic complications: controlled trial of ketanserin. Br Med J 298:424–30

Reunanen A, Takkunen H, Aromaa A (1982) Prevalence of intermittent claudication and its effect on mortality. Acta Med Scand 211: 249–56

Strandness Jr DE, Stahler C (1966) Arteriosclerosis obliterans. JAMA 196:1–4

Tillgren C (1965) Obliterative disease of the lower limbs. II. A study of the cause of disease. Acta Med Scand 178:103–20

32 Secondary Prevention of Ischaemic Heart Disease

P.C. ELWOOD

Secondary prevention is the reduction of death or infarction in subjects who already have clinical signs of ischaemic heart disease. In this context "heart disease" will certainly include previous myocardial infarction, it may include angina and may occasionally also include myocardial ischaemia detected as a screening examination or on electrocardiogram. Consideration of secondary prevention does not simply relate to patients with clinical evidence of disease but is also relevant to population screening and possibly other manifestations of atherosclerosis such as peripheral vascular disease.

The burden of morbidity from ischaemic heart disease in the community is very considerable. Among older men approximately 20%–25% have clinical disease and amongst these, two-thirds or about 15% of such men in the community have symptoms (Table 32.1). Mortality is such subjects is around three times higher than in those without such evidence (Table 32.2). This means that within the community a substantial proportion of deaths and infarctions occur in subjects who already have signs of disease and on whom secondary prevention efforts could be targeted.

The potential for saving lives by secondary prevention is therefore high and the benefits, if prophylaxis in these were effective, would be apparent beyond these patients themselves and could have an appreciable effect on mortality and morbidity in the total community. Prophylaxis is also likely to be far more

Table 32.1. Prevalence (%) of ischaemic heart disease in older men in the United Kingdom

	Age range	Past myocardial infarction	Angina	ECG ischaemia	Any ischaemic heart disease
Representative population samples					
Regional heart study[1]	40–59	9.1	4.8	14.5	23%
Caerphilly[2]	45–59	10.1	7.6	14.1	25%
Speedwell[2]	45–59	6.4	7.9	13.5	21%
Samples of employed men					
Whitehall[3]	40–59	6.5	4.3	5.5	14%
UK Heart Disease[4]	40–59	6.6	3.6	7.8	?

[1] Shaper et al. (1985); [2] Bainton et al. (1988); [3] Rose et al. (1977); [4] WHO European Collaborative Group (1980).

Table 32.2. Numbers of major ischaemic heart disease (IHD) events and per cent per annum (p.a.) in men aged 40–59 years followed for 4 years in the Regional Heart Study (Shaper et al. 1985) and 45–59 years followed for 5 years in the Caerphilly study (MRC Epidemiology Unit 1991)

	Men with prevalent IHD	Men with no prevalent IHD	The total cohort	Events in men with pre-existing IHD	Increased incidence in men with pre-existing IHD
Regional Heart Study	105 (1.3% p.a.)	96 (0.4% p.a.)	202 (0.6% p.a.)	50%	3.3-fold
Caerphilly	41 (2.4% p.a.)	96 (1.0% p.a.)	137 (1.2% p.a.)	30%	2.4-fold

acceptable in subjects who have symptoms than in healthy persons and therefore "effective" prophylaxis is likely to be a much more easily attained goal in these, than in the rest of the community.

The ultimate test of a possible prophylactic measure is its effect on disease incidence. Trials which examine changes in some biochemical or haematological marker of disease, or even some factor which is known to be predictive, are of value and give leads to research. But benefit to health must not be assumed, it must be tested directly. In this connection one is intrigued to note that in a review of studies of fish oils, evidence from 18 animal feeding trials and 47 trials in human subjects were summarised. In their conclusion these authors call for further research to gain greater knowledge of the mechanisms by which n-3 polyunsaturated fatty acids may act, but no need is expressed for trials to test benefit on health or survival (Herold and Kinsella 1986).

The criterion by which the effectiveness of prophylaxis is judged is important. Clearly, death or survival is the bottom line in a disease as serious as ischaemic heart disease and it is difficult to make reasonable judgements as to the benefit of a prophylactic measure which does not significantly increase survival, or to make decisions as to its applicability to the community as a whole. This line of reasoning has been challenged and it has been claimed that the main objective of health promotion and related activities should be the reduction of morbidity rather than mortality. One group has termed this the "compression of morbidity" (Fries et al. 1989) and these and other authors recommend a change of focus from quantity to quality of life: the adding of life to years rather than years to life.

The relevance of such arguments to ischaemic disease, in which over half the major events lead to death, is questionable. In fact, in patients who have already had a myocardial infarct at least 80% and probably 90% of deaths are attributable to ischaemic heart disease and it is therefore eminently reasonable to evaluate any measure used in secondary prevention against total mortality.

Having said all that, it has to be stated that although there have been a number of trials of diet and secondary prevention of death and/or re-infarction almost all the trials have been totally unrealistic in size, and a number of them have had serious design faults (Table 32.3).

Fatty Fish

Studies of Greenland Eskimos in the 1920s led to the suggestion that the absence of ischaemic heart disease might be due to the consumption of fish. Early work led to the identification of certain fatty acids of the n-3 series, in particular eicosapentaenoic acid (EPA), which reduce platelet aggregation and prolong the bleeding time.

Since then there have been numerous studies published on fish and on fish oils. On the one hand these have focused on biochemical mechanisms affected by fish oils, on the other, population studies have examined differences between communities and trends over time in fish consumption and in mortality from ischaemic heart disease. Overall, the consistency in the results of these studies is remarkable (Burr 1989).

Unfortunately, however, there has been only one randomised controlled trial of fatty fish in secondary prevention (Burr et al. 1989). In this, 2033 men were randomised factorially to three interventions, one of which was fatty fish. The amount of fish recommended was modest, namely, at least two main meals per week, contributing around 2.4 g of EPA per week.

Mortality in the 1015 men advised to eat fish was 9.3% over the next 2 years, while it was 12.8% in the 1018 men who had been given no advice about fish. This represents a reduction in mortality of about 29% and a saving of lives of about 35 per 1000 men advised. These last are of course derived from the results of a single trial and they therefore carry a considerable uncertainty.

It is notable however that this trial gave no evidence that non-fatal infarctions were reduced by fish eating. Indeed, there were rather more non-fatal infarctions in the men advised (4.8%) than in the other men (3.2%). It is possible therefore that the effect of fish is to reduce the risk of death after infarction rather than to reduce the incidence of infarction. Evidence consistent with this hypothesis comes from work on experimentally induced infarction in animals and is reviewed in Burr (1989).

It is inevitable that work will continue on metabolic pathways of EPA, and on biochemical and other effects of fish oil. However there is now an urgent need for further trials of fish and mortality. Results from a single trial can never be accepted as conclusive. It is important, moreover, that evidence is collected as to whether or not deaths are sudden so that the effect of fish on death and on infarction can be examined more thoroughly.

Dietary Fibre

There have been at least five prospective studies in which fibre intake has been related to subsequent disease incidence and in all of them the subjects with lower intakes had a raised incidence of ischaemic heart disease (Morris et al. 1977; Kromhout and Coulander 1984; Kushi et al. 1985; Khaw and Barrett-Connor 1987; MRC Epidemiology Unit 1991). At the same time dietary fibre is not independent of energy, in that the more food that is eaten, usually, the more fibre is consumed. Therefore, although fibre does not contribute to energy, it does seem reasonable to standardise fibre intake for energy (using energy as a surrogate for bulk). When this is allowed for (using energy as a surrogate for bulk) the predictive power or fibre for incident ischaemic heart disease is reduced and in two of these studies it is lost altogether (Kromhout and Coulander 1984; MRC Epidemiology Unit 1991).

The danger in observational studies of dietary fibre is that to an extent which is probably greater than for any other food item, fibre is likely to be a marker for a life style. Many dietary and environmental factors may be different in subjects who choose high fibre foods – often recognised as "health foods". That this might be the case is suggested by a large cohort study of almost 11 000 subjects whom Burr and his colleagues identified through health food shops and health food magazines (Burr and Sweetnam 1982). The strength of this study is that the 6000 or so subjects who had a high consumption of fibre, could be compared with 5000 or so who seemed to have had a fairly similar life style with regard to smoking, alcohol intake, activity level, etc. During the first 5 years of follow-up there was no evidence of any difference in the heart disease mortality of these two groups despite a marked difference in cereal fibre consumption.

Unfortunately, again there has been only one randomised controlled trial of dietary fibre (Burr et al. 1989). In this, advice given to 1017 men led to an increase in cereal fibre intake to 19 g per day, compared with 9 g in 1016 controls. After 2 years, however, there was no significant difference in the deaths in those given the advice about fibre (12.1%) compared with those not so advised (9.9%).

Interest in fibre has however shifted yet again and attention is now being focused on "soluble" fibre, which is quite a change from the early ideas about "roughage" and insoluble fibre (Trowell and Burkitt 1981). Interest in soluble fibre arises primarily because it has been shown in numerous small feeding trials to reduce serum cholesterol if eaten in sufficient quantities. There is also interest in fruit and vegetables, the main source of soluble fibre, because these are a significant source of antioxidants. Whatever the biochemical effects of these various dietary items, the ultimate need is for randomised controlled trials and, in the first instance, these trials could well be of secondary prevention.

Reduced Dietary Fat

Undoubtedly fat reduction has been more investigated in secondary prevention than any other potential prophylactic. Table 32.3 summarises most of the published trials and shows the fall in cholesterol achieved in each. In most, a fall in cholesterol level which would be regarded as clinically important was achieved, except for the two trials reported most recently (Woodhill et al. 1978; Burr et al. 1989). In these, only a very modest fall in cholesterol was achieved and this was partly due to patients who had been allocated to the control group spontaneously reducing their fat intake.

Table 32.3. Secondary prevention dietary trials

Reference	No. of patients	Duration	Allocation to prophylaxis	Reduction in cholesterol
Morrison (1960)	100	12 years	Alternate	29%*
MRC (1965)	264	4 years	Random	16%
Rose et al. (1965)	80	2 years	Random	17%
Leren (1966, 1970)	412	10 years	Random	14%
Bierenbaum et al. (1973)	200	10 years	Later: matched	10%*
MRC (1968)	393	4 years	Random	16%
Woodhill et al. (1978)	458	2–7 years	Random	4%
Burr et al. (1989)	2033	2 years	Random	4%

Note: Figures shown in the final column are the excess fall in the groups given dietary advice compared to the changes in the control groups. The figures marked * are for the advised group alone, as changes are not given for the control groups in the report of these trials.

Table 32.4. Secondary prevention dietary trials: Summary of outcome of the eight trials shown in Table 32.3

All trials
Advised: 8121 man-years, 350 deaths (4.3%)
Not advised: 8239 man-years, 380 deaths (4.6%)
Equivalent to 3.2 lives saved per 1000 men per year

Randomised trials
Advised: 6521 man-years, 303 deaths (4.7%)
Not advised: 6641 man-years, 302 deaths (4.6%)
Advised: 5416 man-years, 128 non-fatal myocardial infarctions (2.36%)
Not advised: 5454 man-years, 159 non-fatal myocardial infarctions (2.92%)
Equivalent to 5.5 non-fatal myocardial infarctions saved per 1000 men per year

---

Table 32.4 gives a summary of the outcome of the trials listed in Table 32.3. The evidence of benefit is not entirely convincing, particularly if one only accepts evidence from the randomised trials. At the same time there is evidence of a reduction in non-fatal infarctions.

The same conclusions have been reached by others (Rossouw et al. 1990; Yusuf et al. 1988). For example, Yusuf and Furberg (1987) concluded from an overview of all the secondary prevention trials of dietary therapy that "there is a marginally statistically significant reduction in coronary heart disease events, no difference in cardiac deaths and a non-significant excess in non-cardiac deaths".

The reason for an absence on mortality may be that dietary fat reduction works through an effect on atherosclerosis, and this process may be too far advanced in patients who have already had a myocardial infarct, to be effective in trials lasting only 2–12 years.

Exercise

No randomised trial has yet been completed on exercise nor is any likely. On the other hand there is much evidence from prospective and other studies (Eichner 1983; Lichtenstein 1985).

Morris et al. (1980) have reported on a follow-up of 18 000 civil servants in the Whitehall study. The incidence of heart disease events during the following $8\frac{1}{2}$ years in men who engaged in vigorous leisure time activity was less than half that in the men who took no such exercise. The consistency of this difference, within subgroups of men defined by a variety of factors, encouraged these authors to generalise the benefit and suggest that vigorous exercise is a natural defence against ischaemic heart disease. Their data for men who already had had ischaemic heart disease at baseline is limited but they comment that the findings in these men are "as consistent as could be expected" with the overall findings of benefit.

Further evidence on activity and ischaemic heart disease risk comes from dietary studies because calorie intake is probably a good surrogate for exercise level. Virtually every prospective dietary study has shown lower calorie intake in subjects who go on to infarction compared to those who remain free of ischaemic heart disease. This difference overall is around 5% and in individual studies it varies from 79–273 kcal per day or 3%–10% of total calorie intake. Data from the Caerphilly Study enable this to be examined in men with prevalent disease in whom effects are relevant to secondary prevention. Table 32.5 shows these data and suggests that the benefit of physical exercise is very marked in these men, compared with men who had been disease-free.

Table 32.5. Risk of ischaemic heart disease (IHD) and total energy intake in men aged 45–59 years, followed for 5 years in the Caerphilly Study (MRC Epidemiology Unit 1991)

"Fifth" of energy consumption	No symptomatic IHD at baseline		Symptomatic IHD at baseline	
	No. of men	Annual IHD incidence	No. of men	Annual IHD incidence
1 (Lowest)	373	1.07%	94	3.19%
2	394	1.22%	73	1.92%
3	407	0.93%	59	2.37%
4	406	0.74%	61	2.30%
5 (Highest)	405	0.89%	62	1.61%

Note: The data for men who had had no prevalent IHD are relevant to primary prevention while the data for men who had had symptoms are relevant to secondary prevention (see text).

There is thus a remarkable consistency in the evidence on calorie intake and risk of ischaemic heart disease. Although this gives support to the hypothesis that exercise is protective, evidence from directly focused randomised controlled trials is missing.

Drug Prophylaxis

It is not the purpose of this paper to discuss drug therapy in the acute phase after infarction, but mention of drugs in relation to long-term prophylaxis seems to be appropriate.

Aspirin is of particular interest both because of the low dose required, and the size of its effect. Overview analyses of all the completed trials, which to date number over 30 and have involved over 100 000 patients, indicate that the reduction in total deaths is around 15% and a further 30% of non-fatal infarctions are prevented (Antiplatelet Trialists' Collaboraton 1988).

Cholesterol lowering drugs have at present a less certain role in prophylaxis. The evidence of protection from secondary prevention trials suggests that although there is little short-term improvement in overall survival, there might be with certain drugs over a long period of time (Table 32.6).

An overview of secondary prevention by drugs has concluded that a reduction in cardiac morbidity and mortality of about 20% can be achieved (Yusuf and Cutler 1987). There may however be an excess in non-cardiac mortality (Muldoon et al. 1990) and, until more is known about this, attempts to lower cholesterol by diet or drugs should be pursued cautiously, if at all.

Table 32.6. Trials of cholesterol lowering drugs

Trial		Duration (years)	Reduction in cholesterol (%)	No. of Men		All deaths		Non-fatal myocardial infarctions	
				Treated	Control	Treated	Control	Treated	Control
CDP[a]	1973	6	?	1103	2789	281	708	144	385
Newcastle[b]	1971	5	8	244	253	?	?	30	46
Edinburgh[c]	1971	5	9	350	367	30	30	25	41
Stockholm[d]	1977	5	13	279	276	61	82	35	50

Drug: 9763 man-years, 372 deaths (3.8%)
No drug: 19 949 man-years, 820 deaths (4.1%)
Equivalent to 3.0 lives saved per 1000 men per year

Drug: 10 983 man-years, 234 non-fatal myocardial infarctions (2.1%)
No drug: 21 214 man-years, 522 non-fatal myocardial infarctions (2.5%)
Equivalent to 3.3 non-fatal myocardial infarctions saved per 1000 men per year

[a] Coronary Drug Project Research Group (1975).
[b] Group of Physicians of the Newcastle-upon-Tyne Region (1971).
[c] Research Committee of the Scottish Society of Physicians (1971).
[d] Carlson, Rosenhamer (1988).

Need for Further Research

Secondary prevention trials are not easy, but unlike trials of primary prevention they are within the competence and resources of most epidemiologists, whereas primary prevention is not. Suitable patients are common, a high proportion of them are likely to agree to enter a trial and they are likely to comply with dietary changes for prolonged periods.

It may not, however, be entirely reasonable to extrapolate from secondary prevention and apply conclusions to primary prevention. This is probably true for measures which are relevant to atherosclerosis, such as low fat diets. However, it is possible that measures which have little to do with atherosclerosis may still give protection through effects on mechanisms other than atherosclerosis. This last seems likely to be most important in view of the evidence that tests relating to haemostatic mechanisms have a greater predictive power than any lipid level (Yarnell et al. 1991). The point is further emphasised by the success of both aspirin and of fish in reducing total mortality, yet for neither of which has an effect on atherogenesis been postulated.

References

Antiplatelet Trialists' Collaboration (1988) Secondary prevention of vascular disease by prolonged antiplatelet treatment. Br Med J 296:320–31
Bainton D, Baker IA, Sweetnam PM, Yarnell JWG, Elwood PC (1988) Prevalence of ischaemic heart disease: the Caerphilly and Speedwell surveys. Br Heart J 59:201–6

Bierenbaum ML, Fleischman AI, Raichelson RI, Hayton T, Watson PB (1973) Ten-year experience of modified-fat diets on younger men with coronary heart disease. Lancet i:1404–7

Burr ML (1989) Fish and the cardiovascular system. Prog Food Nutr Sci 13:291–316

Burr ML, Sweetnam PM (1982) Vegetarianism, dietary fibre and mortality. Am J Clin Nutr 36:873–7

Burr ML, Fehily AM, Gilbert JF et al (1989) Effects of changes in fat, fish, and fibre intakes on death and myocardial reinfarction: Diet and re-infarction trial (DART). Lancet ii:757–61

Carlson LA, Rosenhamer G (1988) Reduction of mortality in the Stockholm Ischaemic Heart Disease Secondary Prevention Study by combined treatment with clofibrate and nicotinic acid. Acta Med Scand 223:405–18

Coronary Drug Project Research Group (1975) Clofibrate and niacin in coronary heart disease. JAMA 231:360–81

Eichner ER (1983) Exercise and heart disease: epidemiology of the 'exercise hypothesis'. Am J Med 75:1008–23

Fries JF, Green LW, Levine S (1989) Health promotion and the compression of morbidity. Lancet i:481–3

Group of Physicians of the Newcastle-upon-Tyne Region (1971) Trial of clofibrate in the treatment of ischaemic heart disease. Br Med J 4:767–75

Herold PM, Kinsella JE (1986) Fish oil consumption and decreased risk of cardiovascular disease: a comparison of findings from animal and human feeding trials. Am J Clin Nutr 43:566–98

Khaw KT, Barrett-Connor E (1987) Dietary fibre and reduced ischaemic heart disease mortality rates in men and women: a 12-year prospective study. Am J Epidemiol 126:1093–102

Kromhout D, De Lezenne Coulander C (1984) Diet, prevalence and 10-year mortality from coronary heart disease in 871 middle-aged men: the Zutphen study. Am J Epidemiol 119:733–41

Kushi LH, Lew RA, Stare FJ et al. (1985) Diet and 20-year mortality from coronary heart disease: the Ireland – Boston diet – heart study. N Engl J Med 312:811–18

Leren P (1966) The effect of plasma cholesterol lowering diet in male survivors of myocardial infarction. Acta Med Scand (Suppl) 466:1–92

Leren P (1970) The Oslo Diet – Heart Study. Eleven-year report. Circulation 40:935–43

Lichtenstein MJ (1985) Jogging in middle age. J R Coll Gen Pract 35:341–4

MRC Epidemiology Unit (1991) Studies of cardiovascular disease: Progress Report VIII. Medical Research Council, London

Medical Research Council, Research Committee (1965) Low fat diet in myocardial infarction: a controlled trial. Lancet ii:501–4

Medical Research Council, Research Committee (1968) Controlled trial of soya-bean oil in myocardial infarction. Lancet ii:693–700

Morris JN, Marr JW, Clayton DG (1977) Diet and heart: a postscript. Br Med J ii:1307–14

Morris JN, Everitt MG, Pollard R, Chave SPW (1980) Vigorous exercise in leisure-time: protection against coronary heart disease. Lancet ii:1207–10

Morrison LM (1960) Diet in coronary atherosclerosis. JAMA 173:884–8

Muldoon MF, Manuck SB, Matthews KA (1990) Lowering cholesterol concentrations and mortality: a quantitative review of primary prevention trials. Br Med J 301:309–14

Research Committee of the Scottish Society of Physicians (1971) Ischaemic heart disease: A secondary prevention trial using clofibrate. Br Med J iv:775–84

Rose GA, Thomson WB, Williams RT (1965) Corn oil in treatment of ischaemic heart disease. Br Med J i:1531–3

Rose GA, Reid DD, Hamilton PJS, McCartney P, Keen H, Jarrett RJ (1977) Myocardial ischaemia, risk factors and death from coronary heart disease. Lancet i:105–9

Rossouw JE, Lewis B, Rifkind BM (1990) The value of lowering cholesterol after myocardial infarction. N Engl J Med 323:1112–1119

Shaper AG, Pocock SJ, Walker M, Phillips AN, Whitehead TP, Macfarlane PW (1985) Risk factors for ischaemic heart disease: the prospective phase of the British Regional Heart Study. J Epidemiol Community Health 39:197–209

Trowell HC, Burkitt DP (eds) (1981) Western diseases: Their emergence and prevention. Edward Arnold, London

Woodhill JM, Palmer AJ, Leelarthaepin B, McGilchrist C, Blacket RB (1978) Low fat, low cholesterol diet in secondary prevention of coronary heart disease. Adv Exp Med Biol 109:317–30

WHO European Collaborative Group (1980) Multifactorial trial in the prevention of coronary heart disease. I. Recruitment and critical findings. Eur Heart J 1:3–80

Yarnell JWG, Baker IA, Sweetnam PM et al (1991) Fibrinogen, viscosity, and white blood cell count are major risk factors for ischaemic heart disease. The Caerphilly and Speedwell Collaborative Heart Disease Studies. Circulation 83:836–44

Yusuf S, Cutler J (1987) Single factor trials: Drug studies. In: Olsson AG et al. (eds) Atherosclerosis. Churchill Livingstone, New York, pp 393–8

Yusuf S, Furberg CD (1987) Single factor trials: Control through lifestyle changes. In: Olsson AG et al. (eds) Atherosclerosis. Churchill Livingstone, New York, pp 839–92

Yusuf S, Wittes J, Friedman L (1988) Overview of results of randomized clinical trials in heart disease: unstable angina, heart failure, primary prevention with aspirin, and risk factor modification. Clin Cardiol 260:2259–63

Subject Index